T0180100

Lecture Notes in Computer Science 12844

More information about this subseries at http://www.springer.com/series/7409

Andreas Holzinger · Peter Kieseberg ·
A Min Tjoa · Edgar Weippl (Eds.)

Machine Learning and Knowledge Extraction

5th IFIP TC 5, TC 12, WG 8.4, WG 8.9, WG 12.9
International Cross-Domain Conference, CD-MAKE 2021
Virtual Event, August 17–20, 2021
Proceedings

 Springer

Editors
Andreas Holzinger ⓘD
Institute for Medical Informatics, Statistics
and Documentation and Institute for
Information Systems and Computer Media
Medical University Graz and Graz
University of Technology
Graz, Austria

xAI Lab, Alberta Machine Intelligence
Institute
University of Alberta
Edmonton, AB, Canada

A Min Tjoa ⓘD
Institute of Software Technology
and Interactive Systems
Technische Universität Wien
Vienna, Austria

Peter Kieseberg ⓘD
St. Pölten University of Applied Sciences
St. Pölten, Austria

Edgar Weippl
SBA Research
Vienna, Austria

Research Group Security and Privacy
University of Vienna
Vienna, Austria

ISSN 0302-9743   ISSN 1611-3349   (electronic)
Lecture Notes in Computer Science
ISBN 978-3-030-84059-4   ISBN 978-3-030-84060-0   (eBook)
https://doi.org/10.1007/978-3-030-84060-0

LNCS Sublibrary: SL3 – Information Systems and Applications, incl. Internet/Web, and HCI

This Springer imprint is published by the registered company Springer Nature Switzerland AG
The registered company address is: Gewerbestrasse 11, 6330 Cham, Switzerland

Preface

The International Cross Domain Conference for Machine Learning & Knowledge Extraction (CD-MAKE) is a joint effort of IFIP TC 5, IFIP TC 12, IFIP WG 8.4, IFIP WG 8.9, and IFIP WG 12.9 and is held in conjunction with the International Conference on Availability, Reliability, and Security (ARES). This fifth installment was organized virtually by SBA Research due to the ongoing COVID-19 pandemic, and took place during August 17–20, 2021.

The letters CD in CD-MAKE stand for "Cross-Domain" and describe the integration and appraisal of different fields and application domains to provide an atmosphere to foster different perspectives and opinions. The conference fosters an integrative machine learning approach, considering the importance of data science and visualization for the algorithmic pipeline with a strong emphasis on privacy, data protection, safety, and security. It is dedicated to offering an international platform for novel ideas and a fresh look on methodologies to put crazy ideas into business for the benefit of humans. Serendipity is a desired effect, which may lead to the cross-fertilization of methodologies and the transfer of algorithmic developments.

The acronym MAKE stands for "MAchine Learning & Knowledge Extraction", a field of artificial intelligence (AI) that, while quite old in its fundamentals, has just recently begun to thrive based on both novel developments in the algorithmic area and the availability of vast computing resources at a comparatively low cost.

Machine learning (ML) studies algorithms that can learn from data to gain knowledge from experience and to generate decisions and predictions. A grand goal is in understanding intelligence for the design and development of algorithms that work autonomously (ideally without a human-in-the-loop) and can improve their learning behavior over time. The challenge is to discover relevant structural and/or temporal patterns ("knowledge") in data, which is often hidden in arbitrarily high dimensional spaces, and thus simply not accessible to humans. Knowledge extraction is one of the oldest fields in AI and is seeing a renaissance, particularly in the combination of statistical methods with classical ontological approaches. AI is currently undergoing a kind of Cambrian explosion and is the fastest-growing field in computer science today thanks to the usable successes in machine learning. There are many application domains, e.g., in medicine, etc., with many use cases from our daily lives, e.g., recommender systems, speech recognition, autonomous driving, etc. The grand challenges lie in sensemaking, in context understanding, and in decision-making under uncertainty, as well as solving the problem of explainability. Our real world is full of uncertainties and probabilistic inference enormously influences AI generally and ML specifically. The inverse probability allows us to infer unknowns, to learn from data, and to make predictions to support decision-making. Whether in social networks, recommender systems, health applications, or industrial applications, the increasingly complex data sets require a joint interdisciplinary effort involving the human-in-control to foster a better understanding of the ethical and social issues accountability, retractability, explainability, causability, and privacy, safety and security!

A few words about IFIP: IFIP – the International Federation for Information Processing—is the leading multi-national, non-governmental, apolitical organization in information and communications technologies and computer sciences; it is recognized by the United Nations (UN) and was established in 1960 under the auspices of the UNESCO as an outcome of the first World Computer Congress held in Paris in 1959.

IFIP is incorporated in Austria by decree of the Austrian Foreign Ministry (September 20, 1996, GZ 1055.170/120-I.2/96) granting IFIP the legal status of a non-governmental international organization under the Austrian Law on the Granting of Privileges to Non-Governmental International Organizations (Federal Law Gazette 1992/174). IFIP brings together more than 3500 scientists without boundaries from both academia and industry, organized in more than 100 Working Groups (WGs) and 13 Technical Committees (TCs).

To acknowledge all those who also contributed to the organizational effort and stimulating discussions at CD-MAKE 2021 would be impossible in a preface like this. Many people also contributed to the development of this volume, either directly or indirectly, so, again, it would be impossible to list all of them. We herewith thank all local, national, and international colleagues and friends for their positive and supportive encouragement. Finally, yet importantly, we thank the Springer management team and the Springer production team for their professional support.

Thank you to all! Let's MAKE it!

June 2021

Andreas Holzinger
Peter Kieseberg
Edgar Weippl
A Min Tjoa

Organization

Conference Organizers

Andreas Holzinger   Medical University of Graz and Graz University
of Technology, Austria, and University of Alberta,
Canada
Peter Kieseberg   FH St.Pölten, Austria
Edgar Weippl   SBA Research and University of Vienna, Austria
A Min Tjoa   TU Vienna, Austria

Program Committee

Frantisek Babic   Technical University of Košice, Slovakia
Smaranda Belciug   University of Craiova, Romania
Elisa Bertino   Purdue University, USA
Chris Biemann   Universität Hamburg, Germany
Jiang Bian   University of Florida, USA
Malin Bradley   Vanderbilt University, USA
Ivan Bratko   University of Ljubljana, Slovenia
Guido Bologna   Université de Genève, Switzerland
Francesco Buccafurri   Universita Mediterranea di Reggio Calabria, Italy
Federico Cabitza   University of Milano-Bicocca, Italy
Andre Calero-Valdez   RWTH Aachen University, Germany
Andrea Campagner   University of Milano-Bicocca, Italy
Angelo Cangelosi   University of Manchester, UK
Mirko Cesarini   University of Milano-Bicocca, Milan, Italy
Krzysztof J. Cios   Virginia Commonwealth University, USA
Carlo Combi   University of Verona, Italy
Beatriz De La Iglesia   University of East Anglia, UK
Gloria Cerasela Crisan   Vasile Alecsandri University of Bacau, Romania
Alexiei Dingli   University of Malta, Malta
Josep Domingo-Ferrer   Universitat Rovira i Virgili, Spain
Isao Echizen   National Institute of Informatics, Japan
Massimo Ferri   University of Bologna, Italy
Ulrich Furbach   University of Koblenz, Germany
Hugo Gamboa   PLUX Wireless Biosensors and Universidade Nova de
Lisboa, Portugal
Barbara Hammer   University of Bielefeld, Germany
Pitoyo Hartono   Chukyo University, Japan
Barna Laszlo Iantovics   "George Emil Palade" University of Medicine,
Pharmacy, Sciences and Technology of Targu
Mures, Romania

Contents

Digital Transformation for Sustainable Development Goals (SDGs) - A Security, Safety and Privacy Perspective on AI

Andreas Holzinger[1,2]($\boxtimes$) (iD), Edgar Weippl[3,4] (iD), A Min Tjoa[5] (iD),
and Peter Kieseberg[6] (iD)

[1] Human-Centered AI Lab, Medical University Graz, Graz, Austria
andreas.holzinger@medunigraz.at
[2] xAI Lab, Alberta Machine Intelligence Institute, Edmonton, Canada
[3] SBA Research, Vienna, Austria
[4] Research Group Security and Privacy, University of Vienna, Vienna, Austria
[5] Vienna University of Technology, Vienna, Austria
[6] St. Pölten University of Applied Sciences, St. Pölten, Austria

Abstract. The main driver of the digital transformation currently underway is undoubtedly artificial intelligence (AI). The potential of AI to benefit humanity and its environment is undeniably enormous. AI can definitely help find new solutions to the most pressing challenges facing our human society in virtually all areas of life: from agriculture and forest ecosystems that affect our entire planet, to the health of every single human being. However, this article highlights a very different aspect. For all its benefits, the large-scale adoption of AI technologies also holds enormous and unimagined potential for new kinds of unforeseen threats. Therefore, all stakeholders, governments, policy makers, and industry, together with academia, must ensure that AI is developed with these potential threats in mind and that the safety, traceability, transparency, explainability, validity, and verifiability of AI applications in our everyday lives are ensured. It is the responsibility of all stakeholders to ensure the use of trustworthy and ethically reliable AI and to avoid the misuse of AI technologies. Achieving this will require a concerted effort to ensure that AI is always consistent with human values and includes a future that is safe in every way for all people on this planet. In this paper, we describe some of these threats and show that safety, security and explainability are indispensable cross-cutting issues and highlight this with two exemplary selected application areas: smart agriculture and smart health.

Keywords: Artificial intelligence · Digital transformation · Robustness · Resilience · Explainability · Explainable AI · Safety · Security · AI risks · AI threats · Smart agriculture · Smart health

1   Introduction and Motivation

Often referred to as buzzwords, such as AI, Blockchain, Big Data, Internet of Things (IoT), ..., these technology trends of the last decades are the actual drivers of the digital transformation that is actually taking place [71,75]. Thus, these technologies no longer have just an additional support function, rather these technologies are changing complete process chains and permeate almost all our areas of life and work, from Smart Agriculture to Smart Health to name just two application areas. The main driver of digital transformation is undoubtedly the broad field of artificial intelligence (AI).

AI has gained a lot of attraction in the last decade. Many of the basic concepts date back to the middle of the last century, however the right combination and synergies of three approaches has led to a revolution that now brings AI to everyone's attention: (1) powerful, cost-effective, and available hardware (2) successful methods from statistical machine learning (e.g., Deep Learning), and (3) a growing amount of available data. AI-related components now permeate all sorts of labour, industries and applications, e.g.

- *Autonomous AI systems* that automate decisions without *any* human intervention (e.g., fully autonomous self driving cars [44], autonomous medical diagnosis [2], autonomous drones [19], ...).
- *Automated AI systems* that perform labor-intensive tasks requiring certain intelligence, and complete them automatically within a certain domain and given tasks (e.g., industrial robotic process automatization [1], automated medical workflows [48], automated forest management [49], ...).
- *Assisted AI systems* that help humans perform repetitive routine tasks faster and both quantitatively and qualitatively better (e.g., ambient assisted smart living [69], weather forecasting, ...).
- *Augmented AI systems* that help people understand complex and uncertain future events (e.g., Explainable AI in Digital Pathology [28], Simple Augmented Reality applications [56], Augmented AI in agriculture [68], ...).

This widespread adoption also lowers the barrier to entry for other players in the domain, whether they are scientists from entirely different domains (e.g. health, farming, climate research, ...) using AI technology to solve problems or companies by adding intelligent components to existing tools. While this trend undeniably brings tremendous benefits, opportunities, and possibilities in terms of new capabilities and applications, this rising trend can also lead to problems, especially when it comes to the security, trustworthiness, and privacy of these systems.

In recent years, the topic of sustainability has gained a lot of attention, especially with the declaration of the Sustainable Development Goals (SDGs) by the UN [65] with its 17 core goals. As such, several ideas and approaches have been put forward for using AI-related technologies to support these goals. While this is obviously a very valuable approach, we still need to understand the shortcomings of AI to make these approaches inherently sustainable. While many researchers reduce the problems of AI to purely theoretical aspects, we will define some key issues in this paper. Nevertheless, we believe that AI technologies can significantly improve our lives and support the SDGs through improved digital transformation, however much additional work is needed to actually make a difference.

This paper is organized as follows: Sect. 2 provides an overview of the SDGs as well as some background and related work on how AI can support these goals. Here we also present two specific examples that affect virtually everyone in their daily lives: smart health, with the goal of precision medicine, and smart agriculture, with the goal of precision farming. In Sect. 3, we analyze a selection of topics that we believe need to be considered in supporting the goals outlined in the SDGs through AI technology. While we have derived some of these issues from the relevant literature, we have added some new ones based on our many years of practical experience in developing AI-based systems - always with a focus on safety, security, and privacy. In the Conclusion section, we summarize the paper.

2   Background and Related Work

In this section, we provide an overview of the background of the Sustainable Development Goals, as well as an outline of related work that shows how AI-based systems can support them.

2.1   The UN Sustainable Development Goals (SDGs)

The idea of "sustainable development" was already discussed by the United Nations Brundtland Commission in 1987: "Sustainable development is development that meets the needs of the present without com-promising the ability of future generations to meet their own needs" [13].

The fundamental concept of sustainability originated at the United Nations (UN) Conference on Environment and Development (UNCED) in Rio de Janeiro in 1992 (the so-called "Earth Summit" [63]), where a *Declaration of Principles and Desired Action on International Agreements on Climate Change and Biodiversity and a Declaration of Principles on Forests* were presented. Subsequently, in 2002, the commitment to sustainable development was reaffirmed at the World Summit on Sustainable Development in Johannesburg, South Africa.

The concept was intentionally not clearly defined to allow a way to address the very different challenges: from planning sustainable cities to sustainable livelihoods, from sustainable agriculture to smart health, and the efforts to develop common business standards in the UN Global Compact and the World Business Council for Sustainable Development [61].

The adoption of the Sustainable Development Goals (SDGs) in 2015 [70] signaled the commitment of world leaders to a more sustainable path to inclusive and equitable growth. Also known as the 2030 Agenda, the 17 SDGs cover a wide range of development-related issues and include 169 targets and 304 indicators [16], see Fig. 1 for an overview on the 17 SDGs.

Fig. 1. An overview on the UN Sustainable Development Goals (SDGs) [65]

2.2 Smart Farming and Precision Agriculture - An Example for Supporting SDGs Through AI

As a first application example, we choose a topic that concerns every inhabitant of our planet: Smart Farming, i.e. the use of AI in cyber-physical applications for versatile support of process chains in agriculture. Cyber-physical systems (CPS) have been established for some time [8] and although CPS are very versatile, their engineering is most challenging due to the high degree of heterogeneity. Moreover, the importance of CPS for smart farming is often underestimated, but they form the basis for future precision farming, for better crop management and resource use. In this context, massive amounts of data are already being generated in great variety, which can be collected, analyzed and used for decision making. The goal is to develop smart agriculture that will help address the current major socio-economic challenges worldwide (Goal: "zero hunger"). This domain is essential as it extends from agriculture itself far beyond primary production to virtually the entire food supply chain. Here, AI can help in many ways to gain predictive insights into agricultural operations, make real-time operational decisions, and redesign business processes. This has to incorporate often conflicting economic interests, leading to new business models which are developed this way. Clearly, there are all sorts of challenges of a technical nature here, however there are also tensions and shifts in roles and power relationships between different players in the current food supply chain networks. One example is the need of an evidence-based and critical investigation on the observed shift, from individual small farmers to powerful high-tech "agriculture factories". At the same time, there are public institutions that are now publishing data and where, of course, individual privacy must be ensured. The future of smart agriculture could play out on a continuum of two extreme scenarios: 1) closed, proprietary systems in which the farmer is part of a highly integrated food supply chain, or 2) open, collaborative systems in which the farmer and all other actors and stakeholders in the chain's network need the flexibility in choosing business partners for both technology and food production [17,79]. Closely related to smart agriculture is sustainable forest management. The management of forest ecosystems is underestimated, but of eminent importance for the survival of our planet [24] as the restoration of forested land could help to capture carbon and thus mitigate climate change [6]. Computer-based support tools have been in use in this domain for some time [66]. However, monitoring of forest areas is far from trivial, facing major challenges such as high forest density, complexity and diversity of forest structure, complex topography and climatic conditions, and difficult access for human researchers. Here, unmanned aerial vehicles are already making a valuable contribution by enabling the classification of forest types based on Convolutional Neural Networks [49].

2.3 Smart Health and Precision Medicine - An Example for Supporting SDGs Through AI

As a second example, we choose a topic that also affects each and every one of us individually and is also accompanied by a large number of non-trivial problems: health, and here in particular the emerging domain of smart health. The trend toward higher life expectancy together with the increasing complexity of medicine and health services is causing healthcare costs to rise dramatically worldwide. As a result, enormously high expectations are being placed on AI for health worldwide.

The concept of smart health [30] has huge potential to support a future P4 medicine (preventive, participative, predictive and personalized) or in short *personalized medicine* [23]. The goal of this a.k.a. future *precision medicine* [14] is in modeling the complexity of patients health in order to tailor medical decisions, health practices and therapies to the individual patient - for example, a drug precisely designed for a patient's individual needs and specific background in a given context. This trend towards personalized medicine produces huge amounts of data, which makes manual analysis difficult and almost impossible for a human being [26]. For example massive amounts of sensors produce large amounts of high-dimensional, weakly structured data sets and massive amounts of unstructured information. In the medical domain, many different modalities contribute to an outcome. Consequently, the smart health principle makes medicine a truly data-intensive science. To keep up with these growing volumes of complex data AI approaches are mandatory, however, in the medical domain there is always a need for a human-in-the-loop [27], at least the human-in-control - because of legal aspects [64]. However, the synergies between AI and precision medicine promises to revolutionize healthcare: whilst precision medicine methods help identify phenotypes of patients with less frequent treatment response or special healthcare needs, AI is being used to support clinician decision making through augmented intelligence. Translational research exploring this convergence can help solve precision medicine's most difficult challenges, particularly those where non-genomic and genomic determinants combined with information from patients' symptoms, clinical history, and lifestyle habits will enable personalized diagnosis and prognosis [35].

By deploying complex AI systems in physical-digital ecosystems, future physicians will be supported by their AI assistants in managing their flood of data from different modalities, which requires explainability and causability [29]. At the same time, patients will be supported by their on-line health assistants, and moreover, these technologies will support the *preventive medicine* nature to enable healthier living, wellness and well-being, which will also lead to enormous amounts of private data.

In the medical field, the issues of transparency, accountability, and trust are prerequisites for the integration of AI into daily practice. As the importance of medical AI will certainly continue to grow strongly in the coming years, it is imperative that legal and ethical issues are always considered together [53,54].

2.4   Impact of AI on SDGs

There has been ample publications indicating the benefits of certain AI-based systems on SDGs, as e.g. The academic literature also includes analysis on related subjects, like in [73], where the authors give a comprehensive study on the utilization of artificial intelligence in the development of sustainable business models. In another paper [3], the authors discuss the impact of the ongoing Covid pandemic on the SDGs and the efforts taken into reaching them by 2030. They provide an interesting approach on looking for synergies between different targets and approaches in order to prioritize them. The publication [25] on the other hand gives a very good overview on important topics in this area and also provides a first insight into the challenges associated with using AI for supporting SDGs. To the best of our knowledge the most comprehensive study can be found in [77]. In this work, the authors analyzed all 17 SDGS with their 169 targets with respect to whether AI is beneficial or detrimental for a target, more specifically, they scored, how beneficial and how detrimental AI is for any given target. The measurement was made by conducting a consensus-based expert elicitation process based on results from previous studies. While the paper added a short discussion on security-related issues of utilizing AI for achieving the SGDs, the discussion was rather short and did not go into details.

3   Open Issues on Using AI for SDGs

In this section, we discuss security related issues of intelligent systems, some of them being typically overlooked and having received little attention in the academic literature. In this discussion, we use the term *security* rather loosely for any technical issue that has a detrimental effect on users. Still, we typically limit ourselves to the purely technical IT domain and did not include social and economic issues like increasing job loss, even though these also might have an impact on the overall security of a nation. We have loosely grouped these topics in terms of overarching themes to make the work more stringent, even though several topics might belong to more than one of these themes.

3.1 Data and Models

Many AI techniques, especially advanced techniques of statistical machine learning, including neural networks (deep learning) that gathered a lot of attention during the past years, heavily rely on a key resource: Data. Data is used to train these networks, i.e. the whole AI system is not only defined by the pure mathematical/algorithmic side, but to a large extent by the data that was used for training the system. Thus, a lot of problems in the adoption of intelligent systems lie in the trustworthiness of data sets, ranging from unintentional bias that leads to discrimination of people to malicious attacks trying to interfere with the system and manipulate the decision making process in subtle ways. In the following we have gathered a selection of issues that we see as key concerns when trying to solve SGDs through AI-based systems.

Acquiring Training Data. Many aspects of AI hinge on training data, especially related to issues of trust and quality. Still, acquiring such data is a big problem for many applications and due to various reasons ranging from Privacy related issues in the existing medical data to simply no useful data being available at all. The acquisition of data is especially important, as many other issues can be related to it, e.g. the problem of bias, backdoors and providing a good ground truth. Open data might be a solution for this problem in some applications, e.g. forestation related issues, but it will be hard to find widespread acceptance for such a bold step in the medical domain.

Providing a Ground Truth. Related to the issue of generating suitable sets of training data lies the issue of finding a ground truth. This is especially critical in highly complex research questions, where either it is impossible to gather an oversight of the complete data available or the interpretation of the data is depending on other parameters that are hard to impossible to objectify, like e.g. political opinions. Gathering an oversight on a topic can be difficult either due to the sheer amount of information available, but also due to the information being stored in secluded data vaults. e.g. sensitive medical information. There are an ample number of projects that aim at connecting and joining these data faults in a secure and privacy respecting way. As an example for the latter, finding ground truths in fake news detection is especially difficult [10], as news can never cover a complex topic with a lot of subjective decisions and opinions as a whole, thus making it hard to decide, when information was cut maliciously. Furthermore, many legit new outlets also put some kind of spin at the news they are reporting, consciously or unconsciously, by using different terms for the same people (e.g. "freedom fighters" versus "terrorists") or things. Furthermore, when looking at long-standing conflicts, reports have to cut somewhere in time in order to not become history books - thus often removing important background information. Still, finding a ground truth can also be very challenging in purely technical applications for AI, e.g. in intrusion detection systems (IDS) [15] based on collecting "normal behavior" in order to later detect suspicious traffic.

Bias and Data Quality. Often it is difficult to even find enough data on a subject in order to train a neural network, so the topic of data quality is currently overlooked in many cases [33]. Still, in the recent past, several issues surrounding deficient training data have emerged [52], most prominently regarding racial bias in sensitive applications like predictive policing. Bias is an especially important issue to tackle, as bias in AI applications, e.g. decision support, can become self-reinforcing: If e.g. a certain population is over-proportionally included in a training set the algorithm could advise to look deeper in said population - thus to find even more suitable examples reinforcing the original bias resulting in a spiral of bias reinforcement [57]. There have been several discussions regarding racial bias in predictive policing (see e.g. [57,67]), even though a structured study painted a more complex situation when looking at arrest rates [11]. Still, even besides issues of bias, assessing data quality is a difficult task which requires a lot of further research [5].

Data Preparation and Cleansing. A topic that is typically overlooked in purely theoretical papers, is that data often needs to be prepared in order to be useful. For example, data is often incomplete or contains erroneous records [50]. This is not only true for training data, but also for the processing data. Thus data cleansing is typically applied to the data streams which works on different levels with different techniques [5], by e.g.deleting faulty records, assigning defaults or trying to guess the most probable correct version. This, of course, introduces changes into the data and, in case of training data, into the subsequently calculated model, resulting in different models, i.e. a direct impact of the data cleansing process into the very definition of the neural network at worst. Currently, there is not much discussion on this issue in the scientific community, especially not regarding the legal and organisational implications. It is also a very hard question, which model is more correct when facing two different cleansing strategies that result into different models. The same problems arise in other data preparation steps, e.g. reduction of noise in audio files [74] or pictures, which of course need to be done with respect to the state of the art, but often use heuristic algorithms that sport different results depending on various side parameters (see e.g. [36]). The impact of these data preparation steps need to be analyzed carefully regarding their impact. Selecting an AI technique that is sufficiently stable against the expected level of instability in the data sets is an absolute must, unfortunately, stability has often been pushed into the background of the tool development process.

Sharing Models and Training Data. Sharing data and even trained models is an interesting approach in order to battle the problems of acquiring training data (see [80] for a solution platform). When sharing training data in order to enable other parties to train their models, questions of privacy, but also regarding the intrinsic value of the data, need to be kept in mind [81]. Thus, it might be advisable to use fingerprinting technologies in order to be able to detect data leaks [9,39]. There might also be laws and regulations that need to be complied with in the sharing process. Furthermore, there is always the problem of assessing the quality of the shared data with respect to e.g. sample selection, data preparation and overall data quality (see e.g. [34] for a novel blockchain based solution). Still, pooling data and sharing them with other player might be a viable strategy in many application areas related to SDGs in order to circumvent training data shortage. Another related strategy lies in sharing the models, also called pre-trained systems. Here, the original training data is not given away, but solely the trained model, which, of course, hast to adhere to certain prerequisites specific to the system it is later used in. Pre-trained models are often claimed to solve many of the issues associated with the sharing of training data, especially to those related to the GDPR [60] and privacy, still, this needs to be taken with a grain of salt: On the one hand, there have been attacks against pre-trained models [42,46] devised in the past that had some success in recreating information on the original training data of a model, depending on the models complexity and the availability of side information. Still, what is even more a problem, is the amount of trust that has to be placed into the generator of the model in question:

- It is typically impossible to infer anything on the raw training data used for building the model like using sanity checks or verifying data records. The user must therefore trust the generator that the training data had sufficient quality and is unbiased.
- It is very hard for the user to control, whether backdoors were included into the model [22], e.g. a model trained on the impact of emissions could lead to correct results in all cases, except when a special emission pattern typical for e.g. a specific car brand is encountered. In this special case, the model suddenly calculates a far better result.

Especially the backdoor issue is a major trust problem, especially when a lot of risk and investments are involved. The generator of the model must be extremely trustworthy, still another problem derives from the closed nature of a pre-trained model: It is extremely difficult to expand this model with new data in a controlled way, i.e. even in cases of algorithms that feed new information back into the model and thus expand on the pre-trained model, a lot of transparency is lost due to the unknown nature of the original model training data set. As a result, pre-trained models result in a lot of additional issues, still, for several applications they will be the only feasible solution in the near future.

3.2 Providing Trustworthy Systems

Trust is one of the major issues when it comes to IT-systems that are deployed in critical environments, resulting in the notion of trustworthy artificial intelligence [78]. While this is not related to the SDGs outlined in Sect. 2.1 at first glance, it must be taken into account that (i) some of them relate to people and their data making them sensible data driven applications and (ii) others relate to barriers for big industrial corporations, again, requiring measures to be put in place in order to make people trust them not to be manipulated. On a side note, the topic of explainability is of high importance for providing trust, as outlined in [31].

Security Testing. Testing systems for security is a major aspect in finding vulnerabilities in existing systems and providing remedies in the form of patches. Several testing methods are employed, ranging from code-reviews, where the analyst is in the possession of the systems source code, over architecture analysis to black-box testing methods like digital twins or penetration testing. All of these techniques become increasingly difficult in the presence of AI (see [45] for a survey): For the architecture review, while it certainly stays useful in order to find fundamental flaws like problematic access control, insecure system design and so on, the AI component is typically a black box: The training data virtually defines large parts of the AI behavior, while not being represented in the architecture. The same holds true for source code reviews, since the training data is not within the source code, the very aspects that define the AI are not included. But even when including the training data and the trained model in the review, in general not much useful information can be deduced from them due to the explainability problem [21]. As for penetration testing, AI components add an additional layer of complexity as outlined in [72], where the authors also provided some first attempts for a methodological approach towards the topic. Also with respect to digital twins, side-effects and internal workings of such complexity typically cannot be simulated within reasonable time and financial limits. When using pre-trained models, security testing becomes even more cumbersome, as already outlined in the paragraphs on pre-trained models and bias in Sect. 3.1.

Privacy and Profiling. When utilizing AI for enhancing the targets of SDGs it can be tempting to use as fine-granular data as possible in order to gather good results. This can be a problem with respect to end user privacy [37], especially when dealing with personal data, e.g. when analysing issues of gender equality (SDG 5) or health (SDG 3). While certainly beneficial for the results of the AI component [51], this can be detrimental to end user interests, especially in cases, where the users in question are within a suppressed minority, i.e. it must be made sure that the means that are planned to support the SDGs are not misused in order to hurt them. User profiling [18] can be seen as an extreme form of privacy violation, as digital models of users and their interests, as well as preferences are generated and subsequently exploited, e.g. in order to well-placed targeted marketing. Still, the big companies like Facebook and Google can derive much more from this data. Keeping privacy in mind is thus a fundamental step for designing any AI-based application supporting SDGs.

Data Manipulation Detection. Manipulation can take place at many steps inside a data driven workflow (without guarantee of completeness):

- At the collection phase by only collecting data suitable to the attacker.
- Whenever data resides in a data store, e.g. a database.
- Through the introduction of faulty data into the data streams (see also adversarial machine learning).
- During the calculation and processing stages, especially within complex enrichment workflows, often including external enrichment data of varying volatility.
- When the results are sent to the human decision makers (in case they are still foreseen in the system).
- In case of feedback loops in expert systems in the mechanisms that report the feedback from the (human) expert to the machine learning entity.

Detecting such manipulations, which can often be carried out trivially, especially by an internal attacker like a disgruntled database administrator, a strict integrity providing process like a chain of custody [20] has to be put in place in order to mitigate these threats. It must be kept in mind though that these mechanisms must be secure against very potent internal attackers, like e.g. put forward in [38].

Adversarial Machine Learning. In adversarial machine learning [32], an attacker tries to change the underlying decision model of an AI component by feeding it with specifically crafted data. Often, this feeding needs to be done slowly in order to go undetected. While current attacks are rather quite low-level in nature, their effects can be stunning and might even allow attackers to introduce backdoors into existing systems. Due to the explainability problem, the resulting changed models are often hard to detect, and detection typically focuses on the feeding process though. See publication [62] for an in-depth description of this issue.

Resilience and Stability. When using AI-based systems in order to tackle targets derived from the SDGs, a certain level of resilience [47] is direly needed, i.e. the system needs to be able to adapt to successful attacks and maybe even change. This is especially important, as an unreliable system will lose acceptance rather quickly. The same holds true for the topic of stability, where we use two different meanings for the term *stability*: (i) A system that is running stable and uninterrupted and (ii) The utilization of algorithms that do not behave chaotic, i.e. the output should not change too drastically when making small changes to the input. While the reason for the importance of the first meaning is rather straightforward and can be seen as a part of resilience, the second one is required in order to deal correctly with rounded and/or inaccurate inputs in a correct manner: Since input data, especially concerning natural processes, can never exceed a certain, sometimes quite low, margin for accuracy, an algorithm that reacts very strong on such differences might be useless in a practical context.

3.3 Control

Control in this context means: Who runs the system, who is responsible for the code, for the data, who can change the software - all these elements are vital to clarify when tackling the grand challenges put forward by the SDGs, especially as a lot of financial and political impact is caused by many of them.

Control over Data. Perhaps one of the most important aspects often overlooked in supporting the SDGs with AI-based systems is the issue of control over the data that is processed. This does not only refer to the training data sets, but also the actual processing data that is analyzed. For example, car companies have been found out to change their motor software in order to detect test settings and adjust the exhaust accordingly. Large companies trying to game such systems needs to be taken into account in many measures, especially regarding climate related SDGs.

Control over Systems. What has been said about the data can also be put forward for the system - the one who controls the system can exert a lot of power over the important topics put forward in the SDGs and the methods used to support them.

Control over Rules. Even more overlooked, control over measures to support SDGs can be achieved quite elegantly and simple by being in charge of making the rules: By being able to specify side or target parameters, big companies might try to seemingly fulfill targets set out through the SGDs, but rather than changing root causes just working around them.

3.4   Transparency

Transparency is a very problematic topic in AI [43], especially due to the problem of providing explainable artificial intelligence for systems exceeding quite a low level of complexity. Thus, while this issue has been the root cause for many of the problems already outlined before, we want to discuss some issues very specific to different notions of transparency in data driven systems, ranging from transparency regarding the internal workings of the AI system to issues of reproducibility.

Functional Testing. Functional testing typically involves testing a system for its proper working, i.e. identifying that all features have been implemented, the correct results are calculated and so forth. This is typically done by providing test cases, but also incorporates more advanced techniques like fuzzying or combinatorial testing [55]. For an AI system, it might be hard to determine the actual test cases, i.e. it might not be simple to define the correct function set of the system, especially when thinking of systems in the area of decision support: Did the IDS not report because of an active decision, or because it simply did not call the respective analysis routines at all (a slightly exaggerated example).

Process Transparency. When using data in cascades of intelligent systems, and especially when training neural networks with said data, transparency becomes rather difficult, especially being able to answer questions on the actual sources of information particles that were later on aggregated. This can be a very problematic attack vector for processes that aggregate data from sources of different sensibility [12], be it patient data or vital information on secret business processes of a company. Thus, in order to mitigate a large amount of attacks, transparency over the whole process chain needs to be provided. In case of utilizing sensible personal information, this is also a requirement derived from the GDPR, but related issues can also be encountered in other applications domain, especially within (military) intelligence. In case of AI-supported SDGs, such information can e.g. incorporate sensitive internal technical details of machines, where a successful attack going for extraction of (parts of) this information can cause great damage to the original data owner. Process transparency is also vital in case of re-processing of data.

Reproducibility. In many cases it can be important to exactly reproduce a result (i) in order to proof that it was actually produced the first time or (ii) in order to learn from the calculation process and maybe challenge and adapt it:

- Reproducing a state of knowledge of the human decision maker: AI systems are currently often seen in a supporting role for human decision makers, i.e. the AI analysis the data for patterns and provides a human with the results who is then in charge of the decision. Especially in sensitive and time critical environments, the human decision maker has to take a decision under a lot of stress based on incomplete information. If the decision was wrong, amply time will be dedicated to the subsequent blame game. Thus, it is vital for the human decision maker to be able to reproduce the exact state of knowledge at the point of decision making [41]. This is also very important in order to learn from mistakes and improve on the decision making process as a whole, including the human, as well as the AI component.
- Re-processing data: In many applications, data needs to be re-processed, i.e. the analytical workflows have to be redone on data that has already been processed once. This can have implications in case the process data is fed back into the model, as it would increase the impact of re-processed data, since it would be included into the model again as often as it is re-processed.
- Post-processing data: Sometimes time-sensitive data (e.g. call detail records in telecommunications) arrives late in the analysis process, but still needs to be processed as if it had arrived on the correct time. This is very difficult with respect to the versioning of models and enrichment data, especially in case of feeding back results into the model. The difference between post-processing and re-processing is that re-processing uses the models and enrichment data current at the time of re-processing, while in post-processing the original state at the time the processing should have originally happened needs to be provided.

There are a set of problems surrounding the topic of reproducibility, with the following being most important from our perspective with respect to security:

- Changes in the model: Especially in algorithms that continuously change the model (e.g. self learning algorithms [4]), it can be very hard to (i) track the impact of changes to the model on the decision making process and (ii) go back in time for post-processing. Here, a very fine-granular and still manageable solution for model versioning needs to be provided.

- Heuristic approaches: In case of e.g. random values introduced into the algorithm, running the same algorithm on the same data set using the same model and enrichment information can (and typically will) result in differences in the results. Thus, in order to provide a high level of reproducibility and transparency, all internal values need to be logged in order to be able to redo the whole process.

- Volatile enrichment data: Also enrichment information can be tricky, especially when it is not hosted by the AI system but externally and only invoked through limited interfaces. Managing and versioning this information is vital in order to provide a decent level of reproducibility in many data driven systems.

Deletion and Rectification. Sometimes it becomes necessary to delete or rectify data inside a workflow. Reasons can be different, but especially within the legal domain of the GDPR, persons have the right to revoke their consent to voluntary data processing and having their data deleted from the databases. Changing data in AI processes can be hard, especially when the information had been used in order to train a neural network [7]. While removing the deleted data from the actual trained network might not be required from a legal point of view [76], it can become important in cases where the data is wrong and has an impact on the decision making process, e.g. by introducing a class of cases not actually existing, by introducing bias into the model or a backdoor. Thus, mechanisms on a technical and also on an organizational level must be put into place in order to be able to deal with such requests in an ordered and timely manner.

3.5   Other Issues

In this section, we have gathered some other issues that need to be discussed when planning to support a SDG-target with an intelligent system.

Liability. There is still a much debate inside the legal academic world, as to who is going to be held responsible for damages caused by AI systems [59], especially in expert in the loop systems [58]. In the case of using AI for SDG targets, the topics addressed can be very complex (e.g. climate models) when compared to end user apps, with a high impact of the resulting recommendations on society, the economy and other fields directly affecting millions to billions of citizens in the world. Thus, since these big questions ought to be solved on an international level headed by the UN, liability needs to be solved on an international basis too.

Over-Engineering Due to Ubiquitous AI. This is a very new issue that we did not encounter in any literature, still, we believe it is a big issue with respect to security: In many recent technologies it can be seen that the development path leads from the technology being exotic and expensive first to a fast decline of costs, thus making the technology available to virtually anyone at low costs. This can also be seen in the realm of AI, where AI-based applications become increasingly ubiquitous with many new applications targeting ever increasing customer basis. In addition it can be seen that implementations tend to gravitate towards the use of standard technologies and frameworks, i.e. many implementations go back to the same basic technologies. In order for such a basic technology to survive, it must provide the capabilities required by most implementations, else other frameworks will take its place. This means that even for implementations only requiring a small an primitive subset of technique, typically rather powerful frameworks are used. We can also see this trend in the hardware world, where even for very primitive sensors, standard chips are deployed that run a full UNIX with many advanced features. This is rather problematic from a security standpoint. While it can certainly be argued in many other technologies that using the same fundamental frameworks is in contrary beneficial to the overall security of the system due to the high amount of security analysis received by a single framework, powerful systems also allow for sophisticated attack vectors and typically result in bigger attack surfaces [40]. With respect to AI, additionally the explainability problem must be considered. Using powerful off-the-shelf frameworks can thus result in the utilization of very powerful and highly complex systems for very simple tasks, e.g. (as an exaggerated example) using a trained deep neural network instead of a simple rule set in a decision support system. This problem, combined with the trend of providing AI almost anywhere results in huge amounts of (critical) systems that can only be tested for security at a very high price, thus introducing a huge uncharted attack surface.

4 Conclusion

Artificial intelligence permeates almost all areas of life and work. In this paper, we have developed a brief overview on the topic of supporting the targets of the UN Sustainable Development Goals (SDGs) from a security, safety, and privacy perspective. To this end, we have identified potential problems and threats that AI-based systems are causing now and will cause in the future - in particular, novel threats that are not even thought of in the initial euphoria of planning, developing, or even deploying AI. We discuss this using two selected application areas, Smart Agriculture and Smart Health, both of which are of eminent importance to each and everyone of us. These findings are for scientists, developers and policy makers when considering the impact such solutions to the SDGs will have on industry and society. This inevitably leads to many conflicting interests and strong attacker motivation by powerful entities. Moreover, the goal of this paper is to provide starting points for future work. While we strongly believe that artificial intelligence will play an important role in supporting the goals articulated by the SDGs, implementation must be done carefully to reduce collateral damage and/or not to undermine the original intent by creating tools to the detriment of the supported goals.

Acknowledgements. The authors declare that there are no conflict of interests. This work does not raise any ethical issues. Parts of this work have been funded by the Austrian Science Fund (FWF), Project: P-32554, and the Austrian Research Promotion Agency through grant 866880 (Big Data Analytics).

References

1. Van der Aalst, W.M., Bichler, M., Heinzl, A.: Robotic process automation. Bus. Inf. Syst. Eng. **60**, 269–272 (2018). https://doi.org/10.1007/s12599-018-0542-4
2. Abràmoff, M.D., Lavin, P.T., Birch, M., Shah, N., Folk, J.C.: Pivotal trial of an autonomous ai-based diagnostic system for detection of diabetic retinopathy in primary care offices. NPJ Digit. Med. **1**(1), 1–8 (2018). https://doi.org/10.1038/s41746-018-0040-6
3. Asadikia, A., Rajabifard, A., Kalantari, M.: Systematic prioritisation of SDGs: Machine learning approach. World Dev. **140**, 105269 (2021)
4. Auer, P., Cesa-Bianchi, N., Gentile, C.: Adaptive and self-confident on-line learning algorithms. Journal of Computer and System Sciences 64(1), 48–75 (2002)
5. Azeroual, O., Saake, G., Abuosba, M.: Data quality measures and data cleansing for research information systems. arXiv preprint arXiv:1901.06208 (2019)
6. Bastin, J.-F., et al.: The global tree restoration potential. Sci. **365**(6448), 76–79 (2019). https://doi.org/10.1126/science.aax0848
7. Baumhauer, T., Schöttle, P., Zeppelzauer, M.: Machine unlearning: linear filtration for logit-based classifiers. arXiv preprint arXiv:2002.02730 (2020)
8. Bennaceur, A., et al.: Modelling and analysing resilient cyber-physical systems. In: IEEE/ACM 14th International Symposium on Software Engineering for Adaptive and Self-Managing Systems (SEAMS). IEEE (2019)
9. Boneh, D., Shaw, J.: Collusion-secure fingerprinting for digital data. IEEE Transactions on Information Theory 44(5), 1897–1905 (1998)
10. Bozarth, L., Saraf, A., Budak, C.: Higher ground? How groundtruth labeling impacts our understanding of fake news about the 2016 us presidential nominees. In: Proceedings of the International AAAI Conference on Web and Social Media, vol. 14, pp. 48–59 (2020)
11. Brantingham, P.J., Valasik, M., Mohler, G.O.: Does predictive policing lead to biased arrests? results from a randomized controlled trial. Stat. Public Policy **5**(1), 1–6 (2018)
12. Böhm, C., et al.: GovWILD: integrating open government data for transparency. In: Proceedings of the 21st International Conference on World Wide Web, pp. 321–324 (2012)
13. Bundtland, G.H.: Report of the World Commission on Environment and Development: Our common future. Uni. Nations Gen. Assembly Doc. A **42**(427), 1–300 (1987)
14. Chen, R., Snyder, M.: Promise of personalized omics to precision medicine. Wiley Interdisciplinary Reviews: Systems Biology and Medicine **5**(1), 73–82 (2013). https://doi.org/10.1002/wsbm.1198
15. Du, P., Sun, Z., Chen, H., Cho, J.H., Xu, S.: Statistical estimation of malware detection metrics in the absence of ground truth. IEEE Trans. Inf. Forensics Secur. **13**(12), 2965–2980 (2018)
16. ElMassah, S., Mohieldin, M.: Digital transformation and localizing the sustainable development goals (sdgs). Ecol. Econ. **169**, 106490 (2020). https://doi.org/10.1016/j.ecolecon.2019.106490

17. Eyhorn, F., Muller, A., Reganold, J.P., Frison, E., Herren, H.R., Luttikholt, L., Mueller, A., Sanders, J., Scialabba, N.E.H., Seufert, V.: Sustainability in global agriculture driven by organic farming. Nature Sustainability **2**(4), 253–255 (2019). https://doi.org/10.1038/s41893-019-0266-6

18. Fawcett, T., Provost, F.J.: Combining data mining and machine learning for effective user profiling. In: KDD, pp. 8–13 (1996)

19. Floreano, D., Wood, R.J.: Science, technology and the future of small autonomous drones. Nature **521**(7553), 460–466 (2015). https://doi.org/10.1038/nature14542

20. Giannelli, P.C.: Chain of custody and the handling of real evidence. Am. Crim. L. Rev. 20, 527 (1982)

21. Goebel, Randy, Chander, Ajay, Holzinger, Katharina, Lecue, Freddy, Akata, Zeynep, Stumpf, Simone, Kieseberg, Peter, Holzinger, Andreas: Explainable AI: the new 42? In: Holzinger, Andreas, Kieseberg, Peter, Tjoa, A Min, Weippl, Edgar (eds.) CD-MAKE 2018. LNCS, vol. 11015, pp. 295–303. Springer, Cham (2018). https://doi.org/10.1007/978-3-319-99740-7_21

22. Gu, T., Liu, K., Dolan-Gavitt, B., Garg, S.: Badnets: Evaluating backdooring attacks on deep neural networks. IEEE Access **7**, 47230–47244 (2019)

23. Hamburg, M.A., Collins, F.S.: The path to personalized medicine. New England Journal of Medicine 363(4), 301–304 (2010). doi: 10.1056/NEJMp1006304

24. Hasenauer, H.E.: Sustainable forest management: growth models for Europe. Springer, Heidelberg (2006)

25. Herweijer, C., Waughray, D.: Harnessing Artificial Intelligence for the Earth. Fourth Industrial Revolution for the Earth Series. World Economic Forum (January 2018), p. 52 (2018)

26. Holzinger, A.: Trends in interactive knowledge discovery for personalized medicine: Cognitive science meets machine learning. IEEE Intelligent Informatics Bulletin **15**(1), 6–14 (2014)

27. Holzinger, A.: Interactive machine learning for health informatics: When do we need the human-in-the-loop? Brain Inf. **3**(2), 119–131 (2016). https://doi.org/10.1007/s40708-016-0042-6

28. Holzinger, A., et al.: Towards the augmented pathologist: challenges of explainable-AI in digital pathology. arXiv:1712.06657 (2017)

29. Holzinger, A., Malle, B., Saranti, A., Pfeifer, B.: Towards multi-modal causability with graph neural networks enabling information fusion for explainable ai. Information Fusion **71**(7), 28–37 (2021). https://doi.org/10.1016/j.inffus.2021.01.008

30. Holzinger, A., Röcker, C., Ziefle, M.: From smart health to smart hospitals. In: Smart Health: State-of-the-Art and Beyond. Springer Lecture Notes in Computer Science, LNCS 8700, pp. 1–20. Springer, Heidelberg, Berlin (2015)

31. Holzinger, K., Mak, K., Kieseberg, P., Holzinger, A.: Can we trust machine learning results? Artificial intelligence in safety-critical decision support. Ercim News **2018**, 42-43 (2018)

32. Huang, L., Joseph, A.D., Nelson, B., Rubinstein, B.I., Tygar, J.D.: Adversarial machine learning. In: Proceedings of the 4th ACM Workshop on Security and Artificial Intelligence, pp. 43–58 (2011)

33. Jain, A., et al.: Overview and importance of data quality for machine learning tasks. In: Proceedings of the 26th ACM SIGKDD International Conference on Knowledge Discovery & Data Mining, pp. 3561–3562 (2020)

34. Jiang, X., Yu, F.R., Song, T., Ma, Z., Song, Y., Zhu, D.: Blockchain-enabled cross-domain object detection for autonomous driving: A model sharing approach. IEEE Internet Things J. **7**(5), 3681–3692 (2020)

35. Johnson, K.B., Wei, W., Weeraratne, D., Frisse, M.E., Misulis, K., Rhee, K., Zhao, J., Snowdon, J.L.: Precision medicine, AI, and the future of personalized health care. Clin. Transl. Sci. **14**(1), 86–93 (2021). https://doi.org/10.1111/cts.12884

36. Khan, W.U., Ye, Z., Altaf, F., Chaudhary, N.I., Raja, M.A.Z.: A novel application of fireworks heuristic paradigms for reliable treatment of nonlinear active noise control. Appl. Acoust. **146**, 246–260 (2019)

37. Kieseberg, P., Hobel, H., Schrittwieser, S., Weippl, E., Holzinger, A.: Protecting anonymity in data-driven biomedical science. In: Interactive knowledge discovery and data mining in biomedical informatics, pp. 301–316. Springer (2014)

38. Kieseberg, P., Schantl, J., Frühwirt, P., Weippl, E., Holzinger, A.: Witnesses for the doctor in the loop. In: International Conference on Brain Informatics and Health. pp. 369–378. Springer (2015)

39. Kieseberg, P., Schrittwieser, S., Mulazzani, M., Echizen, I., Weippl, E.R.: An algorithm for collusion-resistant anonymization and fingerprinting of sensitive microdata. Electron. Mark. **24**(2), 113–124 (2014)

40. Kieseberg, P., Weippl, E.R.: Security challenges in cyber-physical production systems. In: International Conference on Software Quality, pp. 3–16 (2018)

41. Kieseberg, P., Weippl, E.R., Holzinger, A.: Trust for the doctor-in-the-loop. Ercim News **2016**(1), 32–33 (2016)

42. Kurita, K., Michel, P., Neubig, G.: Weight poisoning attacks on pre-trained models. arXiv preprint arXiv:2004.06660 (2020)

43. Larsson, S., Heintz, F.: Transparency in artificial intelligence. Internet Policy Review **9**(2), 1–16 (2020)

44. Levinson, J., et al.: Towards fully autonomous driving: systems and algorithms. In: 2011 IEEE Intelligent Vehicles Symposium (IV). IEEE (2011). https://doi.org/10.1109/IVS.2011.5940562

45. Li, J.H.: Cyber security meets artificial intelligence: a survey. Front. Inf. Technol. **19**(12), 1462–1474 (2018)

46. Li, Q., Guo, Y., Chen, H.: Practical no-box adversarial attacks against DNNs. arXiv preprint arXiv:2012.02525 (2020)

47. Linkov, I., Kott, A.: Fundamental concepts of cyber resilience: Introduction and overview. In: Cyber resilience of systems and networks, pp. 1–25. Springer (2019)

48. Liu, C., Xiong, H., Papadimitriou, S., Ge, Y., Xiao, K.: A proactive workflow model for healthcare operation and management. IEEE transactions on knowledge and data engineering **29**(3), 586–598 (2016). doi: 10.1109/TKDE.2016.2631537

49. Liu, T., et al.: Unmanned aerial vehicle and artificial intelligence revolutionizing efficient and precision sustainable forest management. J. Clean. Prod. 127546 (2021). https://doi.org/10.1016/j.jclepro.2021.127546

50. Maletic, J.I., Marcus, A.: Data cleansing: A prelude to knowledge discovery. In: Maimon, O., Rokach, L. (eds.) Data Mining and Knowledge Discovery Handbook, pp. 19–32. Springer, Heidelberg (2009). https://doi.org/10.1007/978-0-387-09823-4_2

51. Malle, B., Kieseberg, P., Holzinger, A.: Do not disturb? classifier behavior on perturbed datasets. In: International Cross-Domain Conference for Machine Learning and Knowledge Extraction. pp. 155–173. Springer (2017)

52. Mehrabi, N., Morstatter, F., Saxena, N., Lerman, K., Galstyan, A.: A survey on bias and fairness in machine learning. arXiv preprint arXiv:1908.09635 (2019)

53. Mueller, H., Mayrhofer, M.T., Veen, E.-B.V., Holzinger, A.: The ten commandments of ethical medical AI. IEEE COMPUT. **54**(7), 119–123 (2021). https://doi.org/10.1109/MC.2021.3074263

54. Nebeker, C., Torous, J., Ellis, R.J.B.: Building the case for actionable ethics in digital health research supported by artificial intelligence. BMC med. **17**(1), 1–7 (2019). https://doi.org/10.1186/s12916-019-1377-7

55. Nie, C., Leung, H.: A survey of combinatorial testing. ACM Computing Surveys (CSUR) **43**(2), 1–29 (2011)

56. Nischelwitzer, A., Lenz, F.J., Searle, G., Holzinger, A.: Some aspects of the development of low-cost augmented reality learning environments as examples for future interfaces in technology enhanced learning. In: Stephanidis, C. (ed.) Universal Access to Applications and Services. Lecture Notes in Computer Science (LNCS, vol. 4556), pp. 728–737. Springer, Berlin, Heidelberg, New York (2007)

57. O'Donnell, R.M.: Challenging racist predictive policing algorithms under the equal protection clause. NYUL Rev. **94**, 544 (2019)

58. Price, W.N., Gerke, S., Cohen, I.G.: Potential liability for physicians using artificial intelligence. JAMA 322(18), 1765–1766 (2019)

59. Reed, C., Kennedy, E.J., Silva, S.N.: Responsibility, autonomy and accountability: Legal liability for machine learning. Soc. Sci. Res. Netw. **243**, 1–31 (2016)

60. Regulation, G.D.P.: Regulation EU 2016/679 of the european parliament and of the council of 27 April 2016. Off. J. Eur. Union (2016)

61. Robert, K.W., Parris, T.M., Leiserowitz, A.A.: What is sustainable development? Goals, indicators, values, and practice. Environ. Sci. Policy Sustain. Dev. 47(3), 8–21 (2005)

62. Salem, A., Wen, R., Backes, M., Ma, S., Zhang, Y.: Dynamic backdoor attacks against machine learning models. arXiv preprint arXiv:2003.03675 (2020)

63. Schlosser, P., Pfirman, S.: Earth science for sustainability. Nature Geoscience 5(9), 587–588 (2012). doi: 10.1038/ngeo1567

64. Schneeberger, D., Stoeger, K., Holzinger, A.: The european legal framework for medical ai. In: International Cross-Domain Conference for Machine Learning and Knowledge Extraction, Springer LNCS 12279, pp. 209–226. Springer, Cham (2020). DOI: https://doi.org/10.1007/978-3-030-57321-8-12

65. SDG, U.: Sustainable development goals (2018)

66. Shao, G., Reynolds, K.M., Shao, G.: Computer applications in sustainable forest management. Springer, London (2006)

67. Shapiro, A.: Reform predictive policing. Nat. news **541**(7638), 458 (2017)

68. Silva, S., Duarte, D., Valente, A., Soares, S., Soares, J., Pinto, F.C.: Augmented intelligent distributed sensing system model for precision agriculture. In: 2021 Telecoms Conference (ConfTELE). IEEE (2021). https://doi.org/10.1109/ConfTELE50222.2021.9435498

69. Singh, D., Merdivan, E., Hanke, S., Kropf, J., Geist, M., Holzinger, A.: Convolutional and recurrent neural networks for activity recognition in smart environment. In: Holzinger, A., Goebel, R., Ferri, M., Palade, V. (eds.) Towards Integrative Machine Learning and Knowledge Extraction: BIRS Workshop, Banff, AB, Canada, July 24–26, 2015, Revised Selected Papers, pp. 194–205. Springer International Publishing, Cham (2017)

70. Stafford-Smith, M., Griggs, D., Gaffney, O., Ullah, F., Reyers, B., Kanie, N., Stigson, B., Shrivastava, P., Leach, M., O'Connell, D.: Integration: the key to implementing the sustainable development goals. Sustain. Sci. **12**(6), 911–919 (2017). https://doi.org/10.1007/s11625-016-0383-3

71. Tang, D.: What is digital transformation? EDPACS - The EDP Audit, Control, and Security Newsletter **64**(1), 9–13 (2021). https://doi.org/10.1080/07366981.2020.1847813

72. Tjoa, S., Buttinger, C., Holzinger, K., Kieseberg, P.: Penetration testing artificial intelligence. ERCIM News **2020**(123), 36–37 (2020)

73. Vaio, A.D., Palladino, R., Hassan, R., Escobar, O.: Artificial intelligence and business models in the sustainable development goals perspective: a systematic literature review. Journal of Business Research **121**, 283–314 (2020)

74. Vaseghi, S.V.: Advanced digital signal processing and noise reduction. Wiley, Hoboken (2008)

75. Verhoef, P.C., Broekhuizen, T., Bart, Y., Bhattacharya, A., Dong, J.Q., Fabian, N., Haenlein, M.: Digital transformation: a multidisciplinary reflection and research agenda. J. Bus. Res. **122**, 889–901 (2021). https://doi.org/10.1016/j.jbusres.2019.09.022

76. Villaronga, E.F., Kieseberg, P., Li, T.: Humans forget, machines remember: Artificial intelligence and the right to be forgotten. Computer Law & Security Review **34**(2), 304–313 (2017)

77. Vinuesa, R., et al.: The role of artificial intelligence in achieving the sustainable development goals. Nat. Commun. **11**(1), 1–10 (2020)

78. Wing, J.M.: Trustworthy AI. arXiv preprint arXiv:2002.06276 (2020)

79. Wolfert, S., Ge, L., Verdouw, C., Bogaardt, M.J.: Big data in smart farming-a review. Agricultural systems 153, 69–80 (2017)

80. Zhao, S., Talasila, M., Jacobson, G., Borcea, C., Aftab, S.A., Murray, J.F.: Packaging and sharing machine learning models via the acumos AI open platform. In: 2018 17th IEEE International Conference on Machine Learning and Applications (ICMLA), pp. 841–846. IEEE (2018)

81. Zhou, X., et al.: A secure and privacy-preserving machine learning model sharing scheme for edge-enabled iot. IEEE Access **9**, 17256–17265 (2021)

When in Doubt, Ask: Generating Answerable and Unanswerable Questions, Unsupervised

Liubov Nikolenko and Pouya Rezazadeh Kalehbasti[✉]

Stanford University, Stanford, CA 94305, USA
liubov@alumni.stanford.edu, pouyar@stanford.edu

Abstract. Question Answering (QA) is key for making possible a robust communication between human and machine. Modern language models used for QA have surpassed the human-performance; however, these models require large amounts of human-generated training data which are costly and time-consuming to create. This paper studies augmenting human-made datasets with synthetic data as a way of surmounting this problem. A state-of-the-art model based on deep transformers is used to inspect the impact of using synthetic answerable and unanswerable questions to complement a well-known human-made dataset. The results indicate a tangible improvement in the performance of the language model (measured in terms of F1 and EM scores) trained on the mixed dataset. Specifically, unanswerable question-answers prove more effective in boosting the model: the F1 score gain from adding to the original dataset the answerable, unanswerable, and combined question-answers were 1.3%, 5.0%, and 6.7%, respectively.
Repository of this paper: github.com/lnikolenko/EQA.

Keywords: Extractive question answering · Unsupervised learning · Natural language processing

1 Introduction

1.1 Problem Statement

Question Answering (QA) is essential to enabling effective communication between humans and machines. As a computer science discipline, it falls under information retrieval and natural language processing (NLP), and it concerns building machines able to answer questions asked by humans in natural languages [5]. Recent years have seen significant progress in Question Answering owing to novel comprehensive public datasets, e.g. SQuAD [19], TriviaQA [10], HOTPOTQA [25], and modern deep-learning models, most notably *BERT* [6]. As an instance of these advancements in Extractive Question Answering (EQA), state-of-the-art models trained on SQuAD dataset have surpassed human performance [6], while BERT model has a performance on par with human's on the updated version of the dataset, SQuAD 2.0 [6].

© IFIP International Federation for Information Processing 2021
Published by Springer Nature Switzerland AG 2021
A. Holzinger et al. (Eds.): CD-MAKE 2021, LNCS 12844, pp. 21–33, 2021.
https://doi.org/10.1007/978-3-030-84060-0_2

Yet, the mentioned achievements come at a price: *massive human-made training datasets*. Generating this massive training data is typically crowd-sourced, and the process takes considerable time and resources [15]. Further, training models on these large datasets is time-consuming and computationally expensive. The problem exacerbates when models are trained on languages other than English which is well-researched and has an abundance of training data available for different tasks. To tackle these challenges, some researchers have tried to create more effective models which can perform better than current models on existing datasets, while others have developed more complex datasets or devised methods of synthesizing training data to get better results from current EQA models. The following section briefly reviews some of these efforts during the past few years.

1.2 Prior Research

Researchers have adopted three major approaches for creating more effective QA systems: (1) create more effective models to better leverage existing datasets, (2) generate more actual training/test data using crowd-sourcing or (3) generate synthetic training/test data to improve the performance of existing models. These approaches are briefly explored below.

Model Development. Qi et al. [17] focus on the task of QA across multiple documents which requires multi-hop reasoning. Their main hypothesis is that the current QA models are too expensive to scale up efficiently to open-domain QA queries, so they create a new QA model called GOLDEN (Gold Entity) Retriever, able to perform iterative-reasoning-and-retrieval for open-domain multi-hop question answering. They train and test the proposed model on HOTPOTQA multi-hop dataset. One highlight of Qi et al. model is that they avoid computationally demanding neural models, such as BERT, and instead use off-the-shelf information retrieval systems to look for missing entities. They show that the proposed QA model outperforms several state-of-the-art QA models on HOTPOTQA test-set.

In another work, Wang et al. [21] develop an open-domain QA system, called R^3, with two innovative features in its question-answering pipeline: a ranker to rank the retrieved passages (based on the likelihood of retrieving the ground-truth answer to a query), and a reader to extract answers from the ranked passages using Reinforcement Learning (RL). Modern deep learning models for open-domain QA use large text corpora as training sets, and use a two-step process to answer questions: (1) Information Retrieval (IR) to select the relevant passages, and (2) Reading Comprehension (RC) to select candidate phrases containing the answer [2,7]. The model proposed in this paper follows this same structure: Ranker module acts as the IR while Reader module acts as the RC. Wang et al. use SGD/backprop to train their Reader and to maximize the probability that the selected span contains the potential answer to the query. They train the Ranker using REINFORCE [23] RL algorithm with a reward function evaluating the quality of the answers extracted from the passages Ranker sends to Reader. They show that this configuration is robust against semantical differences between the question and the passage.

In another study, Lee et al. [13] implemented a recurrent network (called RASOR) on SQuAD dataset for question answering, which resulted in a model with higher EM and F1 score compared to the most successful models up to the date [Match-LSTM]. In analysis, Lee et al. state that a recurrent net enables sharing computation for shared substructures across candidate spans for answering the asked question, and this has resulted in the improved performance of their model compared to the baseline models studied in the paper.

Actual Data Generation. Lewis et al. [14] took on the challenge of 'cross-lingual EQA' by developing a multi-lingual benchmark dataset, called *MLQA*. This dataset covered seven languages including English and Vietnamese with more than 12k instances in English and 5k in the other six languages. They also managed to make each instance included in the benchmark to be paralleled across at least four of their chosen languages. Lewis et al. aimed to reduce the overfit observed in cross-lingual QA models. As their baseline models, Lewis et al. used BERT and XLM models [14]. The dataset they developed only included development and testing set, so for training baseline models, they used the SQuAD v1.1 dataset. Using their test/dev dataset, Lewis et al. finally showed that the transfer results for state-of-the-art models (in terms of EM and F1 score) largely lag behind the training results; hence, more work is required to reduce the variance of high-performance models in EQA.

In a recent paper, Reddy et al. [20] develop a dataset focused on Conversational Question-Answering, called CoQA. They hypothesize that machine QA systems should be able to answer questions asked based on conversations, as humans can do. Their dataset includes 127k question-answer pairs from 8k conversation passages across 7 distinct domains. Reddy et al. show that the state-of-the-art language models (including Augmented DrQA and DrQA+PGNet) are only able to secure an F1 score of 65.4% on CoQA dataset, falling short of the human-performance by more than 20 points. The results of their work shows a huge potential for further research on conversational question answering which is key for natural human-machine communication. Previously, Choi et al. [4] had conducted a similar study on conversational question answering, and using high-performance language models, they obtained an F1 score 20 points less than that of humans on their proposed dataset, called QuAC.

In another work, Rajpurkar et al. [18] focus on augmenting the existing QA datasets with unanswerable questions. They hypothesize that the existing QA models get trained only on answerable questions and easy-to-recognize unanswerable questions. To make QA models robust against unanswerable questions, they augment SQuAD dataset with 50k+ unanswerable questions generated through crowd-sourcing. They observe that the strongest existing language models struggle to achieve an F1 score of 66% on their proposed update to SQuAD dataset (called SQuAD 2.0), while achieving an F1 score of 86% on the initial version of the dataset. Rajpurkar et al. state that this newly developed dataset may spur research in QA on stronger models which are robust against unanswerable questions.

Synthetic Data Generation. Lewis et al. [15] take on the challenge of expensive data-generation for Question Answering task by generating data and training QA models on synthetic datasets. They propose an unsupervised model for question-generating which powers the training process for an EQA model. Lewis et al. aim to make possible training effective EQA models with scarce or lacking training data, especially in non-English contexts. Their question-generation framework generates training data from Wikipedia excerpts. Training data in this work is generated as follows:

1. A paragraph is sampled from English Wikipedia
2. A set of candidate answers within that context get sampled using pre-trained models, such as Named-Entity Recognition (NER) or Noun-Chunkers, to identify such candidates
3. Given a candidate answer and context, "fill-in-the-blank" cloze questions are extracted
4. Cloze questions are converted into natural questions using an unsupervised cloze-to-natural-question translator.

The generated data is then supplied to question-answering model as training data. BERT-LARGE model trained on this data can achieve 56.4% F1 score, largely outperforming other unsupervised approaches. Before this paper, (i) generating training data for SQuAD question-answering and (ii) using unsupervised methods [instead of supervised methods] to generate training data directly on question-answering task were not explored as thoroughly.

In a similar work, Zhu et al. [27] propose a model to automatically generate unanswerable questions based on paragraph-answerable-question pairs for the task of machine reading comprehension. They use this model to augment SQuAD 2.0 dataset and achieve improved F1 scores, compared to the non-augmented dataset, using two state-of-the-art QA models. To create the model for generating unanswerable-questions, Zhu et al. adopt a pair-to-sequence architecture which they show outperforms models with a typical sequence-to-sequence question-generating architecture.

In an earlier work from 2017, Duan et al. [8] propose a question-generator which can use two approaches for generating questions from a given passage (in particular, Community Question Answering websites): (1) a Convolutional Neural Network model for a retrieval-based approach, and (2) a Recurrent Neural Network model for a generation-based approach. They show that the questions synthesized by their model (based on data from YahooAnswers) can outperform the existing generation systems (based on BLEU metric), and it can augment several existing datasets, including SQuAD and WikiQA, for training better language models.

1.3 Objective and Contributions

This paper hypothesizes that for the task of Question Answering (QA), augmenting real data with synthesized data can help train models with a better performance compared to models trained only on real data. This work validates this on the task of Extractive Question Answering (EQA) using BERT language model [6] trained on different combinations of real and artificial data, based on SQuAD 2.0 [18] dataset (as the source of real data) and machine-generated answerable and unanswerable question-answer pairs (as the source of synthetic data). We will use F1 and Exact Match (EM) metrics to measure the performance of the developed models. We use an unsupervised generator-discriminator model based on cloze translation to generate answerable questions, following the work by Lewis et al. [15], and then alter the model to enable it to generate unanswerable questions. We expect the language model trained on augmented data to outperform the model trained on vanilla real data. We also expect models trained on synthetic data composed of both ANS and UNANS questions to yield better results than those trained on synthetic data composed of only ANS or only UNANS questions.

2 Method

2.1 Proposed Model

BERT model trained on 20% of SQuAD 2.0 dataset will act as our baseline model. Improved models will be created by training BERT model on SQuAD 2.0 augmented with (1) answerable questions (ANS) from the work by Lewis et al. [15], (2) UNANS questions (UNANS) generated by the authors of this paper, and (3) a mixture of ANS and UNANS questions. Section 3 provides more details on the experiment designs. The following paragraphs describe the models used to generate the ANS and UNANS datasets.

The model generating synthetic answerable questions was developed by Lewis et al. [15]. It takes as its input a paragraph from English Wikipedia, and uses a Named Entity Recoginition (NER) system - a system that locates and classifies named entities in the text into predefined categories such as people, date, location, organization, etc. - to identify a set of potential answers which it then uniformly samples from. Next, an answer a is generated by identifying a subclause around the named entity using an English syntactic parser. To generate the maximum likelihood question $p(q|a,c)$ from the context c (the paragraph) and answer a, the model produces a cloze statement – i.e. a statement with a masked answer – from the identified sub-clause. An example would be "I ate at McDonald's" which maps to "I ate at [MASK]". Then the system uses unsupervised Neural Machine Translation (NMT) [12] to translate the cloze question into a natural question, and it finally outputs the generated question-answer pair. Figure 1 provides an illustration of question generation pipeline.

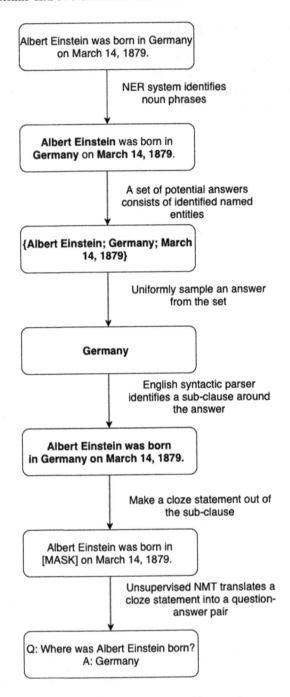

Fig. 1. A pipeline for generating answerable question-answer pairs

We plan to enhance Lewis et al.'s model by enabling it to generate both ANS and UNANS questions. To do this, we will first refactor the model as follows:

1. Model will produce outputs in the same form as SQuAD 2.0 dataset
2. Instead of treating each Wikipedia paragraph as a standalone article, the model will preserve information regarding a paragraph or multiple paragraphs belonging to an article.

To produce UNANS questions we will run the modified model and get ANS synthetic questions. ANS synthetic questions will consist of (paragraph, question, answer) tuples, grouped by article. Afterwards we will remove the generated answers from the output, split each paragraph from its corresponding question in the tuple and randomly pair the question with another paragraph within the same article. This ensures that the questions are indeed unanswerable, since they will be detached from their original context, while staying relevant to the original paragraphs. Sustaining this relevance also helps make the unanswerable questions resilient against word-overlap heuristic [26] because the paragraphs will belong to the same article.

At the end, we will evaluate how well the synthetic training examples complement the SQuAD 2.0 human-labeled data: We will use EM and F1 scores to assess the performance of BERT model (implemented by HuggingFace[1]) on EQA among models trained only on human-generated data and models trained on human-generated data combined with the two sets of synthetic datasets, i.e. ANS and UNANS examples. The GitHub repository of this work contains the entire modeling pipeline required for replicating the results.

2.2 Dataset

In this paper, we train the language models for EQA on the renowned Stanford Question Answering Dataset (SQuAD) 2.0 [18]. This dataset is an updated version of SQuAD 1.0 [19] which was a reading comprehension dataset comprised of 100k+ questions built around Wikipedia articles. SQuAD 2.0 was created by adding 50k crowd-sourced (adversarial) unanswerable questions to the initial dataset.

As the source of answerable synthetic questions, we use the dataset generated by Lewis et al. [15][2]. See Fig. 2 for a synthetic question-answer example. The dataset contains 3.9M answerable question-answer pairs created using a cloze-translating generator. This data is generated in SQuAD 1.0 standard format: we will convert the data into SQuAD 2.0 format to be able to merge it with human-generated question-answer pairs from SQuAD 2.0 dataset. The dataset Lewis et al. [15] generated with their model includes *only* answerable questions. To generate the required unanswerable data, we modify their data-generation pipeline. We have used pre-processed Wikipedia dump[3] as an input

---

[1] https://huggingface.co/ (Last access: 07/01/2020).
[2] github.com/facebookresearch/UnsupervisedQA (Last access: 07/10/2020).
[3] https://dumps.wikimedia.org/ (Last access: 07/10/2020).

to the updated/modified question-answer generation model to generate around 80k unanswerable training examples in SQuAD 2.0 format. Figure 3 contains an instance of the generated unanswerable question-answer pair. All the data required to replicate this paper can be obtained and generated using the instructions from the GitHub repository of this paper.

Context: As the "Bad Boys" era was fading, they were eliminated in five games in the first round of the playoffs by the New York Knicks. The Pistons would not return to the playoffs until 1996. Following the season, Chuck Daly left to coach the New Jersey Nets, and John Salley was traded to the Miami Heat. Meanwhile, the Bulls-Pistons rivalry took another ugly turn as Thomas was left off the Dream Team coached by Daly, reportedly at the request of Michael Jordan.
Question: Who left to coach the New Jersey Nets ?
Answer: Chuck Daly

Fig. 2. An example of a synthetic answerable question-answer pair.

Context: A fiscal deficit is often funded by issuing bonds, such as Treasury bills or consols and gilt-edged securities. These pay interest, either for a fixed period or indefinitely. If the interest and capital requirements are too large, a nation may default on its debts, usually to foreign creditors. Public debt or borrowing refers to the government borrowing from the public.
Question: Who can argue that fiscal policy can still be effective , especially in a liquidity trap where , they argue , crowding out is minimal ?
Answer: N/A

Fig. 3. An example of a synthetic unanswerable question-answer pair.

2.3   Evaluation Metrics

We will use macro-averaged F1 score and EM to evaluate the performance of the models trained in this work. F1 score shows the precision and recall for the words selected as part of the answer actually being part of the correct answer. We first compute the F1 score of the model's predictions against the ground-truth answer represented as bags of tokens, then take the maximum F1 score across all possible answers for a given question, and finally average over all of the questions. EM, on the other hand, indicates the number of exactly correct answers with the same start and end indices.

3 Experiments

We expect to achieve more reliable models for the task of EQA when augmenting actual training data with synthetic data. The synthetic data (question-answers) used for augmenting the actual data in this project has two types: answerable and unanswerable questions. Lewis et al. [15] observed that synthetic answerable questions can boost the performance of QA models when added to actual data from SQuAD 1.1 dataset. Also, Zhu et al. [27] observed that mixing synthetic unanswerable questions derived from human-generated training examples into actual data from SQuAD 2.0 can improve the performance of EQA models. We hence expect that augmenting an actual dataset, *i.e.* SQuAD 2.0 in this work, with a mix of ANS and UNANS can yield an even better performance than using each of them alone to enhance the dataset.

We have devised several experiments to test the mentioned hypothesis with training examples described below:

1. Experiment 0 [Baseline]: Using 26,063 examples from SQuAD 2.0 dataset [the entire dataset was not selected to make the training tractable]
2. Experiment 1-1 [ANS Augmentation]: Using 26,063 examples from SQuAD 2.0, and 391,549 from ANS (from [15])
3. Experiment 1-2 [UNANS Augmentation]: Using 26,063 examples from SQuAD 2.0, and 76,818 from UNANS
4. Experiment 2 [ANS+UNANS Augmentation]: Using 26,063 examples from SQuAD 2.0, 314,731 from ANS, and 76,818 from UNANS

Experiment 0 provides a baseline to compare the other experiment results against. Experiment 1 looks into the impact of exclusive ANS or UNANS data augmentation. Finally, Experiment 2 will show the results of mixing the two approaches of augmentation together.

BERT model adapted to EQA was used to run the mentioned experiments, and the results were evaluated on a set of held-out human-generated data points consisting of 3,618 question-answer pairs. We have tuned the hyper-parameters of the model (number of training epochs, maximum sequence length, etc.) based on our observations from Experiment 0, since it involves a relatively small dataset and is easy to experiment with. We will use these obtained optimal hyper-parameters for the rest of the experiments. During our initial experimentation, we observed that training BERT on the full SQuAD 2.0 dataset takes 9 h on a 1480 MHz 3584 core NVIDIA 1080 TI GPU, so to avoid excessive training times, we decided to use only 20% of the SQuAD dataset and accordingly use a limited portion of the synthetic questions generated by Lewis et al. [15].

Table 1. Results of the three experiments

Experiment	F1 (%)	EM (%)
0	57.61	61.27
1-1	58.90	62.56
1-2	62.56	65.81
2	64.28	66.36

4   Results and Discussion

Table 1 shows the results of the experiments. A few observations can be made based on these results:

- Experiments 1-1 and 1-2 demonstrate that, as expected, adding either ANS or UNANS questions to the human-generated training examples boosts F1 and EM scores of the BERT model for both cases compared to Baseline.
- The results further show that adding the UNANS data to the original dataset (experiment 1-2) has a stronger impact than adding the ANS data (experiment 1-1). Table 2 indicates this point: the normalized impact of adding a single example from the UNANS dataset is almost four-times larger than that of the ANS dataset on the F1 and EM scores compared to the baseline. This can be justified with the following: the original training set has a small portion (only around 1/3) of unanswerable questions, so our synthetic dataset increases the proportion of unanswerable questions and makes the training data more balanced in this regard.
- Finally, the results of experiment 2 show that augmenting the SQUAD 2.0 dataset with both ANS and UNANS at the same time leads to an even greater performance compared to using either of the two datasets to enhance the human-made data, *i.e.* compared to experiments 1-1 and 1-2.

These results confirm our hypothesis mentioned in Sect. 1.3, and show a potential for our novel synthesized unanswerable dataset to further boost the performance of language models similar to BERT for the task of EQA.

Table 2. Comparison between the relative impact of each dataset on the model scores: ANS vs. UNANS

	ANS	UNANS
Gain in F1 (%/example)	0.022	0.086
Gain in EM (%/example)	0.021	0.074

5   Conclusions

This paper studies the impact of augmenting human-made data with synthetic data on the task of Extractive Question Answering by using BERT [6] as the language model and SQuAD 2.0 [18] as the baseline dataset. Two sets of synthetic data are used for augmenting the baseline data: a set of answerable and another set of unanswerable questions-answers. Conducted experiments show that using both these synthetic datasets can tangibly improve the performance of the selected language model for EQA, while the UNANS data, generated by the authors, has a more pronounced impact on improving the performance. Adding the UNANS dataset to the original data yields a gain of 5% in both F1 and EM scores, whereas the ANS dataset yields around a quarter of this gain. Enhancing the original data with a combination of the two synthetic datasets improves the F1 score of BERT on the test-set by 7% and the EM score by 5% which are sizable improvements compared to the performance of the baseline models and similar efforts in the literature. The obtained results indicate the great potential of using synthetic data to complement the costly human-generated datasets: This augmentation can help provide the massive data required for training the modern language models at a very low cost.

6   Limitations

The presented approach has limitations similar to [15]: Although we tried to avoid using any human-labeled data for generating the synthetic question-answers, the question-generating models rely on manually-labeled data from OntoNotes 5 (for NER system) and Penn Treebank (for extracting subclauses). Further, the question-generation pipeline of this work uses English language-specific heuristics. Hence, the applicability of this approach is limited to languages and domains that already have a certain amount of human-labeled data for question generation, and porting this model to another language would require extra preparatory efforts.

An extensive amount of training examples are required to achieve tangible performance gains, and this results in substantial training times and compute costs for both generating synthetic data and training the BERT model. These high training times and resource costs prevented us from performing the experiments on the full SQuAD 2.0 dataset. Nonetheless, given the homogeneity of the original dataset, we expect the synthetic training examples to bring similar performance improvements if added to the full dataset with similar proportions.

7   Future Work

The work presented in this manuscript can be extended in several ways:

- Developing a more sophisticated unsupervised model for unanswerable question generation can be a great extension of this work. Some potential approaches include devising heuristics such as word/synonym overlap for filtering the generated questions and employing the pair-to-sequence model by Zhu et al. [27] on the synthetic training data.

- The computational power available to the authors limited the size of the data used for running the experiments in this work: future efforts can run more extensive experiments to further examine the synthetic data augmentation studied here.
- Breaking down the question types into how, what, where, when, etc. and studying the individual impacts of each question-answer type can also clarify the individual impact of each question type on the performance of the language model. The insights gained from such experiment can help fine-tune the generated data to achieve more effective synthetic datasets.
- Using new language models, e.g. FPNet [24], with the proposed method and using new specialized hardware for training (optimizing) the models, e.g., quantum and digital annealers [11, 16, 22], would also be interesting extensions of this work.
- It would also be interesting to apply the proposed method to other QA tasks such as visual QA [1] and counterfactual QA [3, 9].

Acknowledgments. We would like to thank the CS224N and CS224U course staff from Stanford University, especially Professor Chris Potts, for their guidance and feedback on this research.

Authorship Statements. Liubov implemented unanswerable question generation pipeline and the scripts to process and partition the data. Pouya worked on designing the experiments and composing the paper.

References

1. Antol, S., et al.: VQA: visual question answering. In: Proceedings of the IEEE International Conference on Computer Vision, pp. 2425–2433 (2015)
2. Chen, D., Fisch, A., Weston, J., Bordes, A.: Reading Wikipedia to answer open-domain questions. arXiv preprint arXiv:1704.00051 (2017)
3. Chen, L., Yan, X., Xiao, J., Zhang, H., Pu, S., Zhuang, Y.: Counterfactual samples synthesizing for robust visual question answering. In: Proceedings of the IEEE/CVF Conference on Computer Vision and Pattern Recognition, pp. 10800–10809 (2020)
4. Choi, E., et al.: QuAC: question answering in context. arXiv preprint arXiv:1808.07036 (2018)
5. Cimiano, P., Unger, C., McCrae, J.: Ontology-Based Interpretation of Natural Language. Morgan & Claypool Publishers (2014)
6. Devlin, J., Chang, M.W., Lee, K., Toutanova, K.: BERT: pre-training of deep bidirectional transformers for language understanding. arXiv preprint arXiv:1810.04805 (2018)
7. Dhingra, B., Liu, H., Yang, Z., Cohen, W.W., Salakhutdinov, R.: Gated-attention readers for text comprehension. arXiv preprint arXiv:1606.01549 (2016)
8. Duan, N., Tang, D., Chen, P., Zhou, M.: Question generation for question answering. In: Proceedings of the 2017 Conference on Empirical Methods in Natural Language Processing, pp. 866–874 (2017)

9. Holzinger, A., Malle, B., Saranti, A., Pfeifer, B.: Towards multi-modal causability with graph neural networks enabling information fusion for explainable AI. Inf. Fusion **71**, 28–37 (2021)
10. Joshi, M., Choi, E., Weld, D.S., Zettlemoyer, L.: TriviaQA: a large scale distantly supervised challenge dataset for reading comprehension. arXiv preprint arXiv:1705.03551 (2017)
11. Kalehbasti, P.R., Ushijima-Mwesigwa, H., Mandal, A., Ghosh, I.: Ising-based louvain method: clustering large graphs with specialized hardware. arXiv preprint arXiv:2012.11391 (2020)
12. Lample, G., Ott, M., Conneau, A., Denoyer, L., Ranzato, M.: Phrase-based & neural unsupervised machine translation. In: Proceedings of the 2018 Conference on Empirical Methods in Natural Language Processing (EMNLP) (2018)
13. Lee, K., Salant, S., Kwiatkowski, T., Parikh, A., Das, D., Berant, J.: Learning recurrent span representations for extractive question answering. arXiv preprint arXiv:1611.01436 (2016)
14. Lewis, P., Oğuz, B., Rinott, R., Riedel, S., Schwenk, H.: MLQA: evaluating cross-lingual extractive question answering. arXiv preprint arXiv:1910.07475 (2019)
15. Lewis, P.S.H., Denoyer, L., Riedel, S.: Unsupervised question answering by cloze translation. CoRR abs/1906.04980 (2019). http://arxiv.org/abs/1906.04980
16. Meichanetzidis, K., Toumi, A., de Felice, G., Coecke, B.: Grammar-aware question-answering on quantum computers. arXiv preprint arXiv:2012.03756 (2020)
17. Qi, P., Lin, X., Mehr, L., Wang, Z., Manning, C.D.: Answering complex open-domain questions through iterative query generation. arXiv preprint arXiv:1910.07000 (2019)
18. Rajpurkar, P., Jia, R., Liang, P.: Know what you don't know: unanswerable questions for SQuAD. In: Association for Computational Linguistics (ACL) (2018)
19. Rajpurkar, P., Zhang, J., Lopyrev, K., Liang, P.: Squad: 100,000+ questions for machine comprehension of text. arXiv preprint arXiv:1606.05250 (2016)
20. Reddy, S., Chen, D., Manning, C.D.: CoQA: a conversational question answering challenge. Trans. Assoc. Comput. Linguist. **7**, 249–266 (2019)
21. Wang, S., et al.: R 3: reinforced ranker-reader for open-domain question answering. In: Thirty-Second AAAI Conference on Artificial Intelligence (2018)
22. Wiebe, N., Bocharov, A., Smolensky, P., Troyer, M., Svore, K.M.: Quantum language processing. arXiv preprint arXiv:1902.05162 (2019)
23. Williams, R.J.: Simple statistical gradient-following algorithms for connectionist reinforcement learning. Mach. Learn. **8**(3–4), 229–256 (1992)
24. Yang, Y., Wang, C., Gong, L., Zhou, X.: FPNet: customized convolutional neural network for FPGA platforms. In: 2019 International Conference on Field-Programmable Technology (ICFPT), pp. 399–402. IEEE (2019)
25. Yang, Z., et al.: HotpotQA: a dataset for diverse, explainable multi-hop question answering. arXiv preprint arXiv:1809.09600 (2018)
26. Yih, W.T., Chang, M.W., Meek, C., Pastusiak, A.: Question answering using enhanced lexical semantic models. In: Proceedings of the 51st Annual Meeting of the Association for Computational Linguistics (Volume 1: Long Papers), pp. 1744–1753. Association for Computational Linguistics, Sofia, Bulgaria, August 2013). https://www.aclweb.org/anthology/P13-1171
27. Zhu, H., Dong, L., Wei, F., Wang, W., Qin, B., Liu, T.: Learning to ask unanswerable questions for machine reading comprehension. arXiv preprint arXiv:1906.06045 (2019)

Self-propagating Malware Containment via Reinforcement Learning

Sebastian Eresheim[1,2(✉)] and Daniel Pasterk[3]

[1] St. Poelten University of Applied Sciences, Saint Pölten, Austria
sebastian.eresheim@fhstp.ac.at
[2] University Vienna, Vienna, Austria
[3] Technical University Vienna, Vienna, Austria
daniel.pasterk@tuwien.ac.at

Abstract. We introduce a reinforcement learning based containment system for self-propagating malware in local networks. The system is trained with real-world software and malware and leverages a network of virtual machines for execution and propagation. Instead of relying on labels as is common with supervised learning, we follow a trial-and-error approach in order to learn how to link network traffic to malware infections.

Keywords: Reinforcement learning · Machine learning · Network security

1 Introduction

In 2017, two computer viruses – *WannaCry* and *NotPetya* – emerged and together were responsible for an estimated damage of about 14 billion dollars worldwide. These viruses were typical examples of ransomware, where the malicious software encrypts valuable information of the victim and demands a ransom in exchange for the decryption key. One particularly dangerous behaviour of these two viruses was inherent in their method of dissemination: Both exploited a vulnerability in the file sharing protocol SMB implemented in older versions of Microsoft Windows. This allowed an infected host to write and execute new files on any computer within the local network that accepted the vulnerable protocol version. Before the malware presented the user its ransom note, it had already distributed itself to other hosts on the local network. For each newly infected host the ransomware again tried to infect other hosts on the network, multiplying the threat to the network. By the time the first ransom note was displayed, many more hosts had already been affected. Therefore, a counteracting entity needed to react fast and also keep track of potentially infected hosts, since one unrecognised infection would threaten the whole network all over again. For human security experts such scenarios are a tough challenge and can often only be resolved by inflicting collateral damage to the rest of the operative network.

© IFIP International Federation for Information Processing 2021
Published by Springer Nature Switzerland AG 2021
A. Holzinger et al. (Eds.): CD-MAKE 2021, LNCS 12844, pp. 35–50, 2021.
https://doi.org/10.1007/978-3-030-84060-0_3

The field of reinforcement learning (RL) achieved tremendous accomplishments in recent years. The combination of traditional RL methods and deep learning techniques managed to outperform even some of the best human specialists when executing complicated tasks. Starting at the old games of the Atari 2600 [15], RL agents also outplayed human experts in the board game Go (first based on domain knowledge [19], later entirely through self-play [17]) and the complex real-time computer games Dota 2 [4] and Starcraft 2 [24]. These agents showed that reinforcement learning is capable of finding (nearly) optimal policies in environments that are high-dimensional (in states as well as in actions), (quasi) time-continuous, scarce of reward, and partially observable.

Due to the ability of RL agents to make quick decisions in difficult situations that may have long-term impacts, we show that RL is well suited to prevent self-propagating malware from spreading in local networks. Traditional approaches for solving this problem comprise Intrusion Detection-(IDS) or Intrusion Prevention Systems (IPS), which are built upon supervised learning techniques [23, 28]. Such systems act in accordance to predefined responses, if any, and require a lot of labelled data which is time consuming and/or expensive to gather. Threat response also relies heavily on expert domain knowledge and usually suffers from a certain degree of bias. We argue that an RL approach has 3 key advantages over traditional supervised learning solutions:

1. RL is able to counteract threats not just in a timely manner, but also tailored individually to different situations, in contrast to the predefined reactions of supervised learning. The granularity of individual reactions is defined by the action space.
2. Given a rich action space, RL is able to decrease uncertainty by interacting with the object in question before reacting further. A supervised learning system can be understood as an observer who is restricted to images of an object in order to determine what it might be and how to interact with it. Conversely, RL enables the observer to directly interact with the object and to observe its immediate reactions. For example, the observer could twist and turn the object to change its point of view before making a final decision on a future reaction. In the specific context of self-propagating malware, an RL agent could send specific network packets to a potentially infected host and look for anomalies in the response of the host.
3. RL does not require labelled data. An RL agent requires an environment with which it can interact. Once the environment exists, the amount of data generated by agent interactions can be arbitrarily large. In the scenario of self-propagating malware, labelling all data packets that indicate an infection of a host is a very time-consuming task. RL on the other hand does not need to know initially which data packets are linked to an infection. It simply rewards an agent if it contained an infection and the agent learns the distinguishing factors by itself.

To the best of our knowledge, we present the first RL based containment system for self-propagating malware in local networks. We do so by leveraging virtual machines in combination with real-world software and malware and without the need of labels. Instead, it follows a trial-and-error paradigm.

The rest of this paper is structured as follows: Sect. 2 discusses related work, Sect. 3 introduces the methodical approach to implementing our system, Sect. 4 describes the experiments and presents the results, Sect. 5 discusses these results in detail, and Sect. 6 concludes this paper.

2   Related Work

As mentioned before, current IDS/IPS are usually based on supervised learning [23] and typically consist of multiple components [11]:

- a monitoring component, which can either be host-based or network-based,
- an analysis and recognition component,
- and an alarm component that provides further procedural instructions in case of detection.

In contrast, RL systems only consist of two components – an agent and its environment, whereas most of the intelligence relies on the agent. IDS/IPS can be distinguished into signature-based and anomaly-based systems [5,7,16]. The former are only able to detect known attacks, the later suffer from a high false positive rate. Both require labelled data in order to operate. RL does not rely on labelled data and, if paired with a function approximation method, is capable of generalising from observed states. The fundamental difference between supervised learning and RL is also the reason why benchmark data sets for intrusion detection or malware classification like [10,12,21] are not applicable.

Only few, very recent works in the realm of network security are based on RL. In [29] the authors use RL for automatic feature selection in order to later apply them to a traditional supervised learning approach. [18] focuses on anomalies that are injected into a simulated network and the reward is output based on a correct detection. [2,25,26] deploy tabular temporal-difference learning methods to live sequences of traffic. In [27], a method based on temporal difference learning together with a kernel approximation is used to solve a Markov chain prediction problem. In [9], an actor critic approach is used to apply intrusion detection in the specific case of cyber-physical systems. [14] introduces different deep RL methods for intrusion detection. This approach does use classical RL algorithms, but depends heavily on a deliberately small discounting factor. This results in the overvaluing of the immediate reward while not taking long-term effects into consideration.

In contrast to above related work, we focus on self-propagating malware for conventional client computers by leveraging virtual machines running real-world software and operating systems, instead of simulating the data. Even though it makes sense to use standardised problems and performance metrics, we decided to use a setup with virtual machines to show a functional real-world proof of concept for the proposed method. This way, we can make solid statements about practical applicability, since "secondary variables" such as latency and the actual interactions in the defence against threats are rarely captured by existing data sets.

3   Methods

In order to let an RL agent learn to quarantine (potentially) infected hosts within a computer network, we decided to use an episodic approach, because a network in which every host is infected marks a clear terminal state in the environment. During one episode a host gets infected at a random point in time and the agent needs to learn to intervene only in that case. Each episode is divided into equidistant time-steps where the agent can choose an action. For a small proportion of episodes no infection happens within the network, enabling the agent to learn when not to intervene.

3.1   Environment

We consider a Markov decision process $\mathcal{M} = (\mathcal{S}, \mathcal{A}, \mathcal{R}, p, \chi, \gamma)$, where $\mathcal{S}$ is a state space, $\mathcal{A}$ is an action space, $R : \mathcal{S} \times \mathcal{A} \to \mathbb{R}$ is a reward function, $p : \mathcal{S} \times \mathcal{A} \times \mathcal{S} \to [0, 1]$ is a transition probability function, $\chi \in \Delta(\mathcal{S})$ is the initial state distribution, and $\gamma \in [0, 1)$ is a discount factor. The initial state distribution χ is deterministic, because the environment starts each episode from the same clean initial state. We set the discount factor to $\gamma = 0.99$, because the agent should consider long-term effects.

State Space. We require the state space to contain all relevant information for the agent to base its decision on. In our setup this includes information about all network packets sent within a reasonable amount of time. Each packet is encoded via a feature vector $\phi \in \mathbb{N}^d$, which is a one-hot-encoding where a 1 stands for one data packet, and the dimension the 1 is placed in determines the contextual information. The feature selection is more thoroughly explained in Subsect. 3.2. The state of the environment is then defined as

$$ s := (\sum_{i=0}^{N_t} \phi_i, \sum_{j=t-10}^{t} \sum_{i=0}^{N_j} \phi_i), \tag{1} $$

where N_j is the number of network packets in time-step j, and t is the current time-step. The first component of the state vector is a short term depiction of the current situation, summing up all feature vectors of network packets from the previous time-step. The second component adds more context as it additionally sums up all feature vectors of packets from the previous 10 time-steps. Having the agent only look at the network traffic of the previous time-step enables it to react to short term events, but makes it oblivious to longer-lasting trends. For this reason, the second component was added. This definition results in $\mathcal{S} = \mathbb{N}^d \times \mathbb{N}^d$.

Action Space. The action space consists of only two actions $\mathcal{A} = \{continue, cut\}$ in order to record and replay single episodes. This is due to execution time reasons and is explained in more detail in Sect. 4. The action *continue* lets the environment continue without interruption and the action *cut* disconnects a host from the rest of the network and ends the episode. Figure 1 depicts the backup diagram of the environment, which displays the connections between states and actions.

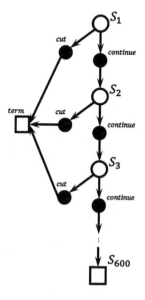

Fig. 1. A backup diagram of the environment. White circles represent states, black circles represent actions and white squares represent terminal states.

Reward Function. The desired behaviour the agent should learn is to a) cut the network connection to a host as soon as an infection has happened, and b) to not intervene otherwise. In order to accomplish these goals, the reward function is defined as: let $I \in \mathbb{N}$ be the time-step an infection happens on a host and let $T \in \mathbb{N}$ be the final time-step. Also let $I \leq T$ denote that during an episode a host is infected and $T < I$ that there is no infection during an episode. Finally let $t \in \mathbb{N}$ be the current time-step. The reward function $R : \mathbb{N} \times \mathbb{N} \times \mathcal{A} \to \mathbb{R}$ is defined as

$$R(I, t, \text{continue}) := \begin{cases} 0 & \text{if } t < T \\ 1 & \text{if } t = T < I \\ -1 & \text{if } I < t = T \end{cases}$$

$$R(I, t, \text{cut}) := \begin{cases} -1 & \text{if } t \leq I \\ \frac{1}{t-I} & \text{if } I < t \end{cases}$$

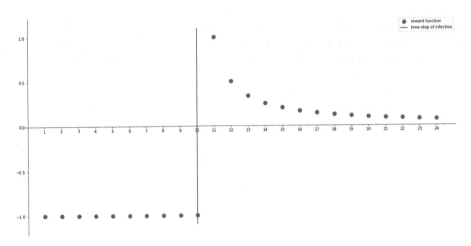

Fig. 2. An example reward function for the action *cut*. In this scenario the infection of the host happens in time-step 10. If the action *cut* is applied before or at the same time-step as the infection, then a reward of −1 is given. If the action is applied after the infection, then the reward is $\frac{1}{t-10}$, where t is the current time-step.

The rapid decrease in reward for past infections ensures that the agent prefers to disconnect the network link earlier than later. On the other hand, the negative reward punishes the agent if it acts too early or not at all when it should have done so. Figure 2 graphically explains the reward function for the action *cut*. Note that the reward is also −1 if the agent cuts the network connection to an infected host in the same time-step as the infection happens, because there couldn't have been a prior indication of that infection. In such a scenario, the agent was simply lucky, but its decision was not based on learned behaviour.

3.2   Feature Selection

For feature selection we only consider metadata, since the corresponding features already resulted in a very high-dimensional space. In our feature selection method, data packets are distinguished by their type and by their destination. Because all of the relevant data that determines the type and destination is of categorical nature, our approach uses one dimension in the feature vector for each combination of these categorical values. This results in a very high, yet mutually exclusive, number of dimensions.

The packet destination (within the LAN) is determined via the IP address. This address is split in a network- and a host part; in our specific setup, we distinguish 3 networks consisting of 256 hosts each. Thus, there are 768 different addresses in this setup. Encoding all addresses, even though only 6 of them are used, is necessary, because the malware actively enumerates all IP addresses in the network and listens for replies. Not encoding all possible addresses would lead to either not being able to model such packets at all, or to not being able to distinguish between these addresses. Also all IPv6 packets are grouped together, regardless of their target, because although it is desired to distinguish IPv6 from IPv4 traffic, IPv6 is not the main focus of this work (due to its enormous address space).

The type of the data packet is defined by the network protocols *arp*, *icmp*, *tcp*, and *udp*. For *tcp* and *udp*, the destination port further identifies the likely type of service – at least for well-known ports ranging from 1 to 1023. Above this threshold, all ports are grouped together for convenience reasons. Finally there is one dimension in the feature vector for each combination of address and type previously mentioned, resulting in 3,154,436 mutually exclusive dimensions. Lastly, 4 additional dimensions for a coarser distinction were added, counting all *ip*, *arp*, *tcp* and *udp* packets, regardless of their ports or destination addresses.

3.3   Agent

The agent uses the formalism of generalised policy iteration [20] in the setting of value-based RL to find an optimal policy π^*. We compare SARSA and Q-Learning as a learning algorithm for the action value function Q. It also uses an ϵ-greedy policy, where different values for ϵ are considered and a memory-based function approximation is applied. To be specific, we apply the k-nearest neighbours algorithm in combination with locality sensitive hashing [6] in order to decrease the query time of the k-NN. This means whenever the value of a state is queried, the average value of the k-nearest neighbours is calculated and used as approximation. The buffer size for the k-NN algorithm comprises 100,000 states.

4   Experiments

In order to return from a terminal state where at least one host is infected with malware to the initial state, where no infection has yet happened, we deployed *virtual machines* (VMs) connected via a virtual network. Utilising VMs makes it possible to use real software as well as malware, which is somewhat rare, due to the fact that many environments in the RL field are abstract games [3,13] or simulations of real world scenarios [1,8,22]. Furthermore, VMs are typically used in companies alongside physical computers, which is why they are representative of a real-world scenario rather than being a simulation/abstraction thereof. As previously mentioned, VMs allow operators to create so-called snapshots – an image of the virtual machine's state at a specific point in time – which the VM can be reverted to. This makes it possible to quickly and reliably undo the damage a malware infection has caused.

In our experiments, the virtual network that connects the VMs corresponds to a star-shaped topology, meaning there is a central VM all other machines are directly connected to. We use 4 VMs in total: the central one, called *agent VM*, and three VMs that can be infected, called *vulnerable VMs*. Figure 3 shows a schematic representation of this setup. The software component that represents the agent is located on the *agent VM*, which also contains some components of the environment, like feature extraction. Network traffic that originates from *vulnerable VMs* is routed via the *agent VM*, where each connection to another *vulnerable VM* can be independently blocked by its firewall. The malware we selected for our experiments, (*NotPetya*), is a Windows executable and therefore only capable to infect Windows hosts. In order to guarantee that the *agent VM* cannot be infected by the malware, it was set up to run Linux. Scenarios where malware actively attacks the agent or the agents' learning process are not within the scope of this work. However, such attacks might be an interesting subject of research for future research.

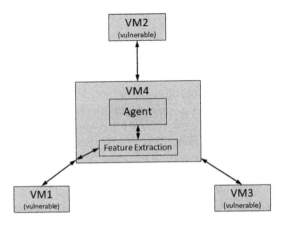

Fig. 3. A schematic representation of the environment. The green elements depict the environment, the light brown element represents the agent. VMs 1–3 (the *vulnerable* VMs) are only connected to VM4 (the *agent* VM), which acts as a router and forwards network data packets. Besides forwarding, the agent VM also captures the data packets and passes them on to a feature extraction unit. The output of this process is then buffered and handed to the agent in the environments state representation. Based on this information the agent then chooses an action. (Color figure online)

Each episode is limited to 10 min of real time, which is divided into equal time-steps of 1 s. During an episode, malware is released on one machine at a random time-step, which then tries to spread to all machines on the local network. At each time-step the agent can interact with the network. This results in a maximum of 600 actions per episode. In order to prevent the agent from interfering with benign hosts, baseline episodes are introduced, where no malware is released. In these episodes the agent should learn to not to interfere with the network. At the beginning of each episode, there is a 20% chance for it to become a baseline episode. The reason behind this particular probability threshold is that finding the earliest time-step after an infection is the critical part of the agents abilities, whereas not interfering in baseline episodes is comparatively simple, especially in our setup. Nevertheless, such baseline episodes are necessary to prevent the agent from learning to always interfere with the network. However, we are aware that the resulting ratio of baseline-to-infected episodes does not reflect the real world, where infection-less episodes are much more likely to occur.

In order to save time during the experiment the environment is adapted to record and replay episodes faster than real-time during the learning process. The downside of this approach is a more restricted action space, since not every consequence of every action at each time-step can be recorded. Therefore the agent's observations are limited to a single network interface instead of all three. Additionally, its action space is reduced to two distinct actions, one no-operation action called *continue*, and one for blocking all network traffic called *cut*. When the agent selects the *continue* action, the episode continues normally and without any agent interaction. By blocking all network traffic, the host is effectively placed in quarantine where it remains until a human expert has finished inspecting it. Since this block cannot be lifted during an episode, the host stays in the same state for the rest of the episode, where it is incapable of sending network messages to other hosts. Note that the agent cannot cut a host's network connection a second time. Since there is no benefit in continuing until the episode runs its predetermined course, the *cut* action effectively ends the episode as soon as it is applied. For a recorded episode, this means that in the case of *continue* action, all recorded messages are replayed in the order they originally appeared. In the case of *cut*, all further recorded messages are suppressed and the episode ends immediately.

Network traffic is captured using *tshark*[1], a well-established tool used for live network data monitoring and analysis. For the record/replay functionality, the recorded network packets are stored in JSON format. During replay, only certain features are extracted from the JSON data. These features are gathered in a feature vector, which serves as the input to the state representation of the environment. More details on the feature selection process are provided in Subsect. 3.2.

In a real-world scenario the initial infection on the network often requires user interaction. A malicious link on a website or a malicious attachment in an e-mail are common initial incident vectors. In our environment this initial user behaviour is simulated. The file containing the actual malware is already in place on every vulnerable VM, but is not executed by default. A small script that automatically runs at startup listens on a specific port for the instruction to infect the machine. This port is not included in the data collection of the *agent VM*, as it would enable the agent to learn to look out for this very instruction. As soon as such an instruction comes from the *agent VM* the script executes the stored malicious file and releases the malware on the host and into the network. The *agent VM* is the only VM that sends such infection instructions to the vulnerable VMs.

For the experiments we created a data set of 20 recorded episodes. Due to the 20% probability of an episode to become a baseline episode, this resulted in 3 episodes in which no infection happened and 17 with a random infection

---

[1] https://www.wireshark.org/docs/man-pages/tshark.html.

time-step. This data set is then extended by two of similar size in order to see how representative the generated data is and how the transfer of a trained model between the data sets performs. Please note that this is not considered transfer learning as we are not using a trained model for different problems, but rather have several sets of samples from the same distribution. We conducted a hyper-parameter search for the three parameters step-size α, greedyness ϵ and amount of nearest neighbours k via a grid search. Values considered in the search were for α: $[0.1, 0.2, 0.4, 0.6, 0.8]$, for ϵ: $[0.1, 0.001, 0.0001]$ and for k: $[1, 5]$. The best performing hyper-parameters were $\alpha = 0.4$, $\epsilon = 0.0001$ and $k = 1$. Figure 4 depicts the average reward for an agent with these hyper-parameters. The data sets can be fount at https://github.com/seresheim/self-propagating-malware-containment-via-reinforcement-learning.

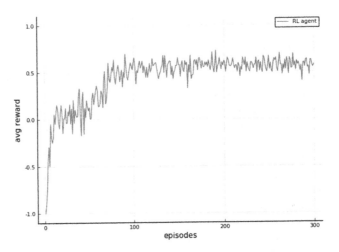

Fig. 4. Average reward of an agent with the hyper-parameters $\alpha = 0.4$, $\epsilon = 0.0001$ and $k = 1$. A reward of 1 means the agent cut the network connection exactly one time-step after an infection, a reward of 0.5 means it cut the connection two time-steps after an infection, 0.33 after 3, etc. A reward of -1 is given in a bad outcome (no intervention or intervened too early). The results are averages over 50 runs.

Besides the initial data set (data set 1), two additional data sets of similar size were recorded (data set 2 and data set 3) to further evaluate the setup. Figure 5a shows a comparison of all 3 data sets, when the agent is only trained on a single data set. The sub-figures Fig. 5b, Fig. 5c, and Fig. 5d show comparisons between the data sets. In each of these sub-figures in the first 50 episodes all 3 agents were trained on one of the 3 different data sets, each agent on a separate one. After 50 episodes, all agents were switched to the same data set. Thus these 3 sub-figures display how beneficial it is to transfer knowledge based on one particular data set to an environment based on another data set. If the data distributions of the data sets are the same, then an agent with transferred knowledge should perform as well as an agent that was trained only on the switched to data set.

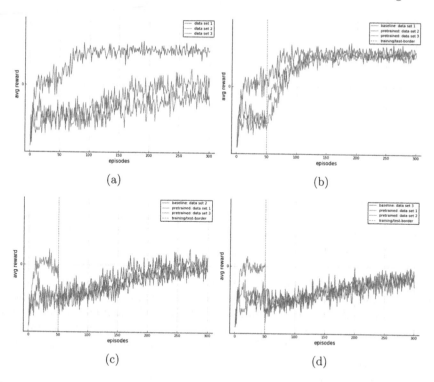

Fig. 5. Besides the initial data set (data set 1), two additional data sets were recorded. Figure (a) shows a comparison of the 3 data sets with agents performing on the sets separately. Figures (b)–(d) show a comparison of transferred knowledge. For the first 50 episodes, each agent trains on its separate data set. After 50 episodes all agents are transferred to the same baseline data set. This shows how good a trained knowledge base performs on a different data set. Each line in all 4 subplots is averaged over 50 runs.

Finally, Fig. 6, Fig. 7, and Fig. 8 show the results when these three data sets are each split into a training- and test set. This split is done in a ratio of 3:1. Each figure uses an agent that only learns from the test set as an upper bound. Such an agent is clearly overfit to the test set, but also shows an estimate of the maximum possible average reward. The training phase includes 100 episodes of the training set and the following test phase includes 50 episodes of the test set.

5   Discussion

The results of Fig. 4 show that our general concept is feasible and that an agent is capable to learn how to distinguish between malicious and benign network traffic without requiring labels for each data packet or time frame. The figure also shows that the agent is not able to achieve the maximum reward. We suspect the reason behind this is a discrepancy between the time-step the environment

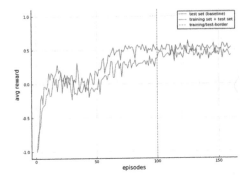

Fig. 6. A train-test comparison on data set 1. Data set 1 is split into a training- and test set (3:1). An agent is then trained on the training set and evaluated on the test set. Another agent solely operating on the test set acts as upper bound. Each line in the graph is averaged over 50 runs.

records the release of the malware and the earliest time-step the malware sends out data packets. The reward calculation heavily depends on the assumption that this time difference is negligible. The theoretical maximum of the reward function (1) is only met if the agent cuts the network connection to the infected host exactly one time-step after the infection has happened. If the malware does not interact with the network for several time-steps, the agent is not able to detect the infection that early. Instead it can only detect the infection when the malware first sends data packets across the network, resulting in a lower actual reward maximum.

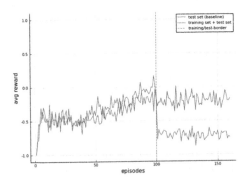

Fig. 7. A train-test comparison on data set 2. Data set 2 is split into a training- and test set (3:1). An agent is then trained on the training set and evaluated on the test set. Another agent solely operating on the test set acts as upper bound. Each line in the graph is averaged over 50 runs.

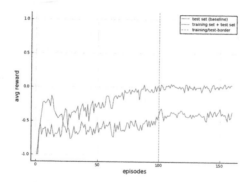

Fig. 8. A train-test comparison on data set 3. Data set 3 is split into a training- and test set (3:1). An agent is then trained on the training set and evaluated on the test set. Another agent solely operating on the test set acts as upper bound. The base line is averaged over 50 runs, the train-test split is averaged over 100 runs.

In our experiments of Fig. 5a, the results indicate that three separate agents with the same hyper-parameters trained on our three data sets do not achieve the same average reward. We suspect the graphs for data set 2 and data set 3 in Fig. 5a perform poorly, because of the previously described effect of latency in the malware's execution. In essence, we assume that the distribution of the latency is different in the three data sets.

Switching from data set 1 to one of the other two results in a performance drop, as depicted in Fig. 5c and Fig. 5d. This is expected, considering the time delay assumption holds. Furthermore, the performance after the switch from data set 1 to 2 or 3 does not significantly decrease in comparison to the respective baseline. Switching from data set 2 or 3 to data set 1 (Fig. 5b) results in a steep increase in average reward in the first few episodes after the change. This reinforces the time delay assumption, since an increase based on new experience would require more episodes in the new environment. On the other hand, the average reward does not increase immediately to the level of the baseline, therefore also suggesting a difference in data distribution.

Figure 6 shows no performance drop after the switch from training data to test data. The results nearly match the ones of the baseline, suggesting the data of train- and test set are of similar distribution and the agent is capable of learning the relevant information from the training set. Figure 7, on the other hand, shows a significant drop in average reward immediately after the change of the data set, and a lower overall performance than the baseline agent thereafter. A likely explanation for this outcome is the agent lacking necessary information it requires in the test set. Since the model (the k-NN algorithm) uses raw data, we assume that the information is generally missing from the training data set and not excluded in the model building process. Figure 8 displays a similar image to Fig. 6, albeit in a lower range of average reward as well as with a larger gap to the baseline. A closer inspection of the raw data of individual runs indicated that the learning success depends on the replay order of the recorded episodes, further indicating that some time-steps are better predictors in the k-NN algorithm than others. We suspect this phenomenon occurs due to a lack of intelligence in the accumulation mechanism of data points for the k-NN buffer. Currently all data is added to the comparison buffer in a first-in, first-out order, regardless of the data present in the buffer. Adding more intelligence to the selection process of which data points to add, which to keep, and which to discard could add more value to the overall structure.

6   Conclusion

In this paper, we demonstrated that it is possible to create a self-propagating malware containment system through trial- and error learning. For that purpose, we applied reinforcement learning algorithms to an environment of virtual machines containing real world software as well as real-world malware. In particular, we compared SARSA and Q-Learning by leveraging a k-nearest neighbours based function approximation approach. This approach compares states via the amount of data packets sent within a certain time-span, grouped by destination and packet type. The agents' action space contains only a non-response action as well as a network connection cut. Our empirical results show that a trained agent is capable of distinguishing when to cut a network connection to an infected host and when to not intervene with the system. Additionally, the RL based approach allows our agent to learn the difference between benign and malicious data packets without the need of labelling each packet.

Acknowledgement. The authors would like to thank Tobias Dam, Robert Luh, Sebastian Schrittwieser and Clemens Heitzinger for their support and guidance throughout the project. This research was funded in whole, or in part, by the Austrian Science Fund (FWF) P 32706-N.

References

1. Anwar, A., Raychowdhury, A.: Autonomous navigation via deep reinforcement learning for resource constraint edge nodes using transfer learning. arXiv e-prints arXiv:1910.05547, October 2019
2. Sukhanov, A.V., Kovalev, S.M., Stýskala, V.: Advanced temporal-difference learning for intrusion detection. IFAC-PapersOnLine **48**(4), 43–48 (2015). 13th IFAC and IEEE Conference on Programmable Devices and Embedded Systems
3. Bellemare, M.G., Naddaf, Y., Veness, J., Bowling, M.: The arcade learning environment: an evaluation platform for general agents. J. Artif. Intell. Res. **47**, 253–279 (2013)
4. Berner, C., et al.: Dota 2 with large scale deep reinforcement learning. arXiv preprint arXiv:1912.06680 (2019)
5. Chandola, V., Banerjee, A., Kumar, V.: Anomaly detection: a survey. ACM Comput. Surv. (CSUR) **41**(3), 1–58 (2009)
6. Datar, M., Immorlica, N., Indyk, P., Mirrokni, V.S.: Locality-sensitive hashing scheme based on p-stable distributions. In: Proceedings of the Twentieth Annual Symposium on Computational Geometry, pp. 253–262 (2004)

7. Dennis, C., Ingram, J., Kremer, H., Rowe, N.: Distributed intrusion detection for computer systems using communicating agents, November 2012
8. Dosovitskiy, A., Ros, G., Codevilla, F., Lopez, A., Koltun, V.: CARLA: an open urban driving simulator. In: Proceedings of the 1st Annual Conference on Robot Learning, pp. 1–16 (2017)
9. Feng, M., Xu, H.: Deep reinforecement learning based optimal defense for cyber-physical system in presence of unknown cyber-attack. In: 2017 IEEE Symposium Series on Computational Intelligence (SSCI), pp. 1–8. IEEE (2017)
10. Habibi Lashkari, A., Abdul Kadir, A.F., Taheri, L., Ghorbani, A.: Toward developing a systematic approach to generate benchmark android malware datasets and classification, pp. 1–7, October 2018
11. Khan, S., Loo, J., Ziauddin, Z.: Framework for intrusion detection in IEEE 802.11 wireless mesh networks. Int. Arab J. Inf. Technol. **7**, 435–440 (2010)
12. Kolias, C., Kambourakis, G., Stavrou, A., Gritzalis, S.: Intrusion detection in 802.11 networks: empirical evaluation of threats and a public dataset. IEEE Commun. Surv. Tutor. **18**, 1 (2015)
13. Lanctot, M., et al.: OpenSpiel: a framework for reinforcement learning in games. CoRR abs/1908.09453 (2019). http://arxiv.org/abs/1908.09453
14. Lopez-Martin, M., Carro, B., Sanchez-Esguevillas, A.: Application of deep reinforcement learning to intrusion detection for supervised problems. Expert Syst. Appl. **141**, 112963 (2019)
15. Mnih, V., et al.: Playing Atari with deep reinforcement learning. arXiv preprint arXiv:1312.5602 (2013)
16. Patcha, A., Park, J.M.J.: An overview of anomaly detection techniques: existing solutions and latest technological trends. Comput. Netw. **51**, 3448–3470 (2007)
17. Schrittwieser, J., et al.: Mastering Atari, go, chess and shogi by planning with a learned model. Nature **588**(7839), 604–609 (2020)
18. Servin, A.: Towards traffic anomaly detection via reinforcement learning and data flow (2007)
19. Silver, D., et al.: Mastering the game of go with deep neural networks and tree search. Nature **529**(7587), 484–489 (2016)
20. Sutton, R.S., Barto, A.G.: Reinforcement Learning: An Introduction. MIT Press, Cambridge (2018)
21. Tavallaee, M., Bagheri, E., Lu, W., Ghorbani, A.: A detailed analysis of the KDD cup 99 data set. In: IEEE Symposium on Computational Intelligence for Security and Defense Applications, CISDA 2, July 2009
22. Todorov, E., Erez, T., Tassa, Y.: Mujoco: a physics engine for model-based control. In: 2012 IEEE/RSJ International Conference on Intelligent Robots and Systems, pp. 5026–5033. IEEE (2012)
23. Verwoerd, T., Hunt, R.: Intrusion detection techniques and approaches. Comput. Commun. **25**, 1356–1365 (2002)
24. Vinyals, O., et al.: Grandmaster level in StarCraft II using multi-agent reinforcement learning. Nature **575**(7782), 350–354 (2019)
25. Xu, X.: Sequential anomaly detection based on temporal-difference learning: principles, models and case studies. Appl. Soft Comput. **10**, 859–867 (2010)
26. Xu, X., Xie, T.: A reinforcement learning approach for host-based intrusion detection using sequences of system calls. In: Huang, D.-S., Zhang, X.-P., Huang, G.-B. (eds.) ICIC 2005. LNCS, vol. 3644, pp. 995–1003. Springer, Heidelberg (2005). https://doi.org/10.1007/11538059_103

27. Xu, X., Xie, T., Hu, D., Lu, X.: Kernel least-squares temporal difference learning. Int. J. Inf. Technol. **11**, 54–63 (2005)
28. Yeo, L.H., Che, X., Lakkaraju, S.: Understanding modern intrusion detection systems: a survey (2017)
29. Zhiyang, F., Wang, J., Geng, J., Kan, X.: Feature selection for malware detection based on reinforcement learning. IEEE Access **7**, 176177–176187 (2019)

Text2PyCode: Machine Translation of Natural Language Intent to Python Source Code

Sridevi Bonthu[1,3]($\boxtimes$) (ID), S. Rama Sree[2] (ID), and M. H. M. Krishna Prasad[3] (ID)

[1] Vishnu Institute of Technology, Bhimavaram, Andhra Pradesh, India
[2] Aditya Engineering College, Surampalem, Andhra Pradesh, India
ramasree_s@aec.edu.in
[3] Jawaharlal Nehru Technological University, Kakinada, Andhra Pradesh, India

Abstract. Natural Language Processing has improved tremendously with the success of Deep Learning. Neural Machine Translation (NMT) has arisen as the most powerful with the power of Deep Learning. The same idea has been recently applied to source code. Code Generation (CG) is the task of generating source code from natural language input. This paper introduces a Python parallel corpus of natural language intent and source code pairs. It also proposes a Code Generation model based on Transformer architecture used for NMT by using code tokenization and code embeddings on the custom parallel corpus. The proposed architecture achieved a good BLEU score of 32.4 and Rouge-L of 82.1, which is on par with natural language translation.

Keywords: Deep learning · Attention · Transformer · Code generation · Python · Neural machine translation

1 Introduction

Code Generation (CG), summarization, and retrieval are the main applications of Natural Language Processing (NLP) in the Software Engineering domain [5]. These three tasks, mostly powered by the data-driven models, rely on a parallel corpus of natural language intent and source code for training and evaluation. CG aims to generate code based on Natural Language Intent i.e., description of the problem. It is an indispensable activity of the programmers while implementing specific intents. Enhancing software quality and improving the productivity of the programmer are dependent on the correct code [7]. However, generating the source code is time-consuming, costly, and error-prone. Therefore Automatic Code Generation process becomes greatly important for software development.

Many researchers these days employing encoder-decoder frameworks for sequence-to-sequence (seq2seq) learning, which are popular for Neural Machine Translation (NMT). In seq2seq models, the encoder takes a sentence and produces another sentence as output [17]. Attention Mechanism can be added to

© IFIP International Federation for Information Processing 2021
Published by Springer Nature Switzerland AG 2021
A. Holzinger et al. (Eds.): CD-MAKE 2021, LNCS 12844, pp. 51–60, 2021.
https://doi.org/10.1007/978-3-030-84060-0_4

help the model to focus on the relevant information [1] and it helped improve the performance of NMT applications. As a continuation, Transformer architecture [18] was proposed by Google to perform seq2seq tasks well. There are a lot of benefits for transformers when compared with traditional seq2seq models like it also employs attention, and lends itself to parallelization which makes proper utilization of GPUs.

```
Write a python function that returns the sum of
n natural numbers

def sum_natural(num):
    if num < 0:
        print("Please enter a positive number!")
    else:
        sum = 0
        while(num > 0):
            sum += num
            num -= 1
        return num
```

```
Write a python function to find the area of a
circle, whose radius is given.

def findArea(r):
    PI = 3.142
    return PI * (r*r)
```

Fig. 1. Two example records from the parallel corpus of natural language intent and source code pairs.

This paper proposes an NMT model to translate a problem description given in the English Language to a Python source code snippet. The proposed model is based on Transformer architecture, and Attention is employed to learn the alignment between the text and the code. The task can be described as follows - Given a parallel corpus of Problem Description and Python Source Code pairs as shown in Fig. 1, create a model that generates a well-indented Python Source Code of the Problem Description. This task involves solving several subproblems because of the associated challenges with programming languages. The model learns a mapping from $\mathcal{X}$ to $\mathcal{Y}$ where $\mathcal{X} = \{x^{(i)}\}$, a set of problem descriptions, $x^{(i)} = \{x_1^{(i)}, x_2^{(i)} \cdots x_n^{(i)}\}$ denotes the token sequence of ith example description and $\mathcal{Y} = \{y^{(i)}\}$, a set of source code snippets, $y^{(i)} = \{y_1^{(i)}, y_2^{(i)} \cdots y_m^{(i)}\}$ denotes the token sequence of ith example source code.

Contributions of this paper

- Creation of a parallel corpus for Python Code Generation from Natural Language description.
- Train a Transformer based NMT model to generate Python Code.
- Study the effect of Python Code Tokenization on the accuracy of the model.
- Study the effect of Pre-trained Code Embeddings on the accuracy of the model.

The remaining part of the paper is organized as follows. Section 2 introduces the proposed approach for python code generation, Sect. 3 gives an insight into the curated dataset and its statistics, Sect. 4 provides a clear explanation about the setting of baseline models on the dataset, and Sect. 5 presents the results obtained by the three models on the dataset followed by the conclusion.

2 Proposed Approach

This paper proposes a novel approach based on Transformer architecture, by incorporating source code tokenization and code embeddings as shown in Fig. 2.

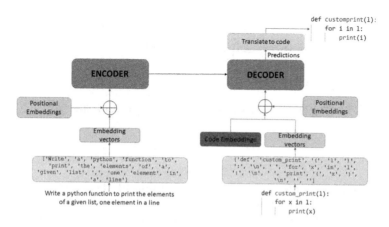

Fig. 2. Architecture of NMT model to translate natural language intent to python source code

The encoder and decoder layers of the proposed architecture are identical to the ones proposed in [18] and it is shown in Fig. 3. The entire Transformer architecture follows stacked self-attention and feed-forward fully connected layers for both the encoder and decoder as shown in Fig. 3. Our model used a stack of 3 such identical layers. The encoder takes Natural Language Intent whereas the decoder works on Python Source Code. Since the Transformer models contain no recurrence and convolution operations, the positional information is provided by learning the positional embeddings and summed with the inputs at encoder and decoder. The learned Code Embeddings are also added along with input embeddings before sending to the decoder to provide more contextual information on python source code. The predictions of the decoder were passed to the test code block, which generates a well-indented source code.

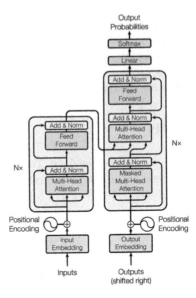

Fig. 3. The transformer - model architecture [18]

3   Dataset

The Dataset is a parallel corpus of a description of the problem in Natural Language (source) and python source code (target). An example record is shown in the Fig. 1. The data is collected from various sources and it is open for public use.

3.1   Extraction and Preparation

The Dataset is extracted from public GitHub repositories labeled as containing Python source code[1], Stackoverflow[2], ProjectEuler[3], and few examples were crowdsourced from graduate students. Crowd sourcing of Python examples helped to reduce class imbalance problem. 12473 files were successfully collected by crowdsourcing and extraction. The dataset has many duplicate source codes, very noisy, and the code examples are very diverse. They contain simple code snippets of a single line and many lines, simple functions, complex functions, code consisting of many functions, classes, etc., and need a lot of preprocessing. The length of the code samples is ranging from 10 to 7199 characters. Preprocessing plays a very crucial role with this data because of the noise and the variance in the code lengths.

Initially, the number of pairs in the corpus is 42473 and it is further processed in the following ways. To build the Python corpus, we removed the source codes

---

[1] https://github.com/trending/python.
[2] https://stackoverflow.com/questions/405374/python-source-code-collection.
[3] https://projecteuler.net/.

which are of other languages. The empty lines and decorative elements do not contribute to training, hence removed. As NMT systems show lower translation quality on very long sequences [10], a simple rule is adopted to filter out all the source code snippets having a length higher than 1000 characters. The Natural Language intent is tokenized using Spacy English tokenizer and the python source code is by a custom tokenizer. The number of pairs in the corpus after preprocessing is 4299.

An example of an extracted Natural Language Description and Python source code is provided in Fig. 1.

3.2 Dataset Description

The final dataset consists of python source codes of varying lengths ranging from 10 characters to 1000 characters. An example extracted source code whose length is 10 characters is

```
a , b = b , a
```

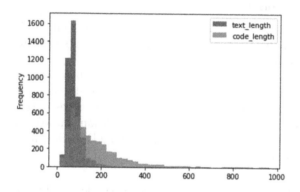

Fig. 4. Distribution of text and code lengths

The distribution of the Text and Source code lengths is shown in the Fig. 4. The majority of the questions have a length of 50 to 150 characters. But there is a lot of variance in the distribution of source code snippets. There is no relation between the Question text of the problem and the respective source code. The corpora summary statistics were reported in Table 1. The dataset is partitioned with a train-test-split of $85\% - 15\%$ into training and validation sets.

4 Experiments

This section discusses the training data preparation, baseline models, training details, and their outcomes. This study used two human-level translation evaluation criteria, which are BLEU and ROUGE scores along with two general evaluation metrics, Cross Entropy Loss and Perplexity. All the experiments given below are performed using GPU, Tesla K80 with 25 GB RAM provided by Google Colab. All the implementations are available for public use[4].

Table 1. Statistics of the corpora

	Mean	Std.	Median	Max	Min
Description	74.48	31.77	68	309	23
Source code	167.0	131.6	128	963	10

The transformer implementation is adopted as a baseline model and implemented with few changes. The models were evaluated in three training settings. First, both the source and target are tokenized using an English spacy tokenizer and fed to the transformers Encoder and Decoder layers. Second, the target source code is tokenized with the help of CuBERT [8] and fed to the decoder layer. Third, Code Embeddings were learned and added before fed to the decoder layer.

4.1 Baseline Model

Since the dataset is completely new, containing source codes only specific to the Python programming language, it is useful to assess its difficulty by providing some baseline results for other researchers to compare and improve upon. To obtain baseline results, we train the NMT model based on the Transformer architecture [18] by following the implementation of the Sockeye [6] toolkit for NMT. The encoder and decoder of the Transformer respectively have 3 sublayers and 8 heads for multi-head attention mechanism. We train our models with Adam optimizer [9] and ReduceLRonPlateau [13] scheduler. The remaining hyperparameters are listed in Table 2.

Table 2. Parameters used for Text2PyCode model training

Parameter	Value
Encoder layers	3
Decoder layers	3
Encoder embedding dimensions	512
Decoder embedding dimensions	512
Encoder attention heads	8
Decoder attention heads	8
Dropout	0.1
Learning rate	1e-3
Epochs	100
Batch size	64

---

[4] https://github.com/sridevibonthu/Text2PyCode.

Initial Model. This model is trained on the inputs tokenized by Spacy[5] English tokenizer which can split compound adjectives and nouns in English. In this setup, both the Encoder and Decoder layers fed the word embeddings created by following the Word2Vec [2,12] algorithm.

Tokenized Model. The Spacy tokenizer has only been evaluated on English, German and Spanish sentences and it has no model for any programming Language. The 16 tokens obtained for a python code snippet with the Spacy tokenizer is shown below. The exponentiation operator $**$ is tokenized into two, whereas $range(1$ into one token.

```
tokenize_text (" print ([ x ** 2] for x in range(1, 10))")

['print', '(', '[', 'x', '*', '*', '2', ']', 'for',
'x', 'in', 'range(1', ',', '10', ')', ')']
```

To address the above issue, a code tokenizer CuBERT is used to tokenize the source code before sending to the decoder layers. The CuBERT is a pre-trained contextual embedding of source code [8]. The above example after tokenization has resulted in 18 tokens as shown below.

```
tokenize_code (" print ([ x ** 2] for x in range(1, 10))")

['print', '(', '[', 'x', '**', '2', ']', 'for', 'x',
'in', 'range', '(', '1', ',', '10', ')', ')', '']
```

Final Model. NLP has improved tremendously after the success of the word embedding techniques [4]. The pre-trained word embeddings like Word2Vec [12], Glove [15], FastText [15] are all trained on natural languages and their context vectors cannot hold the semantics of source code tokens. With the help of the GenSim [16] library of Python, we have trained code embeddings on the CoNala [19] python corpus[6]. These pre-trained vectors are also added while building vocabulary to the model. The final model architecture is shown in the Fig. 2.

4.2   Metrics

To quantify the performance of the baselines, we choose the four performance measures, which are widely used in NMT studies. They are Cross-Entropy Loss, Perplexity, BLEU, and ROUGE scores. For loss, and perplexity lower values are better and for BLEU, and ROUGE the higher the value of the performance measure, the better the performance of the model. Cross Entropy loss is the very common loss function used in NMT and it is detailed in Eq. 1 where $|S|$ is

---

[5] https://spacy.io.
[6] https://conala-corpus.github.io/.

the length of the sentence, $|V|$ is the length of the vocabulary and $\hat{y}_{w,e}$ is the predicted probability of the vocab entry e on word w and $y_{w,e}$ can be either 1 or 0 based on whether the vocab the vocab entry is correct or not.

$$\sum_{w=1}^{|S|}\sum_{e=1}^{|V|} y_{w,e} log(\hat{y}_{w,e} \tag{1}$$

Perplexity of a Language Model is the average branching factor in predicting the next word [3]. Perplexity is calculated using the Eq. 2, where N is the number of words. Lower perplexity value indicates a better model.

$$PP(W) = \frac{1}{P(w_1, w_2, \cdots w_n)^{\frac{1}{N}}}$$
$$= N\sqrt{\frac{1}{P(w_1, w_2, \cdots w_n)}} \tag{2}$$

BLEU [14] calculates the similarity by computing the n-gram precision of a system translated sentence to the reference sentence. Its computation is shown in Eq. 3. where p_n is the geometric average of the modified n-gram precision, using N number of N-grams and w_n are positive weights summing to one. BP stands for brevity penalty. It is computed based on the length of the candidate translation and reference length.

$$BLEU = BP \cdot exp \left(\sum_{n=1}^{N} w_n logp_n \right) \tag{3}$$

ROUGE is a framework for automatic evaluation of summaries [11]. The variants of this metric are ROUGE-N, ROUGE-L, ROUGE-W, ROUGE-S. We reported ROUGE-1, ROUGE-2 and ROUGE-L for our experiments. ROUGE-N is computed as shown in Eq. 4 where n stands for the length of the n-gram, $gram_n$ is the maximum number of n-grams co-occuring in a candidate summary and $Count_{match}(gram_n)$ is the n-grams co-occurring in a set of reference summaries.

$$ROUGE - N = \frac{\sum_{S \in Reference Summaries} \sum_{gram_n \in S} Count_{match}(gram_n)}{\sum_{S \in Reference Summaries} \sum_{gram_n \in S} Count(gram_n)} \tag{4}$$

5   Results

Results of all three experiments on the Python parallel corpus are tabulated. We report Cross Entropy loss and perplexity in Table 3 and BLUE and ROUGE scores in Table 4. Perplexity and loss of both Tokenized and the proposed final model Fig. 2 have a very small difference. BLEU score and ROUGE score are clearly stating that the proposed model which equips both code tokens and code embeddings is performing well.

Table 3. Validation loss and perplexity

Model	Tokenization	Embeddings	Val. loss	Perplexity
Initial model	No	No	1.412	4.105
Tokenized model	Yes	No	1.236	3.443
Final model	Yes	Yes	**1.218**	**3.382**

Table 4. BLEU and ROUGE Scores on the three models

Model	Tokenization	Embeddings	BLEU		ROUGE		
					R1	R2	RL
Initial model	No	No	20.59	P	73.7	60.0	78.9
				R	67.8	57.2	68.8
				F1	67.1	56.5	71.6
Tokenized model	Yes	No	26.74	P	76.6	65.1	84.4
				R	79.3	66.8	79.0
				F1	75.4	64.0	79.9
Final model	Yes	Yes	**32.40**	P	**79.5**	**68.9**	**85.1**
				R	**81.5**	**71.3**	**81.3**
				F1	**78.4**	**68.3**	**82.1**

6 Conclusion

Automatic Code Generation has the potential to make programmers working in software companies or projects more efficient by allowing them to integrate various codes more easily from natural language intents thereby it improves the programmer's productivity. In this paper, we aim to create a parallel corpus and build a model based on Transformer architecture for CG. In order to enhance the model, code tokenization and code embeddings are leveraged. The experimental results of the proposed architecture on the source code dataset are on par with the results of Natural Language translation. This work can be extended to generate comments along with source code for the problem descriptions. It can also be extended to translate source code from one language to another.

References

1. Bahdanau, D., Cho, K., Bengio, Y.: Neural machine translation by jointly learning to align and translate. arXiv preprint arXiv:1409.0473 (2014)
2. Bojanowski, P., Grave, E., Joulin, A., Mikolov, T.: Enriching word vectors with subword information. Trans. Assoc. Comput. Linguist. **5**, 135–146 (2017)
3. Chen, S.F., Beeferman, D., Rosenfeld, R.: Evaluation metrics for language models (1998)
4. Chen, Z., Monperrus, M.: A literature study of embeddings on source code. arXiv preprint arXiv:1904.03061 (2019)

5. Gu, X., Zhang, H., Kim, S.: Deep code search. In: 2018 IEEE/ACM 40th International Conference on Software Engineering (ICSE), pp. 933–944. IEEE (2018)
6. Hieber, F., et al.: Sockeye: a toolkit for neural machine translation. arXiv preprint arXiv:1712.05690 (2017)
7. Iyer, S., Konstas, I., Cheung, A., Zettlemoyer, L.: Summarizing source code using a neural attention model. In: Proceedings of the 54th Annual Meeting of the Association for Computational Linguistics (Volume 1: Long Papers), pp. 2073–2083 (2016)
8. Kanade, A., Maniatis, P., Balakrishnan, G., Shi, K.: Learning and evaluating contextual embedding of source code. In: International Conference on Machine Learning, pp. 5110–5121. PMLR (2020)
9. Kingma, D.P., Ba, J.: Adam: a method for stochastic optimization. arXiv preprint arXiv:1412.6980 (2014)
10. Koehn, P., Knowles, R.: Six challenges for neural machine translation. arXiv preprint arXiv:1706.03872 (2017)
11. Lin, C.Y.: Rouge: a package for automatic evaluation of summaries. In: Text Summarization Branches Out, pp. 74–81 (2004)
12. Mikolov, T., Chen, K., Corrado, G., Dean, J.: Efficient estimation of word representations in vector space. arXiv preprint arXiv:1301.3781 (2013)
13. Mukherjee, K., Khare, A., Verma, A.: A simple dynamic learning rate tuning algorithm for automated training of DNNs. arXiv preprint arXiv:1910.11605 (2019)
14. Papineni, K., Roukos, S., Ward, T., Zhu, W.J.: Bleu: a method for automatic evaluation of machine translation. In: Proceedings of the 40th Annual Meeting of the Association for Computational Linguistics, pp. 311–318 (2002)
15. Pennington, J., Socher, R., Manning, C.D.: Glove: global vectors for word representation. In: Proceedings of the 2014 Conference on Empirical Methods in Natural Language Processing (EMNLP), pp. 1532–1543 (2014)
16. Rehurek, R., Sojka, P., et al.: journal=Retrieved from genism. org, y.: Gensim-statistical semantics in python
17. Sutskever, I., Vinyals, O., Le, Q.V.: Sequence to sequence learning with neural networks. arXiv preprint arXiv:1409.3215 (2014)
18. Vaswani, A., et al.: Attention is all you need. In: Advances in Neural Information Processing Systems, pp. 5998–6008 (2017)
19. Yin, P., Deng, B., Chen, E., Vasilescu, B., Neubig, G.: Learning to mine aligned code and natural language pairs from stack overflow. In: International Conference on Mining Software Repositories, pp. 476–486. MSR, ACM (2018). https://doi.org/10.1145/3196398.3196408

Automated Short Answer Grading Using Deep Learning: A Survey

Sridevi Bonthu[1,3]($\boxtimes$)(iD), S. Rama Sree[2](iD), and M. H. M. Krishna Prasad[3](iD)

[1] Vishnu Institute of Technology, Bhimavaram, Andhra Pradesh, India
[2] Aditya Engineering College, Surampalem, Andhra Pradesh, India
ramasree_s@aec.edu.in
[3] Jawaharlal Nehru Technological University, Kakinada,
Kakinada, Andhra Pradesh, India

Abstract. Automated Short Answer Grading (ASAG) is the task of assessing short answers authored by students by leveraging computational methods. The task of ASAG is investigated for many years, but this task continues to draw attention because of the associated research challenges. One of the core constraints of ASAG is the limited availability of domain-relevant training data. The task of ASAG can be tackled with several approaches and they can be broadly categorized into the traditional approaches based on handcrafted features and the Deep Learning based approaches. Researchers are applying Deep Learning Approaches for the past five years to address this problem owing to the increasing popularity of this area. The paper aims to summarize various existing deep learning approaches researchers followed to address this problem and to investigate whether Deep Learning based techniques are outperforming traditional approaches from the selected 38 papers. The paper also outlines several state-of-the-art datasets that can be used to do this work and the evaluation metrics to be used for both Regression and Classification settings.

Keywords: Short Answer Grading · Natural Language Processing · Deep learning · Evaluation metrics · Corpora · LSTM · Attention

1 Introduction

The application of Natural Language Processing(NLP) techniques to assess the short answers authored by students is referred to as Automatic Short Answer Grading (ASAG). The prime objective of ASAG is to automatically score the free-text answers from the students according to the corresponding reference answers [43], and it has attracted great attention from a variety of research communities. The alarming spread of the novel Corona virus (COVID-19) with its domino effect placed many industries in crisis. This outbreak has not spared the education sector either. Institutions shifted to online teaching mode as is has become a preferred way to facilitate learning during the lockdown period and

© IFIP International Federation for Information Processing 2021
Published by Springer Nature Switzerland AG 2021
A. Holzinger et al. (Eds.): CD-MAKE 2021, LNCS 12844, pp. 61–78, 2021.
https://doi.org/10.1007/978-3-030-84060-0_5

with this approach the assessment has become a major challenge. Assessment can be done through multiple choice questions (MCQ), one-word answers, short answers, and essay answers [12]. Short and essay answers have been recognized as tools to perform a deeper assessment of student's knowledge than MCQs. Because of the powerful technology available now-a-days, many are learning new concepts through MOOCs. Most of the online courses follow peer grading for accurate answer grading [49], but peer graders possess different mindsets thereby, a lot of variation in grades obtained. Grading short answers and essay answers accurately by human graders is becoming tedious as they have to evaluate a soft copy. Automation in the evaluation of short and essay answers is a growing need in the field of education through which evaluation can be done easily, fairly, quickly, and without any bias. There is also a need of providing useful feedback on answers to students which is again a cumbersome job to teachers.

In the learning process, the assessment of knowledge plays a key role in effective teaching [32]. Manual scoring takes a considerable amount of time and the provision of meaningful feedback even takes more. Manual scoring of answers can suffer from inconsistency since the human grader must infer meaning from the candidate's answer which is a free text comprised of the candidate's own words. The human grader may also get strained after evaluating few responses and the way he corrects the other responses may also change. But on the other hand, accepting free text answers from students is a widely preferred assessment tool, and should be used throughout the learning process, due to their effectiveness in developing the cognitive skills of students and also demonstrating knowledge in short texts [28]. Therefore there is a need to develop tools for addressing these challenges in assessment.

ASAG is not a new approach, but it needs to adopt the latest technologies in the current scenarios has become important. The problem of short answer grading has attracted significant attention of researchers over the years. Various approaches, starting from traditional hand-crafted features [32, 45] to more recent deep learning models [41] and their combination [27] were available. As shown in Fig. 1 the method-eras for the ASAG problem was viewed as rule-based and statistical methods [8] earlier. From 2007 people started using Machine Learning and from 2015 using Deep Learning. Due to the increased popularity of Deep Learning in the areas of computer vision, speech and text, researchers started employing it in this domain, to achieve accurate automatic grading systems. Therefore we researched the works that employed deep learning methods to solve the ASAG problem as no survey specific to deep learning methods is present.

The ASAG problem has been modelled as a supervised learning problem. It can be viewed as either a regression task or a classification task based on whether the student answer is assigned a mark/grade or categories like 'correct', 'partially correct', 'incorrect' [44]. Input to this problem is a pair of short answers consisting of Reference Answer (Q_r) and Student Answer (Q_s) to a question Q. Output is a label(classification) or a grade(Regression) based on the extent of similarity between Q_r and Q_s.

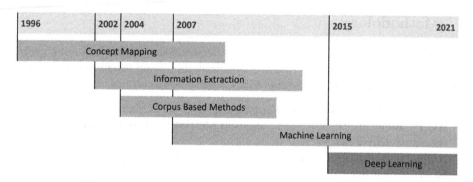

Fig. 1. The era of ASAG system.

For regression models, the objective is to learn

$$Y = f(\vec{X}, \vec{w}) \tag{1}$$

where $\vec{X}$ refers to the input feature vector obtained by finding similarity between Q_r and Q_s. $\vec{x} = x_1, x_2, ...x_n$ is an n-dimensional similarity vector of a pair (Q_r, Q_s). $\vec{w}$ is the model parameters to be estimated.

For classification problems, the objective is to represent the score with a category k for every data instance, like

$$score(X_i, k) = \beta_k \cdot X_i \tag{2}$$

where β_k is the vector of weights corresponding to category k and X_i is feature vector based on similarity for data instance i i.e., the pair (Q_r, Q_s). The final category is obtained by considering the highest scored category.

$$k^* = \underset{i}{argmax} \quad score(X_i, k) \tag{3}$$

The goal of this paper is to study the Deep Neural Network approaches followed for this task and their impact on ASAG when compared with traditional machine learning approaches as they are capable of automatically inducing syntactic and semantic features from the text [26]. To the best of our knowledge, there are three [8,18,42] literature studies on Short Answer Grading without restrictions on the approach. There is no study dedicated to Deep Learning approaches to solve the ASAG task.

The organization of the remaining paper is as follows. In Sect. 2 the methodology of the research is presented, Sect. 3 addressed the existing Corpora to carry out this task, Sect. 4 presents the overview of the Evaluation Metrics, Sect. 5 discusses the various state-of-the-art Deep Learning approaches. Section 6 presents the observations of the authors for the framed Research Questions. Finally, the conclusions of the work can be seen in Sect. 7.

2   Methodology

This survey seeks to explore, analyze and understand the current state-of-the-art of ASAG with a focus on the ASAG works that used deep learning approaches. The elaborated Research Questions to address the objective of the survey are:

- **RQ1:** *"What are the various datasets available to perform ASAG?"*
- **RQ2:** *"What are the various Evaluation Metrics used to measure the performance of ASAG tasks?"*
- **RQ3:** *"Which Deep Learning approaches are used?"*
- **RQ4:** *"What are the results obtained?"*

The search keywords identification is made based on the preliminary research. The identified keywords are *"Automatic, Short Answer Grading, Scoring, Assessment, Natural Language Processing, Deep Learning, Question, Answer, Response, etc."*. Similar words from the keywords are grouped and a search query is created using Boolean operators to search for research contributions. The primary online databases considered as sources for this survey are *Google Scholar, IEEE Xplore, ACM Digital Library, Elsevier-ssrn, EBSCO, ACL Anthology.* ACL Anthology[1] is a great source for this survey as it hosts 64000 papers especially on the study of NLP. The total number of retrieved papers from the databases is 2000 in number. After removing the duplicates, we were left with 676 papers. In the next stage, we removed few papers based on the title, as the survey mainly concentrates on the Deep Learning approaches. The resultant 87 papers were skimmed through the title, abstract, introduction, their contributions, model architecture, novelty, published venue etc., After the filtering process, the number of papers considered for this survey is 38. From these 38 papers, the datasets, metrics, deep learning techniques are further studied concerning the framed Research Questions.

3   Corpora

One of the primary challenges in addressing the ASAG problem is the non-availability of the datasets of natural responses. A great variety of datasets are used in the reviewed papers, they show a lot of variation in terms of the language, the topic of the question, grading scale, number of questions, reference answers. This section describes six majorly used datasets of the English language from the 38 reviewed papers and provides an overview of the dataset details and their pros and cons. A prime observation is that many of the ASAG datasets like ASAP, SemEval-2013, Joint SRA(Beetle & ScientsBank) are released through competitions. Table 1 provides a glimpse of example records from the Texas and SemEval-2013 datasets. Table 2 provides an idea about the usage of these datasets by the Deep Learning community.

ASAP - Kaggle provides a dataset for doing ASAG with the name ASAP-AES, for Hewlett Foundations Automated Assessment Prize competition

---
[1] https://www.aclweb.org/anthology/.

Table 1. Example of model question, reference answers and students answers from SemEval-2013 and Texas datasets.

Texas	
Question	What is the role of a prototype program in problem solving?
Model Answer	To simulate the behaviour of portions of the desired software product
Model Vocabulary	Simulate, behaviour, portion, desire, software, product
Student Answer 1	High risk problems are address in the prototype program to make sure that the program is feasible. A prototype may also be used to show a company that the software can be possibly programmed
Student Answer 2	It simulates the behavior of portions of the desired software product
SemEval-2013	
Question	Lee has an object he wants to test to see if it is an insulator or a conductor. He is going to use the circuit you see in the picture Explain how he can use the circuit to test the object
Reference Answer	If the motor runs, the object is a conductor
Student Answer	He could know if it works

(ASAP) on Kaggle [20]. It can be downloaded from the official website of Kaggle[2]. This dataset contains 10686 samples belong to 8 different sets of essays and each of them is generated from a single prompt. These essays range from a length of 150 to 550 words per response on average. All the essays in the dataset were hand-graded by either two or three instructors. The main challenge with this dataset is every set has a different grading scale.

Beetle and ScientsBank - Students Response Analysis (SRA) was a task of annotating student-authored answers with categories that in turn will help the dialog systems to generate suitable and useful feedback on errors. The SRA corpus mainly consists of two distinct corpora: (1) BEETLE data, which is entirely based on transcripts of students interacting with the BEETLE II tutorial dialogue system [13], and (2) SCIENTSBANK data, which is based on the corpus of student answers to assessment questions collected by Nielsen et al. [36]. The BEETLE corpus mainly comprised of 56 questions in the domain of basic electricity and electronics requiring one or two sentence answers, and it has nearly 3000 student answers to those 56 questions. The SCIENTSBANK corpus contains approximately 10,000 answers to 197 assessment questions in 15 different science domains. Student answers in the BEETLE corpus are annotated manually by trained human using a scheme that straightforwardly maps to SRA annotations. A fine-grained scheme that automatically labels using a set of question-specific heuristics and also manually revising them based on the definition of the class [13], is adopted to convert the labels of SCIENTSBANK corpus into SRA labels. The researchers who want to work on these datasets need to further filter and transform the corpus to produce training and test data sets.

---

[2] https://www.kaggle.com/.

Texas - This dataset consists of 80 questions collected from an undergraduate course titled Data Structures. Questions scattered across ten different assignments and two tests, each on a related set of topics along with $2,273$ student responses. Students Answers were graded by two human graders who consider the model answer provided for each question. The average of the two scores assigned by human graders is used as the final gold score for each student answer [14]. The dataset can be obtained from the archive hosted at the website[3].

Cairo - Cairo University's dataset consists of a total of 610 questions which are 10 answers for 61 questions. These are collected from only one chapter of the official Egyptian curriculum for the Environmental Science course. The average length of a student's answer is 2.2 sentences, 20 words, or 103 characters. The dataset contains a collection of students' responses along with their grades that vary between 0 and 5 according to an assessment of two human evaluators. An English version of the Cairo University data set is also available to research this area. This dataset can be downloaded from the webpage[4].

Powergrading - The Powergrading dataset [3] contains 10 individual prompts from U.S. immigration exams with about 700 responses each. Each prompt is accompanied by one or more reference responses. Responses in this dataset are repetitive as they are very short and the percentage of correct answers is very high. This dataset can be utilized to test the model's ability to perform well on extremely short responses. It was originally used for the task of (unsupervised) clustering [3] so that there are no state-of-the-art scoring results available for this dataset.

Statistics - Stefano Menini et al. released a dataset [30] with the name Statistics to perform short answer grading which is publicly available at the webpage[5]. This data has been partially collected using data from the real statistics exams spanning different years, and the same is partially extended by the authors of this paper. The dataset contains the group of sentences written by students, with a unique sentence ID, the type of statistical analysis it refers to, its degree in a range from 0 to 1, and its fail/pass result, flanked with a manually defined gold standard (i.e., the correct answer).

4   Evaluation Metrics

Evaluation metrics are used to quantitatively measure the performance of the ASAG models or to compare the performance with the baselines. As show in Fig. 2 different metrics can be used based on the way the ASAG system is built i.e., as a classifier or as a regressor. This section provides an insight into the metrics that can be used with ASAG regressor and classifier.

---

[3] http://lit.csci.unt.edu/index.php/Downloads.

[4] http://www.aucegypt.edu/src/datasets.htm.

[5] https://zenodo.org/record/3257363.

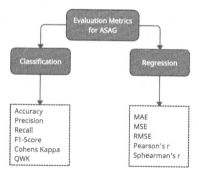

Fig. 2. Evaluation Metrics for ASAG task.

4.1   Regression Metrics

Root Mean Square Error (RMSE). RMSE is a standard way to measure the performance of a regression model. Formally it is defined as "root of the residual sum of squares resulting from comparing the predictions $\hat{y}$ and ground truth y" [19]. RMSE is calculated by following the Eq. 4.

$$\sqrt{\frac{1}{n}\sum_{i=1}^{n} n\frac{(\hat{y}_i - y_i)^2}{n}} \qquad (4)$$

Concerning to ASAG task, in the Eq. 4 $\hat{y}_1, \hat{y}_2, ..., \hat{y}_n$ are predicted grades, and $y_1, y_2, ..., y_n$ are grades assigned by human graders and n is the number of observations. RMSE measure is simple to calculate, but the drawback is the RMSE value depends on the order of magnitude of the observed values [35].

Correlation Co-efficient. The correlation coefficient measures the strength of association between two variables. The value of the coefficient indicates the degree of association and it varies between $+1$ and -1 [5]. A positive value is required for the ASAG task, as it needs a strong correlation between reference answers and student answers. Researchers are using the Pearson Correlation coefficient or Spearman's correlation coefficient for short answer grading tasks. Pearson's r is calculated as shown in the Eq. 5.

$$r = \frac{\sum(x_i - \bar{x})(y_i - \bar{y})}{\sqrt{\sum(x_i - \bar{x})^2 \sum(y_i - \bar{y})^2}} \qquad (5)$$

where r is pearson's correlation coefficient, x_i is the grade assigned by the human grader, $\bar{x}$ is the mean of the values of x-variable, y_i is the grader predicted by the model and $\bar{y}$ is the mean value of the y variable.

The Spearman's ρ is calculated by using the Eq. 6.

$$\rho = 1 - \frac{6\sum d_i^2}{n(n^2 - 1)} \qquad (6)$$

where ρ is the Spearman's correlation coefficient, d_i is the difference between the human assigned grade and the predicted grade and n is the number of observations.

Pearson's r evaluates the linear relation between two variables whereas Spearman's ρ is based on the ranked values for each variable rather than the raw data.

4.2 Classification Metrics

4.3 F1-Score

F1-score is the common performance measure of classifiers. It is the harmonic mean of Precision and Recall and it is calculated using the Eq. 7.

$$F1Score = \frac{2 \times Precision \times Recall}{Precision + Recall} \tag{7}$$

where

$$Precision = \frac{TruePositive}{TruePositive + FalsePositive}$$

$$Recall = \frac{TruePositive}{TruePositive + FalseNegative}$$

F1-score metric is more preferable when the class distribution is imbalanced [16]. The two variants of the F1-score used for the ASAG task are *macro-F1* and *weighted-F1* scores.

Macro-averaged-F1 score or *macro-F1* is an arithmetic mean of the per-class F1 scores, It is used to assess the quality of problems with multiple binary labels or classes, but it gives the same importance to each label/class [34]. Its value varies from 0 to 1, and 1 is the best value. It is computed by using the Eq. 8.

$$Macro - F1 = 2 \times \frac{MacroAveragePrecision \times MacroAverageRecall}{MacroAveragePrecision^{-1} + MacroAverageRecall^{-1}} \tag{8}$$

where k is the generic class and K is the Number of classes

$$MacroAveragePrecision = \frac{\sum_{k=1}^{K} Precision_k}{K}$$

$$MacroAverageRecall = \frac{\sum_{k=1}^{K} Recall_k}{K}$$

$$Precision_k = \frac{TP_k}{TP_k + FP_k}$$

$$Recall_k = \frac{TP_k}{TP_k + FN_k}$$

Weighted-average-F1 score, or *weighted-F1*, is the weighted F1 score of each class by the number of samples from that class [38].

Cohen's Kappa. Cohen's kappa statistic handles both multi-class and imbalanced class problems well and it is developed to account for the possibility that answer graders guess on at least some variable due to uncertainty [29]. It is calculated by following the Eq. 9.

$$\kappa = \frac{p_o - p_e}{1 - p_e} = 1 - \frac{1 - p_o}{1 - p_e} \tag{9}$$

where p_o and p_e are observed and expected agreement respectively. The kappa value can range from -1 to +1. Values of 0 or less, indicate that the trained answer grader is useless. It is the most widely used statistic, but the acceptable level of kappa value is questioned in few areas like health research [29].

Quadratic Weighted Kappa. The QWK metric is used to calculate the level of agreement between two ratings [6]. Concerning the ASAG task, it can be used to find the agreement between the predicted grades and the ground truth. It also considers the *by chance* probability of assigning the same grade to a sample by both raters. Generally, it ranges from 0 to 1 and it can also be negative if there is less agreement. To calculate QWK, first, the weight matrix W is constructed according to Eq. 10.

$$W_{i,j} = \frac{(i - j)^2}{(N - 1)^2} \tag{10}$$

where i is the rating assigned by the human grader, j is the predicted rating and N is the total number of ratings. Next, the QWK is calculated according to the equation.

$$k = 1 = \frac{\sum_{i,j} W_{i,j} O_{i,j}}{\sum_{i,j} W_{i,j} E_{i,j}} \tag{11}$$

Here the matrix O contains the scores observed such that rating i is given by human grader and j is assigned by the model. O_{ij} corresponds to the adoption records that have a rating of i and predicted a rating of j. E is the histogram matrix of expected ratings, obtained by multiplying the histogram vectors of both human grader score and model predicted score.

5 Deep Learning Approaches

The researchers employed various mechanisms like Transfer Learning, Siamese LSTM, clustering, Latent Semantic Analysis, Bidirectional Transformers, Paragraph Embeddings, Deep Autoencoders, Attention Networks, Transformer based pretraining from the past few years. Recent advancements in the domain of deep learning for NLP made it promising to use deep learning architectures, such as the Attention mechanism, Transformer, for increasingly complex NLP tasks. Some of the promising contributions using Deep Learning based experiments from 2016 were shown in Fig. 3 and the Table 2 summarizes their performance. Many of the researchers have done experimentation on LSTM [22,33,39,51] and its variants.

Table 2. Individual Results on ASAG: Outperformed Deep Learning approaches along with their attained scores. Most of the models used variants of LSTM.

	Year	Approach	Corpus	Evaluation							
				Accuracy	Macro-F1	Weighted-F1	QWK κ	Cohen's κ	Pearson's r	Spearman's ρ	RMSE
[33]	2016	LSTM	SICK	–	–	–	–	–	0.88	0.83	–
[1]	2016	LSTM	ASAP	–	–	–	–	0.96	0.96	0.91	2.4
[22]	2017	Bi-LSTM	Sem-Eval	–	–	–	–	–	0.55	–	0.75
			Texas	–	–	–	–	–	0.64	–	0.83
[41]	2017	LSTM	ASAP	–	–	–	0.74	–	–	–	–
			Powergrading	–	–	–	90.36	–	–	–	–
			SRA	–	74.54	–	–	–	–	–	–
[10]	2017	Bi-LSTM	Sem-Eval	0.76	–	–	–	–	–	–	–
[43]	2018	LSTM	Sem-Eval	79.26	78.58	79.1	–	–	–	–	–
			Texas	–	–	–	–	–	0.57	–	0.9
[24]	2019	LSTM	LSI	66.36	63.09	65.58	–	–	–	–	–
			K12	88.9	81.5	–	–	–	–	–	–
[14]	2019	LSTM	Texas	–	–	–	–	–	0.63	–	0.89
			Cairo	–	–	–	–	–	0.79	–	0.92
[50]	2019	Transformers	Custom	80.17	81.5	–	–	–	–	–	–
[46]	2019	Transformers	ScientsBank	75.9	72.0	75.8	–	–	–	–	–
[9]	2020	Transformers	SemEval	79.7	79.1	79.7	–	–	–	–	–
			ScientsBank	76.0	75.0	76.0	–	–	0.65	–	0.88
[40]	2021	LSTM	Ukara	–	–	–	–	–	–	–	–

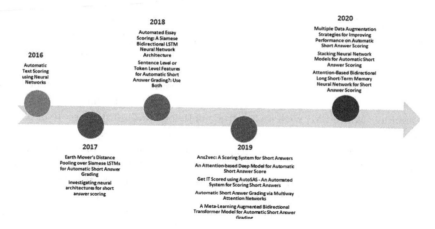

Fig. 3. Prominent contributions using Deep Learning.

LSTM - LSTMs are complex activation units that can selectively remember or forget things. Kumar et al. [22] applied Siamese bidirectional LSTMs, a pooling layer based on Earth-Mover-Distance (EMD) across all hidden states from both the LSTMs of the Siamese Network and a flexible final regression layer to output scores to address the ASAG problem. Authors enhanced training through a task-specific data augmentation strategy and they experimented on a publicly available dataset (SemEval) and scored an RMSE of 0.830 superior to the baselines on LSTM. Riordan et al. [41] explored the effectiveness of multiple architectures on three publicly available datasets: ASAP-SAS, Powergrading, SRA and answered the questions like convolution layer produce useful features or not, can we use smaller hidden layers, the role of bi-directional LSTMs, Attention and concluded that the basic neural architectures of pretrained embedding with LSTM is reasonably effective architecture for short answer grading. Conneau et al. [10] proposed a novel Joint Multi-Domain neural model (JMD-ASAG)for ASAG by using bi-LSTMS and similarity scorers per domain, that learns both generic and domain-specific aspects simultaneously. It achieves this by utilizing multiple domain-specific corpora SemEval-2013, and without requiring a large generic corpus. Prabhudesai et al. [39] proposed a Siamese Bi-directional LSTM Neural Network based Regressor which combines the benefits of both Deep Learning and Feature Engineering. The novel approach, authors followed is the augmentation of the data based on highly rated reference answers to improve the training.

Attention - It is the most influential idea in the areas of Deep Learning for NLP. The central idea behind Attention is to utilize all the states of intermediate encoders to construct the context vectors required by the decoders. Liu et al. [24] proposed a generalized end-to-end ASAG framework that aims at extracting linguistic information from the student and reference answers automatically and to model the semantic relations between student and reference answers accurately. The evaluation of this model is done on a real-world K12 dataset. Their model leveraged the multi-way attention and transformer layers to improve the matching between words in a sentence [47]. Authors adopted AUC and accuracy as metrics and this model outperformed several state-of-the-art baselines like Logistic regression, Gradient boosted decision tree, Multichannel convolutional neural networks, Sentence embedding by Bidirectional Transformer block (Bi-Transformer), Multiway Attention Network (MAN), Manhattan LSTM with max-pooling (MaLSTM) with an accuracy of 0.8899 and an AUC score of 0.9444. Gong et al. [15] also proposed a deep learning based method with an attention mechanism to solve this task. The proposed method combines pretrained embedding word vector and Recurrent Neural Network (RNN) model with attention to learn answer vector and then learns response answer vector and reference answer vector are fed to the logistic regression model to predict response answer score. Authors are claiming that they have achieved a relative 10% increase in performance compared to the baseline model results by over 8% in some of the question prompts as evaluated by Quadratic Weighted Kappa (QWK), showing performance comparable to humans.

Transformers - The paper 'Attention Is All You Need' [48] introduced a novel architecture called Transformer. As the title indicates, it leverages the attention mechanism. BERT(Bidirectional Representation of the Transformer) is the state-of-the-art model for learning textual representations. Wang et al. [50] introduced ml-BERT method for grading short answer questions to advance the existing models in ASAG when the training data is not sufficient. The authors combined BERT, with meta-learning, a training framework that leverages additional data and learning tasks to improve model performance when labeled data is limited. With the incorporation of meta-learning, the model achieved a remarkable accuracy of 80.17% and an F1 score of 0.815 when compared with the baseline BERT model whose accuracy is 77.8%. Camus et al. [9] investigated Transformers for ASAG with fine-tuning different pretrained Transformer based architectures and they also showed that models trained with knowledge distillation are feasible for use in short answer grading. Sung et al. [46] proposed two ways to update the pre-trained BERT language model for the short answer grading. The authors illustrated the utilization of unstructured textbook data and labeled question-answer data for the model update. On the benchmarking dataset of SemEval-2013, they report up to 10% absolute improvement in macro-average-F1 over state-of-the-art results. The authors also addressed the training time to come up with an optimal model and indicated that task-specific transfer takes place within the initial few epochs only.

Some authors experimented skip-thought vectors, various kinds of embeddings with deep learning models. Saha et al. [43] developed novel token-level features that are specifically tuned for understanding partially correct sentences. Gomaa et al. [14] proposed a scoring system for short answers, with the name Ans2vec. This approach utilizes skip-thought vectors to convert model and student's answers into meaningful vectors to measure similarity between them. This model tested against 3 different benchmarking datasets and for the Texas dataset it achieved 0.63 as Pearson correlation value. Yaman, et al. [23] proposed a supervised regression model, AutoSAS which can be used to grade short answers easily in a classroom setting. This work mainly leveraged the features such as Weighted Keywords and Word2Vec/Doc2Vec embeddings. Examples of such features are prompt information, weighted keywords, lemmatized response, and lexical overlap. Authors have also concluded that the additional features like word frequency, difficulty, statistics of the word, and sentence length do not figure highly either in ranking or the accuracy values being affected significantly. Authors claim that AutoSAS obtained state-of-the-art performance and achieves better performance.

6   Results

This study is conducted to help the researchers who want to solve the ASAG problem using deep learning. To fulfil this aim we framed a set of four Research questions and presented in Sect. 2. We have studied a set of recent 38 papers on Deep Learning approaches to solve Short Answer grading. Our findings and observations are summarized below.

RQ1: *"What are the various datasets available to perform ASAG?"* - To address this question, we have studied the strengths and weaknesses of the datasets that are used by a majority of the research community and presented the details in Sect. 3. Deep Learning requires more data than traditional machine learning to train the model [2]. Many of the datasets meant for short answer grading have a minimal number of records. The student answers in few datasets are either well-formed or nonsentential responses. There is a need for good corpora with many training records to apply deep learning. Another observation is that the questions in the datasets are collected from the subjects like Environmental science, Data Structures, etc. None of the datasets cover questions from programming subjects to test the real strengths of Deep Learning approaches.

RQ2: *"What are the various Evaluation Metrics used to measure the performance of ASAG tasks?"* - Evaluation metrics of the ASAG task vary based on the way the problem is solved. Section 2 provides an insight on the metrics for both the Regression and the Classification setting of the ASAG task. Most ASAG challenges specify the metric to measure the performance. It is easy to calculate all ASAG metrics and most of them are available as part of the open source libraries available.

RQ3: *"Which Deep Learning approaches are used?"* - Approaches followed in all the 38 papers were studied. The pros and cons observed in papers that have proposed well performing models are presented in Table 3. Researchers have used various deep neural networks like Fully Connected Networks, Convolutional Neural Networks, Recurrent Neural Networks like LSTM and transformers to address the ASAG task. We have observed that many authors have acquired good performance through LSTM with attention, but LSTM are sequential in nature and cannot benefit the usage of GPUs.

RQ4: *"What are the results obtained?"* - The Deep Learning model will not alone determine the success of the ASAG system pipeline in production. The

Table 3. Strengths and weaknesses of the Deep Models of ASAG

	Pros	Cons
[1]	Learnt and used sentence specific embeddings	Not able to distinguish the occurrence of single word in multiple places as gradients are calculated at the end
[22]	Proposed a framework by cascading 3 neural building blocks. Siamese BiLSTM, pooling layers based on Earth-Mover-Distance, regression layer	
[41]	Experimented neural architectures and shown that they can outperform non-neural using LSTM	Performance is marginal in one datasets out of the tested 3
[39]	Proposed a siamese architecture with both CNN and LSTM	Convolution layer can process entire text at a time on top LSTM is placed, which is sequential
[43]	Proposed a feature encoding based on partial similarities of tokens	
[14]	Proposed an uncomplicated short answer grading model by employing skip-thought vectors	Achieved good accuracy on few but not on Cairo dataset
[15]	Followed new HanLP tokenization	
[23]	Augmentation, Ablation study, feedback provision	Tested on a single dataset
[24]	Parallel computation is possible as the model is based on transformer	Compared baselines are not of ASAG
[50]	A method to augment the BERT with meta-learning to improve its performance is proposed	Tested on their custom dataset
[27]	Used multiple data augmentation strategies, and also done ablation studies	Improvement in the performance is marginal
[40]	Introduced stacking architecture for XGBoost and Neural Networks, and handled class imbalance problems	Tested the approach on only one dataset

latency of the inference time, the ease of fine-tuning, and reproducibility of the model on a similar and smaller dataset also plays role in the success. Many of the ASAG systems proposed using Deep Learning approaches performed well when compared against several state-of-the-art models which used hand-crafted features and the time required for feature extraction is reduced. Researchers discussed the benefit of providing feedback to the students on their responses automatically, but no deep learning model has trained on that. Even though many Deep Learning models proposed and achieved accuracy, they still leave space for new techniques and methods to achieve higher performance. Among the models studied, the ones which employed attention mechanism performed well. Encoding text is at the heart of understanding language. The encoding models like Deep Averaging Network(DAN) which uses multi-task learning [37] are designed to be as general-purpose as possible. This implies that researchers can combine self-attention, co-attention, hierarchical-attention, and multiple-attention in different ways and experiment to achieve better performance. In the last year, the Transformer model was followed by Reformer [21], Longformer [4], GTrXL [11], etc., and BERT was followed by XLNet [54], RoBERTa [25], T5 [53], etc., and new tokenizations like Byte-Pair encoding, Word-Piece Encoding, Sentence-Piece Encoding were introduced [31]. Transformers are quadratic. These all can be tested for better accuracy on existing deep learning models of the ASAG system. GPT3 [7] is pretty huge with 175GB parameters, and because of its capacity GPU inferences can be costly. Quantization, pruning techniques, and usage of onnx can decrease the inference time 10–40 times [17]. We may not need a model as big as BERT-base all the time and the inference latency requirements to push a need to make smaller models [52]. Researchers should work on finding a sweet spot between training the model from scratch and transfer learning.

7 Conclusion and Future Challenges

An ASAG system that takes less amount of time to train, infer, and more accurate is the need of the now-a-days education sector. This is a research problem for many decades and researchers have started adopting Deep Learning approaches for the past 5 years. This paper examines the existing models based on deep neural networks, the corpora, and the evaluation metrics used for this task. Among the studied models, the ones which employed attention mechanism performed well and it is also identified that there are limited corpora to work in this area. There is a need for coming up with new good datasets which also involve recent subjects like programming languages and there is a lot of scopes to continue working in this domain by employing improved encoding mechanisms, GPT, quantization, pruning, etc.

In Sect. 5, we discussed the methodological decisions made by authors in current Deep Learning approaches for ASAG, and in Sect. 6, we also shared our views on the reviewed papers. Based on this study, potential next steps to ameliorate the results are formulated as our research agenda. The major challenges identified and that should be addressed in future research are - (a) A corpus covering technical questions with a good number of training examples is identified as a need, as many of the university campuses are training their students for placements. (2) An ASAG model which can learn from context and takes less training, inference time. (3) A model which provides immediate feedback along with the grade.

References

1. Alikaniotis, D., Yannakoudakis, H., Rei, M.: Automatic text scoring using neural networks. In: Proceedings of the 54th Annual Meeting of the Association for Computational Linguistics (Volume 1: Long Papers), Berlin, Germany, pp. 715–725. Association for Computational Linguistics, August 2016. https://doi.org/10.18653/v1/P16-1068. https://www.aclweb.org/anthology/P16-1068

2. Angelov, P., Sperduti, A.: Challenges in deep learning. In: ESANN (2016)
3. Basu, S., Jacobs, C., Vanderwende, L.: Powergrading: a clustering approach to amplify human effort for short answer grading. Trans. Assoc. Comput. Linguist. **1**, 391–402 (2013)
4. Beltagy, I., Peters, M.E., Cohan, A.: Longformer: the long-document transformer. arXiv preprint arXiv:2004.05150 (2020)
5. Benesty, J., Chen, J., Huang, Y., Cohen, I.: Pearson correlation coefficient. In: Benesty, J., Chen, J., Huang, Y., Cohen, I. (eds.) Noise Reduction in Speech Processing. STSP, vol. 2, pp. 1–4. Springer, Heidelberg (2009). https://doi.org/10.1007/978-3-642-00296-0_5
6. Brenner, H., Kliebsch, U.: Dependence of weighted kappa coefficients on the number of categories. Epidemiology 199–202 (1996)
7. Brown, T.B., et al.: Language models are few-shot learners. arXiv preprint arXiv:2005.14165 (2020)
8. Burrows, S., Gurevych, I., Stein, B.: The eras and trends of automatic short answer grading. Int. J. Artif. Intell. Educ. **25**(1), 60–117 (2015)
9. Camus, L., Filighera, A.: Investigating transformers for automatic short answer grading. In: Bittencourt, I.I., Cukurova, M., Muldner, K., Luckin, R., Millán, E. (eds.) AIED 2020. LNCS (LNAI), vol. 12164, pp. 43–48. Springer, Cham (2020). https://doi.org/10.1007/978-3-030-52240-7_8
10. Conneau, A., Kiela, D., Schwenk, H., Barrault, L., Bordes, A.: Supervised learning of universal sentence representations from natural language inference data. arXiv preprint arXiv:1705.02364 (2017)
11. Dai, Z., Yang, Z., Yang, Y., Carbonell, J., Le, Q.V., Salakhutdinov, R.: Transformer-XL: attentive language models beyond a fixed-length context. arXiv preprint arXiv:1901.02860 (2019)
12. Dwivedi, C.: A study of selected-response type assessment (MCQ) and essay type assessment methods for engineering students. J. Eng. Educ. Transform. **32**(3), 91–95 (2019)
13. Dzikovska, M.O., et al.: Semeval-2013 task 7: the joint student response analysis and 8th recognizing textual embodiment challenge. In: Second Joint Conference on Lexical and Computational Semantics (* SEM): Seventh International Workshop on Semantic Evaluation (SemEval 2013), vol. 2. Association for Computational Linguistics (2013)
14. Gomaa, W.H., Fahmy, A.A.: Ans2vec: a scoring system for short answers. In: Hassanien, A.E., Azar, A.T., Gaber, T., Bhatnagar, R., F. Tolba, M. (eds.) AMLTA 2019. AISC, vol. 921, pp. 586–595. Springer, Cham (2020). https://doi.org/10.1007/978-3-030-14118-9_59
15. Gong, T., Yao, X.: An attention-based deep model for automatic short answer score. Int. J. Comput. Sci. Softw. Eng. **8**(6), 127–132 (2019)
16. Grandini, M., Bagli, E., Visani, G.: Metrics for multi-class classification: an overview. arXiv preprint arXiv:2008.05756 (2020)
17. Guerra, L., Zhuang, B., Reid, I., Drummond, T.: Automatic pruning for quantized neural networks. arXiv preprint arXiv:2002.00523 (2020)
18. Hasanah, U., Permanasari, A.E., Kusumawardani, S.S., Pribadi, F.S.: A review of an information extraction technique approach for automatic short answer grading. In: 2016 1st International Conference on Information Technology, Information Systems and Electrical Engineering (ICITISEE), pp. 192–196. IEEE (2016)
19. Hyndman, R.J., Koehler, A.B.: Another look at measures of forecast accuracy. Int. J. Forecast. **22**(4), 679–688 (2006)

20. Kaggle: The Hewlett Foundation: Automated Essay Scoring—Kaggle. https://www.kaggle.com/c/asap-aes/. Accessed 04 Oct 2021
21. Kitaev, N., Kaiser, L., Levskaya, A.: Reformer: the efficient transformer. arXiv preprint arXiv:2001.04451 (2020)
22. Kumar, S., Chakrabarti, S., Roy, S.: Earth mover's distance pooling over Siamese LSTMs for automatic short answer grading. In: IJCAI, pp. 2046–2052 (2017)
23. Kumar, Y., Aggarwal, S., Mahata, D., Shah, R.R., Kumaraguru, P., Zimmermann, R.: Get it scored using autosas-an automated system for scoring short answers. In: Proceedings of the AAAI Conference on Artificial Intelligence, vol. 33, pp. 9662–9669 (2019)
24. Liu, T., Ding, W., Wang, Z., Tang, J., Huang, G.Y., Liu, Z.: Automatic short answer grading via multiway attention networks. In: Isotani, S., Millán, E., Ogan, A., Hastings, P., McLaren, B., Luckin, R. (eds.) AIED 2019. LNCS (LNAI), vol. 11626, pp. 169–173. Springer, Cham (2019). https://doi.org/10.1007/978-3-030-23207-8_32
25. Liu, Y., et al.: Roberta: a robustly optimized BERT pretraining approach. arXiv preprint arXiv:1907.11692 (2019)
26. Lopez, M.M., Kalita, J.: Deep learning applied to NLP. arXiv preprint arXiv:1703.03091 (2017)
27. Lun, J., Zhu, J., Tang, Y., Yang, M.: Multiple data augmentation strategies for improving performance on automatic short answer scoring. In: AAAI, pp. 13389–13396 (2020)
28. McDaniel, M.A., Anderson, J.L., Derbish, M.H., Morrisette, N.: Testing the testing effect in the classroom. Eur. J. Cogn. Psychol. **19**(4–5), 494–513 (2007)
29. McHugh, M.L.: Interrater reliability: the kappa statistic. Biochemia medica **22**(3), 276–282 (2012)
30. Menini, S., Tonelli, S., De Gasperis, G., Vittorini, P.: Automated short answer grading: a simple solution for a difficult task. In: CLiC-it (2019)
31. Mnasri, M.: Recent advances in conversational nlp: Towards the standardization of chatbot building. arXiv preprint arXiv:1903.09025 (2019)
32. Mohler, M., Bunescu, R., Mihalcea, R.: Learning to grade short answer questions using semantic similarity measures and dependency graph alignments. In: Proceedings of the 49th Annual Meeting of the Association for Computational Linguistics: Human Language Technologies, pp. 752–762 (2011)
33. Mueller, J., Thyagarajan, A.: Siamese recurrent architectures for learning sentence similarity. In: Proceedings of the AAAI Conference on Artificial Intelligence, vol. 30 (2016)
34. Narasimhan, H., Pan, W., Kar, P., Protopapas, P., Ramaswamy, H.G.: Optimizing the multiclass f-measure via biconcave programming. In: 2016 IEEE 16th International Conference on Data Mining (ICDM), pp. 1101–1106. IEEE (2016)
35. Neill, S.P., Hashemi, M.R.: Ocean modelling for resource characterization, Chapter 8. In: Fundamentals of Ocean Renewable Energy, pp. 193–235 (2018)
36. Nielsen, R.D., Ward, W.H., Martin, J.H., Palmer, M.: Annotating students' understanding of science concepts. In: LREC (2008)
37. Perone, C.S., Silveira, R., Paula, T.S.: Evaluation of sentence embeddings in downstream and linguistic probing tasks. arXiv preprint arXiv:1806.06259 (2018)
38. Powers, D.M.: Evaluation: from precision, recall and f-measure to ROC, informedness, markedness and correlation. arXiv preprint arXiv:2010.16061 (2020)
39. Prabhudesai, A., Duong, T.N.: Automatic short answer grading using Siamese bidirectional LSTM based regression. In: 2019 IEEE International Conference on Engineering, Technology and Education (TALE), pp. 1–6. IEEE (2019)

40. Rajagede, R.A., Hastuti, R.P.: Stacking neural network models for automatic short answer scoring. In: IOP Conference Series: Materials Science and Engineering, vol. 1077, p. 012013. IOP Publishing (2021)

41. Riordan, B., Horbach, A., Cahill, A., Zesch, T., Lee, C.: Investigating neural architectures for short answer scoring. In: Proceedings of the 12th Workshop on Innovative Use of NLP for Building Educational Applications, pp. 159–168 (2017)

42. Roy, S., Narahari, Y., Deshmukh, O.D.: A perspective on computer assisted assessment techniques for short free-text answers. In: Ras, E., Joosten-ten Brinke, D. (eds.) CAA 2015. CCIS, vol. 571, pp. 96–109. Springer, Cham (2015). https://doi.org/10.1007/978-3-319-27704-2_10

43. Saha, S., Dhamecha, T.I., Marvaniya, S., Sindhgatta, R., Sengupta, B.: Sentence level or token level features for automatic short answer grading?: use both. In: Penstein Rosé, C., et al. (eds.) AIED 2018. LNCS (LNAI), vol. 10947, pp. 503–517. Springer, Cham (2018). https://doi.org/10.1007/978-3-319-93843-1_37

44. Sahu, A., Bhowmick, P.K.: Feature engineering and ensemble-based approach for improving automatic short-answer grading performance. IEEE Trans. Learn. Technol. 13(1), 77–90 (2019)

45. Sultan, M.A., Salazar, C., Sumner, T.: Fast and easy short answer grading with high accuracy. In: Proceedings of the 2016 Conference of the North American Chapter of the Association for Computational Linguistics: Human Language Technologies, pp. 1070–1075 (2016)

46. Sung, C., Dhamecha, T.I., Mukhi, N.: Improving short answer grading using transformer-based pre-training. In: Isotani, S., Millán, E., Ogan, A., Hastings, P., McLaren, B., Luckin, R. (eds.) AIED 2019. LNCS (LNAI), vol. 11625, pp. 469–481. Springer, Cham (2019). https://doi.org/10.1007/978-3-030-23204-7_39

47. Tan, C., Wei, F., Wang, W., Lv, W., Zhou, M.: Multiway attention networks for modeling sentence pairs. In: IJCAI, pp. 4411–4417 (2018)

48. Vaswani, A., et al.: Attention is all you need. In: Advances in Neural Information Processing Systems, pp. 5998–6008 (2017)

49. Vu, L.: A case study of peer assessment in a composition MOOC: students' perceptions and peer-grading scores versus instructor-grading scores. In: Handbook of Research on Innovative Pedagogies and Technologies for Online Learning in Higher Education, pp. 178–217. IGI Global (2017)

50. Wang, Z., Lan, A.S., Waters, A.E., Grimaldi, P., Baraniuk, R.G.: A meta-learning augmented bidirectional transformer model for automatic short answer grading. In: EDM (2019)

51. Xia, L., Guan, M., Liu, J., Cao, X., Luo, D.: Attention-based bidirectional long short-term memory neural network for short answer scoring. In: Guan, M., Na, Z. (eds.) MLICOM 2020. LNICST, vol. 342, pp. 104–112. Springer, Cham (2021). https://doi.org/10.1007/978-3-030-66785-6_12

52. Xu, C., Zhou, W., Ge, T., Wei, F., Zhou, M.: Bert-of-Theseus: compressing BERT by progressive module replacing. arXiv preprint arXiv:2002.02925 (2020)

53. Xue, L., et al.: mT5: a massively multilingual pre-trained text-to-text transformer. arXiv preprint arXiv:2010.11934 (2020)

54. Yang, Z., Dai, Z., Yang, Y., Carbonell, J., Salakhutdinov, R., Le, Q.V.: XLNet: generalized autoregressive pretraining for language understanding. arXiv preprint arXiv:1906.08237 (2019)

Fair and Adequate Explanations

Nicholas Asher[1(✉)], Soumya Paul[2], and Chris Russell[3]

[1] IRIT, Université Paul Sabatier, 118 route de Narbonne, 31062 Toulouse, France
asher@irit.fr
[2] Telindus, 18 rue du Puits Romain, 8070 Bertrange, Luxembourg
soumya.paul@telindus.lu
[3] Amazon Research, Tübingen, Germany
cmruss@amazon.com

Abstract. Recent efforts have uncovered various methods for providing explanations that can help interpret the behavior of machine learning programs. Exact explanations with a rigorous logical foundation provide valid and complete explanations, but they have an epistemological problem: they may be too complex for humans to understand and too expensive to compute even with automated reasoning methods. Interpretability requires *good* explanations that humans can grasp and can compute.

We take an important step toward specifying what good explanations are by analyzing the epistemically accessible and pragmatic aspects of explanations. We characterize sufficiently good, or fair and adequate, explanations in terms of counterfactuals and what we call the *conundra of the explainee*, the agent that requested the explanation. We provide a correspondence between logical and mathematical formulations for counterfactuals to examine the partiality of counterfactual explanations that can hide biases; we define fair and adequate explanations in such a setting. We then provide formal results about the algorithmic complexity of fair and adequate explanations.

1 Introduction

Explaining the predictions of sophisticated machine-learning algorithms is an important issue for the foundations of AI. Recent efforts [4,19,34,35,38] have shown various methods for providing explanations. Among these, model-based, logical approaches that completely characterise one aspect of the decision promise complete and valid explanations.

Such logical methods are thus *a priori* desirable, but they have an epistemological problem: they may be too complex for humans to understand or even to write down in human-readable form. Interpretability requires *epistemically accessible* explanations, explanations humans can grasp *and compute*. Yet what is a sufficiently complete and *adequate* epistemically accessible explanation, a *good explanation* still needs analysis [30]. We propose to characterize sufficiently good, or fair and adequate, explanations in terms of counterfactuals—explanations,

© IFIP International Federation for Information Processing 2021
Published by Springer Nature Switzerland AG 2021
A. Holzinger et al. (Eds.): CD-MAKE 2021, LNCS 12844, pp. 79–97, 2021.
https://doi.org/10.1007/978-3-030-84060-0_6

that is that are framed in terms of what would have happened had certain conditions (that do not obtain) been the case—and what we call the *conundrum* and *fairness requirements* of the *explainee*, the person who requested the explanation or for whom the explanation is intended). It is this conundrum that makes the explainee request an explanation. Counterfactual explanations, as we argue below, are a good place to start for finding accessible explanations, because they are typically more compact than other forms of explanation.

We argue that a fair and adequate explanation is relative to the cognitive constraints and fairness requirements of an explainee $\mathcal{E}$ [1,5,28]. $\mathcal{E}$ asks for an explanation for why π when she wasn't expecting π. Her not expecting π follows from beliefs that must now be revised—how to specify this revision is the conundrum of $\mathcal{E}$. An adequate explanation is a pragmatic act that should solve the conundrum that gave rise to the request for explanation; solving the conundrum makes the explanation useful to $\mathcal{E}$ [15]. In addition, an adequate explanation must lay bare biases that might be unfair or injurious to $\mathcal{E}$ (the fairness constraint). In effect, this pragmatic act is naturally modelled in a game theoretic setting in which the explainer must understand explainee E's conundrum and respond so as to resolve it. A cooperative explainer will provide an explanation in terms of the type he assigns to E, as the type will encode the relevant portions of E's cognitive state. On the other hand the explainee will need to interpret the putative explanation in light of her model of the explainer's view of his type. Thus, both explainer and explainee have strategies that exploit information about the other—naturally suggesting a game theoretic framework for analysis.

In developing our view of fair and adequate explanations, we will exploit both the logical theory of counterfactuals [26] and mathematical approaches for adversarial perturbation techniques [4,9,23,24,33,40]. We provide a correspondence between logical and mathematical formulations for counterfactuals, and we analyze how counterfactual explanations can hide biases. We then formalize conundra and fair and adequate explanations, and we develop a game theoretic setting for proving computational complexity results for finding fair and adequate explanations in non cooperative settings.

2   Background on Explanations

Following [1,5], we take explanations to be answers to *why* questions. Consider the case where a bank, perhaps using a machine learning program, judges $\mathcal{E}$'s application for a bank loan and $\mathcal{E}$ is turned down. $\mathcal{E}$ is in a position to ask a *why* question like,

(1)   why was I turned down for a loan?

when her beliefs would not have predicted this. Her beliefs might not have been sufficient to infer that she wouldn't get a loan; or her beliefs might have been mistaken—they might have led her to conclude that she would get the loan. In any case, $\mathcal{E}$ must now revise her beliefs to accord with reality. *Counterfactual explanations*, explanations expressed with counterfactual statements, help $\mathcal{E}$ do this by offering an incomplete list of relevant factors that together with unstated properties of $\mathcal{E}$ entail the *explanandum*—the thing $\mathcal{E}$ needs explained, in this case her not getting the loan. For instance, the bank might return the following answer to (1):

(2)   Your income is €50K per year.

(3)   If your income had been €100K per year, you would have gotten the loan.

The counterfactual statement (3) states what given all of $\mathcal{E}$'s other qualities would have been sufficient to get the loan. But since her income is in fact not €100K per year, the semantics of counterfactuals entails that $\mathcal{E}$ does not get the loan. (3) also proposes to $\mathcal{E}$ how to revise her beliefs to make them accord with reality, in that it suggests that she mistakenly thought that her actual salary was sufficient for getting the loan and that the correct salary level is €100K per year.[1]

Counterfactual explanations, we have seen, are *partial*, because they do not explicitly specify logically sufficient conditions for the prediction. They are also *local*, because their reliance on properties of a particular sample makes them valid typically only for that sample. Had we considered a different individual, say $\mathcal{D}$, the bank's explanation for their treatment of $\mathcal{D}$ might have differed. $\mathcal{D}$ might have had different, relevant properties from $\mathcal{E}$; for instance, $\mathcal{D}$ might be just starting out on a promising career with a salary of €50K per year, while $\mathcal{E}$ is a retiree with a fixed income.

The partiality and locality of counterfactuals make them simpler and more epistemically accessible than other forms of explanation. Moreover, the logical theory of counterfactuals enables us to move from a counterfactual to a complete and logically valid explanation. So in principle counterfactual explanations can provide both rigour and epistemic accessibility. But not just any partiality will do, since partiality makes possible explanations that are misleading, that hide injurious or unfair biases. To show how the partiality of counterfactual explanations can hide unfair biases, consider the following scenario. The counterfactual in (2) might be true but it also might be *misleading*, hiding an unfair bias. (1)–(2) can be true while another, more morally repugnant explanation that hinges on $\mathcal{E}$'s being female is also true. Had $\mathcal{E}$ been male, she would have gotten the loan with her actual salary of €50K per year. A fair and adequate explanation should expose such biases.

We now move to a more abstract setting. Let $\hat{f} \colon X^n \to Y$ be a machine learning algorithm, with X^n an n-dimensional feature space encoding data and Y the prediction space. Concretely, we assume that $\hat{f}$ is some sort of classifier. When $\hat{f} = \pi$, an explainee may want an explanation, an answer to the question, "why π?" We will say that an explanation is an event by an *explainer*, the

---

[1] [10] provide a superficially similar picture to the pragmatic one we present, but their aim is rather different, to provide a semantics for argumentation frameworks. For us the pragmatic aspect of explanations is better explained via a game theoretic framework; see below.

provider of the explanation, directed towards the explainee (the person requesting the explanation or to whom the explanation is directed) with a conundrum. An explanation will consist of of an *explanandum*, the event or prediction to be explained, an *explanans*, the information that is linked in some way to the explanandum so as to resolve the explainee's conundrum. When the explanation is about a particular individual, we call that individual the *focal point* of the explanation.

Explanations have thus several parameters. The first is the scope of the explanation. For a *global* explanation of $\hat{f}$, the explainee wants to know the behavior of $\hat{f}$ over the total space X^n. But such an explanation may be practically uncomputable; and for many purposes, we might only want to know how $\hat{f}$ behaves on a selection of data points of interest or focal points, like $\mathcal{E}$'s bank profile in our example.[2] Explanations that are restricted to focal points are *local* explanations.

Explanations of program behavior also differ as to the nature of the *explanans*. In this paper, we will be concerned with *external* explanations that involve an explanatory link between features of input or feature space X and the output in Y without considering any internal states of the learning mechanism [11]. These are attractive epistemically, because unpacking the algorithms' internal states and assigning them a meaning can be a very complicated affair.

A third pertinent aspect of explanations concerns the link between *explanans* and the *explanandum*. [14,18,19] postulate a deductive or logical consequence link between *explanans* and *explanandum*. [19] represent $\hat{f}$ as a set of logic formulas $\mathcal{M}(\hat{f})$. By assuming features with binary values[3], an *instance* is then a set of literals that assigns values to every feature in the feature space. An *abductive* explanation of why π is a subset minimal set of literals $\mathcal{I}$ such that $\mathcal{M}(\hat{f}), \mathcal{I} \models \pi$. Abductive explanations exploits universal generalizations and a deductive consequence relation. They explain why *any instance* $\hat{x}$ that has $\mathcal{I}$ is such that $\hat{f}(\hat{x}) = \pi$ and hence are known as *global* explanations [29].

Counterfactuals offer a natural way to provide epistemically accessible, partial explanations of properties of individuals or focal points. The counterfactual in (3) gives a sufficient reason for $\mathcal{E}$'s getting the loan, *all other factors of her situation being equal* or being as equal as possible (*ceteris paribus*) given the assumption of a different salary for $\mathcal{E}$. Such explanations are often called *local* explanations [8,29], as they depend on the nature of the focal point; they are also partial [38], because the antecedent of a counterfactual are not by themselves logically sufficient to yield the formula in the consequent. Deductive explanations, on the other hand, are invariant with respect to the choice of focal point. But because counterfactual explanations exploit *ceteris paribus* conditions, factors that deductive explanations must mention can remain implicit in a counterfactual explanation. Thus, counterfactual explanations are typically more compact

---

[2] We are implicitly assuming that $\hat{f}$ is too complex or opaque for its behaviour to be analyzed statically.

[3] By increasing the number of literals we can simulate non binary values, so this is not really a limitation as long as the features are finite.

and thus in principle easier to understand.[4] Counterfactuals are also intuitive vehicles for explanations as they also encode an analysis of causation [26].

2.1   Counterfactual Explanations for Learning Algorithms

The canonical semantics for a counterfactual language $\mathcal{L}$, which is a propositional language to which a two place modal operator $\Box\!\!\rightarrow$ is added, as outlined in [26] exploits a possible worlds model for propositional logic, $\mathfrak{A} = \langle W, \leq, [\![.]\!]\rangle$, where: W is a non-empty set (of worlds), $\leq$ is a ternary similarity relation ($w' \leq_w w''$), and $[\![.]\!] : P \rightarrow W \rightarrow \{0,1\}$ assigns to elements in P, the set of proposition letters or atomic formulas of the logic, a function from worlds to truth values or set of possible worlds. Then, where $\models$ represents truth in such a model, we define truth recursively as usual for formulas of ordinary propositional logic and for counterfactuals $\psi \Box\!\!\rightarrow \phi$, we have:

Definition 1. $\mathfrak{A}, w \models \psi \Box \rightarrow \phi$ *just in case:* $\forall w'$, *if* $\mathfrak{A}, w' \models \psi$ *and* $\forall w''(\mathfrak{A}, w'' \models \psi \rightarrow w' \leq_w w'')$, *then:* $\mathfrak{A}, w' \models \phi$

What motivates this semantics with a similarity relation? We can find both epistemic and metaphysical motivations. Epistemically, finding a closest or most similar world in which the antecedent ϕ of the counterfactual $\phi \Box\!\!\rightarrow \psi$ is true to evaluate its consequent ψ follows a principle of belief revision [12], according to which it is rational to make minimal revisions to one's epistemic state upon acquiring new conflicting information. A metaphysical motivation comes from the link Lewis saw between counterfactuals and causation; $\neg\phi\Box\!\!\rightarrow \neg\psi$ implies that if ϕ hadn't been the case, ψ wouldn't have been the case, capturing much of the semantics of the statement ϕ *caused* ψ. The truth of such intuitive causal statements, however, relies on the presence of a host of secondary or enabling conditions. Intuitively the statement that if I had dropped this glass on the floor, it would have broken is true; but in order for the consequent to hold after dropping the glass, there are many elements that have to be the same in that counterfactual situation as in the actual world—the floor needs to be hard, there needs to be a gravitational field around the strength of the Earth's that accelerates the glass towards the floor, and many other conditions. In other words, in order for such ordinary statements to be true, the situation in which one evaluates the consequent of a counterfactual has to resemble very closely the actual world.

Though intuitive, as this logical definition of counterfactuals stands, it is not immediately obvious how to apply it to explanations of learning algorithm behavior. We need to adapt it to a more analytical setting. We will do so by interpreting the similarity relation appealed to in the semantics of counterfactuals as a distance function or norm as in [39] over the feature space X^n, an n-dimensional space, used to describe data points. To fill out our semantics for counterfactuals in this application, we identify instances in X^n as the relevant "worlds" for the semantics of the counterfactuals. We now need to specify a

---

[4] See [18] for some experimental evidence of this.

norm for X^n. A very simple norm assumes that each dimension of X^n is orthogonal and has a Boolean set of values; in this case, X^n has a natural L_1 norm
or Manhattan or edit distance [36].[5] While this assumption commits us to the
fact that the dimensions of X^n capture *all the causally relevant* factors and that
they are all independent—both of which are false for typical instances of learning
algorithms, it is simple and makes our problem concrete. We will indicate below
when our results depend on this simplifying assumption.

A logic of counterfactuals can now exploit the link between logic formulas,
features of points in X^n, and a learning algorithm $\hat{f}$ described in [19,22]. Suppose
a focal point $\hat{x}$ is such that $\hat{f}(\hat{x}) = \eta$. A counterfactual $A \; \square\!\!\rightarrow \pi$ that is true at
the point $\hat{x}$, where π is a prediction incompatible with η, has an antecedent that
is a conjunction of literals, each literal defining a feature value, and that provides
a sufficient and minimal shift in the features of $\hat{x}$ to get the prediction π. Each
counterfactual that explains the behavior of $\hat{f}$ around a focal point $\hat{x} \in X^n$ thus
defines a minimal transformation of the features of $\hat{x}$ to change the prediction.
We now define the transformations on X^n that counterfactuals induce.

Definition 2. *Let $i \subset n$. A fixed transformation Δ_i is a function $\Delta_i : X^n \to X^n$
such that for $x \in X^n$, if $\Delta_i(x) = y$, then x and y differ only in the dimensions
in i. We write $x =_i x'$ to mean that x and x' share the same values along
dimensions i. Given $x \in X^n$, and $\hat{f}(x) = \eta$ and where $\|.\|_{X^n}$ is a natural norm
on X^n, we shall be interested in the following types of transformations.*

(i) *$\Delta_i(x)$ is* appropriate *if $\hat{f}(\Delta_i(x)) = \pi$ where η and π are two incompatible
predictions in Y.*

(ii) *$\Delta_i(x)$ is* minimally appropriate *if it is appropriate and in addition, $\forall x' \in X$
such that $\Delta_i(x) =_i x'$ and $\hat{f}(x') = \pi$, $\|x' - x\|_{X^n} \geq \|\Delta_i(x) - x\|_{X^n}$.*

(iii) *$\Delta_i(x)$ is* sufficiently appropriate *if it is appropriate and in addition, for any
$j \subsetneq i$, $\Delta_j(x)$ is not appropriate.*

(iv) *$\Delta_i(x)$ is* sufficiently minimally appropriate *if it is both sufficiently and minimally appropriate.*

Note that when X is a space of Boolean features, then conditions (ii) and (iv) of
Definition 2 trivially hold. Given a focal point $\hat{x}$ in X^n, minimally appropriate
transformations represent the minimal changes necessary to the features of $\hat{x}$ to
bring about a change in the value predicted by $\hat{f}$.

Let $\hat{f} : X^n \to Y$ and consider now a counterfactual language $\mathcal{L}_{\hat{f}}$ with a set
of formulas Π that describe the predictions in Y of $\hat{f}$.

Definition 3. *A counterfactual model $\mathcal{C}_{X^n, \hat{f}}$ for $\mathcal{L}_{\hat{f}}$ with $\hat{f} : X^n \to Y$ is a triple
$\langle W, \leq, [\![.]\!] \rangle$ with W a set of worlds $W = X^n$, $\leq$ defined by a norm $\|.\|$ on X^n
and $[\![.]\!] : P \cup \{\Pi\} \to W \to \{0,1\}$ such that for $A \in P$, $[\![A]\!]_w = 1$ iff w has feature
A and for $\pi \in \Pi$, $[\![\pi]\!]_w = 1$ iff $\hat{f}(w) = \pi$.*

---

[5] In fact, we only assume a finite set of finitely valued features, since an n-valued
feature is definable with n Boolean valued features. By complicating the language
and logic [7], we can have probability estimates on literals and so encode continuous
feature spaces.

Given a counterfactual model $\mathcal{C}_{X^n,\hat{f}}$ for $\mathcal{L}_{\hat{f}}$ with norm $||.||$ on X^n, we say that $||.||$ is $\mathcal{L}_{\hat{f}}$ definable just in case for worlds $w, w_1 \in X^n$, there is a formula ϕ of $\mathcal{L}_{\hat{f}}$ that separates w_1 from all $w_2 \in X^n$ such that $||w_2 - w|| < ||w_1 - w||$—i.e. for all w_2, $||w_2 - w|| < ||w_1 - w||$, $\mathcal{C}_{X^n,\hat{f}}, w_1 \models \phi$ and $\mathcal{C}_{X^n,\hat{f}}, w_2 \not\models \phi$.

Proposition 1. *Let $\hat{f} \colon X^n \to Y$ and let $\mathcal{C}_{X^n,\hat{f}}$ be a counterfactual model for $\mathcal{L}_{\hat{f}}$ with an $\mathcal{L}_{\hat{f}}$ definable norm. Suppose also that $\hat{f}(w) = \eta$. Then: $\mathcal{C}_{X^n,\hat{f}}, w \models \phi \,\square\!\!\rightarrow \pi$, where $\pi \in \Pi$ and ϕ is a separating formula iff there is a minimally appropriate transformation, $\Delta_i : X^n \to X^n$, where $\hat{f}(\Delta_i(w)) = \pi$, and $\mathcal{C}_{X^n,\hat{f}}, \Delta_i(w) \models A$.*

Proposition 1 follows easily from Definitions 1, 2 and 3.

Proposition 1 is general and can apply to many different norms and languages. We will mostly be concerned here with a special and simple case:

Corollary 1. *Let $\mathcal{L}_{\hat{f}}$ be a propositional language with a set P of propositional letters, where P is the set of Boolean valued features of X^n, and let $\mathcal{C}_{X^n,\hat{f}}$ be a counterfactual model for $\mathcal{L}_{\hat{f}}$ with an L_1 norm. Then: $\mathcal{C}_{X^n,\hat{f}}, w \models A \,\square\!\!\rightarrow \pi$, where $\pi \in \Pi$ and A is a conjunction of literals in P iff there is a minimally appropriate transformation over the dimensions i fixed by A, $\Delta_i : X^n \to X^n$, where $\hat{f}(\Delta_i(w)) = \pi$, and $\mathcal{C}_{X^n,\hat{f}}, \Delta_i(w) \models A$.*

We can generate minimally appropriate transformations via efficient (poly-time) techniques like optimal transport or diffeomorphic deformations [4,9,23, 33,40] for computing adversarial perturbations [24]. In effect all of these diverse methods yield counterfactuals or sets of counterfactuals given Proposition 1. A typical definition of an adversarial perturbation of an instance x, given a classifier, is that it is a smallest change to x such that the classification changes. Essentially, this is a counterfactual by a different name. Finding a closest possible world to x such that the classification changes is, under the right choice of distance function, the same as finding the smallest change to x to get the classifier to make a different prediction.[6]

The great advantage of Proposition 1 is that marries efficient techniques to generate counterfactual explanations with the logical semantics of counterfactuals that provides logically valid (LV) explanations from counterfactual explanations, unlike heuristic methods [27,35,35]. Thus, counterfactual explanations build a bridge between logical rigour and computational feasibility.

Proposition 2. *A counterfactual explanation given by a minimally appropriate $\Delta_i(\hat{x})$ in $\mathcal{C}_{X^n,\hat{f}}$, with an L_1 norm and X^n with Boolean valued features, yields a minimal, LV explanation in at worst a linear number of calls to an NP oracle.*

---

[6] Such minimal perturbations may not reflect the ground truth, the causal facts that our machine learning algorithm is supposed to capture with its predictions, as noted by [25]. We deal with this in Sect. 4.

Proof Sketch. The atomic diagram [6] of $\mathcal{C}_{X^n,\hat{f}}$ in which each world is encoded as a conjunction of literals (Boolean values of the features P of X^n together with predictions from Y), encodes $\mathcal{M}(\hat{f})$. Further, given Corollary 1 and Definition 3, each minimally appropriate Δ_i defines a set of literals $\mathcal{L}_{\Delta_i}$ describing $\Delta_i(\hat{x})$ such that $\Delta_i(\hat{x}), \mathcal{M}(\hat{f}) \models \pi$. [18,19] provide an algorithm for finding a subset minimal set of literals $\mathcal{E} \subseteq \mathcal{L}_{\Delta_i}$ with $\mathcal{E}, \mathcal{M}(\hat{f}) \models \pi$ in a linear number relative to $|\mathcal{L}_{\Delta_i}|$ of calls to an NP oracle [21].   □

3   From Partial to More Complete Explanations

We have observed that counterfactual explanations are intuitively simpler than deductive ones, as they typically offer only a partial explanation. In fact there are three sorts of partiality in a counterfactual explanation. First, a counterfactual explanation is *deductively incomplete*; it doesn't specify the *ceteris paribus* conditions and so doesn't specify what is necessary for a proof of the prediction π for a particular focal point. Second, counterfactual explanations are also partial in the sense that they don't specify all the sufficient conditions that lead to π; they are hence *globally incomplete*. Finally, counterfactuals are partial in a third sense; they are also *locally incomplete*. To explain this sense, we need a notion of *overdetermination*.

Definition 4. *A prediction* $\pi \in Y$ *by* $\hat{f} : X \to Y$ *is* overdetermined *for a focal point* $\hat{x} \in X$ *if the set of minimally sufficiently appropriate transformations of* $\hat{x}$

$$O(\hat{x}, \pi, \hat{f}) = \{\Delta_i : \Delta_i(\hat{x}) \text{ is minimally sufficiently appropriate}\}$$

contains at least two elements.

Locally incomplete explanations via counterfactuals can occur whenever $\hat{f}$'s counterfactual decisions are over-determined for a given focal point. Many real world applications like our bank loan example will have this feature.

Locally incomplete explanations can, given a particular ML model $\mathcal{M}_{\hat{f}}$, hide implicitly defined properties that show $\hat{f}$ to be unacceptably biased in some way and so pose a problem for fair and adequate explanations. Local incompleteness allows for several explanatory counterfactuals with very different *explanans* to be simultaneously true. This means that even with an explanation, $\hat{f}$ may act in ways unknown to the agent $\mathcal{E}$ or the public that is biased or unfair. Worse, the constructor or owner of $\hat{f}$ will be able to conceal this fact if the decision for $\mathcal{E}$ is overdetermined, by offering counterfactual explanations using maps Δ that don't mention the biased feature.

Definition 5. *A prejudicial factor* P *is a map,* $P : X^n \to X^n$ *and* $\hat{f}$ *exhibits a* biased dependency *on prejudicial factor* P *just in case for some* $i \neq 0$, Δ_i, *and for some incompatible predictions* η *and* π,

$$\hat{f}(\hat{x}) = \hat{f}(\Delta_i(\hat{x})) = \eta \quad and \quad \hat{f}(P(\hat{x})) = \hat{f}(P(\Delta_i(\hat{x}))) = \pi$$

Dimensions of the feature space that are atomic formulas in $\mathcal{L}_{\hat{f}}$ can provide examples of a prejudicial factor P. But prejudical factors P may be also implicitly definable in $M_{\hat{f}}$. Assume that $\hat{\cdot}$ is a map from real individuals x to their representation as data points $\hat{x} \in \hat{X}$. Then: P is $M_{\hat{f}}$ implicitly definable just in case: for all x such that $\hat{x} \in \hat{X}$, $x \in \|P\|$ iff for some boolean combination E of atoms of $\mathcal{L}_{\hat{f}}$, $M_{\hat{f}} \models E(\hat{x})$.

We've just described some pitfalls of locally incomplete counterfactual explanations. We now show how to move from a partial picture of the behavior of $\hat{f}$ to a more complete one using counterfactuals. Imagine that at a focal point $\hat{x}$, $\hat{f}(\hat{x}) = \eta$ and we want to know why not π.

Definition 6. *In a counterfactual model $\mathcal{C}_{X^n, \hat{f}}$ with a set of Boolean valued features P, the collection of counterfactuals $\mathbf{S}_{\mathcal{C}, \hat{x}, \pi} = \{\phi \; \Box\!\!\rightarrow \; \pi : \mathcal{C}_{X^n, \hat{f}}, \hat{x} \models \phi \; \Box\!\!\rightarrow \; \pi$ with ϕ a Boolean combination of values for atoms in $P\}$ true at $\hat{x}$ gives the complete explanation for why π would have occurred at $\hat{x}$.*

Appropriate transformations Δ_i on X^n in a counterfactual model $\mathcal{C}_{X^n, \hat{f}}$ to produce π associated with counterfactuals via Proposition 1 can capture $\mathbf{S}_{\mathcal{C}, \hat{x}, \pi}$ and permit us to plot the local complete explanation of $\hat{f}$ around a focal point $\hat{x}$ with regard to prediction π.

Definition 7. $\mathbf{B}_{\mathcal{C}, \hat{x}, \pi} = \{\Delta_i(\hat{x}) : \Delta_i$ *is a minimal appropriate transformation for some $i \subset n\}$*

Proposition 3. *In a counterfactual model $\mathcal{C}_{X^n, \hat{f}}$, $\mathbf{B}_{\mathcal{C}, \hat{x}, \pi} = \{y \in X^n : \exists (\phi \; \Box\!\!\rightarrow \psi) \in \mathbf{S}_{\mathcal{C}, \hat{x}, \pi}$ such that y is a closest ϕ world to $\hat{x}$ where $\mathcal{C}_{X^n, \hat{f}}, y \models \psi\}$.*

For the remainder of this section we will fix a counterfactual model $\mathcal{C}_{X^n, \hat{f}}$ to simplify notation.

We are interested in the space $\mathcal{N}_{\hat{f}, \hat{x}, \pi}$ around $\hat{x}$ with boundary $\mathbf{B}_{\hat{x}, \pi}$.

Definition 8. 1. $\mathcal{N}_{\hat{f}, \hat{x}, \pi}$ *is the subspace of X^n such that (i) $\hat{x} \in \mathcal{N}_{\hat{f}, \hat{x}, \pi}$ and (ii) $\mathcal{N}_{\hat{f}, \pi, \hat{x}}$ includes in its interior all those points z for which $\hat{f}(z) = \hat{f}(\hat{x})$ and (iii) the boundary of $\mathcal{N}_{\hat{f}, \hat{x}, \pi}$ is given by $\mathbf{B}_{\hat{x}, \pi}$.*
2. $\mathcal{N}^d_{\hat{f}, \pi, \hat{x}}$ *is a subspace of $\mathcal{N}_{\hat{f}, \hat{x}, \pi}$ with boundary $\mathbf{B}^d_{\hat{x}, \pi}$, where $\mathbf{B}^d_{\hat{x}, \pi} = \mathbf{B}_{\hat{x}, \pi} \cap B_d(\hat{x})$, where $B_d(\hat{x}) = \{y \in X^n : \|y - \hat{x}\| \leq d\}$.*
3. $\mathbf{S}^d_{\hat{x}, \pi} = \{y : \exists (\phi \; \Box\!\!\rightarrow \psi) \in \mathbf{S}_{\hat{x}, \pi} \wedge \mathcal{C}_{X^n, \hat{f}}, y \models \psi \wedge \|y - \hat{x}\| \leq d\}$.

The set $\mathbf{S}_{\hat{x}, \pi}$ can have a complex structure in virtue of the presence of *ceteris paribus* assumptions. Because strengthening of the antecedent fails in semantics for counterfactuals, the counterfactuals in (4) relevant to our example of Sect. 2 are all satisfiable at a world without forcing the antecedents of (4)b or (4)c to be inconsistent:

(4)   a.   If I were making €100K euro, I would have gotten the loan.
       b.   If I were making €100K or more but were convicted of a serious financial fraud, I would not get the loan.

    c.   If I were making €100K or more and were convicted of a serious financial fraud but then the conviction was overturned and I was awarded a medal, I would get the loan.

The closest worlds in which I make €100k do not include a world w in which I make €100k but am also convicted of fraud. Counterfactuals share this property with other conditionals that have been studied in nonmonotonic reasoning [13, 32]. However, if the actual world turns out to be like w, then by weak centering (4)a turns out to be false, because the *ceteris paribus* assumption in (4)a is that the actual world is one in which I'm not convicted of fraud.

In $\mathbf{S}_{\hat{x},\pi}$ we can count how many times the value of the consequent changes as we move from one antecedent to a logically more specific one (e.g., does the prediction flip from A to $A \wedge C$ or from $A \wedge C$ to $A \wedge C \wedge D$). For generality, we will also include in the number of flips, the flips that happen when we change the Boolean value of a feature—going from A to $\neg A$ for example. We will call the number of flips the *flip degree* of $\mathbf{S}_{\hat{x},\pi}$.

There is an important connection between the flip degree of $\mathbf{S}_{\hat{x},\pi}$ and the geometry of $\mathcal{N}_{\hat{f},\hat{x},\pi}$. In a counterfactual model, the move from one antecedent ϕ_1 of a counterfactual c_1 a to logically more specific antecedent ϕ_2 of c_2, with $c_1, c_2 \in \mathbf{S}_{\hat{x},\pi}$ will, given certain assumptions about the underlying norm yield $\hat{x} < y < z$, with y being a closest to $\hat{x}$ point verifying ϕ_1 and z a closest point verifying ϕ_2. In fact we generalize this property of norms.

Definition 9. *A norm $||.||$ in a counterfactual model $\mathcal{C}_{X^n,\hat{f}}$ respects the logical consequence relation $(\models)$ of the model iff for any $z \in X^n$ such that $\mathcal{C}_{X^n,\hat{f}}, z \models \psi$ and for $\phi_1 \models \phi_2 \models ... \models \phi_n \models \neg\psi$, there are collinear $x_1, ...x_n \in X^n$ such that for each i, x_i is a closest point to z such that $\mathcal{C}_{X^n,\hat{f}}, x_i \models \phi_i$ and $||x_{i+1} - z|| \leq ||x_i - z||$.*

Remark 1. *An L1 norm for a counterfactual model is a logical consequence respecting norm.*

In addition, a flip (move from a point verifying ϕ_1 to a point verifying ϕ_2 corresponds to a move from a transformation Δ_i to a transformation Δ_j with $i \subset j$. Thus, flips determine a partial ordering under $\subseteq$ over the shifted dimensions i: thus $\Delta_i \leq \Delta_j$, if $i \subseteq j$. We are interested in the behavior of $\hat{f}$ with respect to the partial ordering on Δ_i.

Definition 10. *$\hat{f}$ is nearly constant around $\hat{x}$, if for every sufficiently minimally appropriate Δ_i for all $\Delta_j \supset \Delta_i$, $\hat{f}(\Delta_j(\hat{x})) = \hat{f}(\Delta_i(\hat{x}))$.*

A nearly constant $\hat{f}$ changes values only once for each combination of features/dimensions d_i moving out from a focal point $\hat{x}$. So at some distance d, nearly constant $\hat{f}$ becomes constant $\hat{f}$. For a nearly constant $\hat{f}$ around $\hat{x}$, $\mathbf{S}_{\hat{x},\pi}$, has flip degree 1. A complete local explanation for $\hat{f}$'s prediction of π within d, $\mathbf{S}_{\hat{x},\pi}^d$, is a global explanation $\hat{f}$'s behavior with respect to π.

We can generalize this notion to define an n-shifting $\hat{f}$. If $\hat{f}$ flips values n times moving out from $\hat{x}$, $\mathbf{S}_{\hat{x},\pi}$ has flip degree n.

Proposition 4. *Suppose A counterfactual model has a logical consequence respecting norm, then:* $\mathbf{S}_{\hat{x},\pi}$, *has a flip degree* ≤ 2 *iff* $\mathcal{N}_{\hat{f},\hat{x},\pi}$ *forms a convex subspace of* $\hat{f}[X]$.

Proof Sketch. Assume $\mathbf{S}_{\hat{x},\pi}$ has flip degree ≥ 3. Then $\mathbf{S}_{\hat{x},\pi}$ will contain counterfactuals with antecedents ϕ, χ, δ such that $\phi \models \chi \models \delta$ but, say, ϕ and δ counterfactually support π but not χ. As the underlying norm respects $\models$, there are collinear points x, y, and z, where x is a closest point to $\hat{x}$ where ϕ is true, y is a closest χ world, and z is a closest δ world such that $\hat{x} < z < y < x$. But $\hat{x}, y_{\chi} \in \mathcal{N}_{\hat{f},\hat{x},\pi}$, while $x_{\phi}, z_{\delta} \in \mathbf{B}_{\hat{x},\pi}$ and $\notin \mathcal{N}_{\hat{f},\hat{x},\pi}$, which makes $\mathcal{N}_{\hat{f},\hat{x},\pi}$ non convex. Conversely, suppose $\mathcal{N}_{\hat{f},\hat{x},\pi}$ is non convex. Using the construction of counterfactuals from the boundary $\mathbf{B}_{\hat{x},\pi}$ of $\mathcal{N}_{\hat{f},\hat{x},\pi}$ will yield a set with flip degree 3 or higher.   □

The flip degree of $\mathbf{S}_{\hat{x},\pi}$ gives a measure of the degree of non-convexity of $\mathcal{N}_{\hat{f},\hat{x},\pi}$, and a measure of the complexity of an explanation of $\hat{f}$'s behavior. A low flip degree for $\mathbf{S}_{\hat{x},\pi}^{d}$ with minimal overdeterminations provides a more general and comprehensive explanation. With Proposition 4, a low flip degree converts a local complete explanation into a global explanation, which is *a priori* preferable. It is also arguably closer to our prior beliefs about basic causal processes. The size of $\mathbf{S}_{\hat{x},\pi}^{d}$ gives us a measure to evaluate $\hat{f}$ itself; a large $\mathbf{S}_{\hat{x},\pi}^{d}$ doesn't approximate very well a good scientific theory or the causal structures postulated by science. Such a $\hat{f}$ lacks generality; it has neither captured the sufficient nor the necessary conditions for its predictions in a clear way. This could be due to a bad choice of features determining $\hat{f}$'s input X^{n} [9]; too low level or unintuitive features could lead to lack of generality with high flip degrees and numerous overdeterminations. Thus, we can use $\mathbf{S}_{\hat{x},\pi}^{d}$ to evaluate $\hat{f}$ and its input representation X^{n}.

The flip degree of $\mathbf{S}_{\hat{x},\pi}$ and the topology of $\mathcal{N}_{\hat{f},\hat{x},\pi}$ can also tell us about the relation between counterfactual explanations based on some element in X and ground truth instances provided during training. Our learning algorithm $\hat{f}$ is trying to approximate or learn some phenomenon, which we can represent as a function $f : X \rightarrow Y$; the observed pairs $(z, f(z))$ are ground truth points for $\hat{f}$. Ideally, $\hat{f}$ should fit and converge to f—i.e., with the number of data points N $\hat{f}$ is trained on $lim_{N \rightarrow \infty} \hat{f}^{N} \rightarrow f$; in the limit explanations of the behavior of $\hat{f}$ will explain f, the phenomenon we want to understand. Given that we generate counterfactual situations using techniques used to find adversarial examples, however, counterfactual explanations may also be based on adversarial examples that have little to no intuitive connection with the ground truth instances $\hat{f}$ was trained on. While these can serve to explain the behavior of $\hat{f}$ and as such can be valuable, they typically aren't good explanations of the phenomenon f that $\hat{f}$ is trying to model. [25] seek to isolate good explanations of f from the behavior of $\hat{f}$ and propose a criterion of topological connectedness for good counterfactual explanations. This idea readily be implemented as a constraint on $\mathcal{N}_{\hat{f},\hat{x},\pi}$: roughly, $\hat{f}$ as an approximation of f will yield good counterfactual explanations relative to a focal point x only if for any point y outside of $\mathcal{N}_{\hat{f},\hat{x},\pi}$, there is a path of points $y_1, ...y_n$ wholly within C between y and a ground truth data point p such that $f(p) = \hat{f}(p) = \hat{f}(y_i) = \hat{f}(y)$.[7]

4 Pragmatic Constraints on Explanations

While we have clarified the partiality of counterfactual explanations, AI applications can encode data via hundreds even thousands of features. Even for our simple running example of a bank loan program, the number of parameters might provide a substantial set of counterfactuals in the complete local explanation given by $\mathbf{S}_{\hat{x},\pi}$. This complete local explanation might very well involve too many counterfactuals for humans to grasp. We still to understand what counterfactual explanations are *pragmatically relevant* in a given case.

Pragmatic relevance relies on two observations. First, once we move out a certain distance from the focal point, then the counterfactual shifts intuitively cease to be about the focal point; they cease to be counterparts of $\hat{x}$ and become a different case. Exactly what that distance is, however, will depend on a variety of factors about the explainee $\mathcal{E}$ and what the explainer believes about $\mathcal{E}$. Second, appropriate explanations must respond to the particular *conundrum* or cognitive problem that led $\mathcal{E}$ to ask for the explanation [1,5,28]. On our view, the explainee $\mathcal{E}$ requires an explanation when her beliefs do not lead her to expect the observed prediction π. When $\mathcal{E}$'s beliefs suffice to predict $\hat{f}(\hat{x}) = \pi$, she has *a priori* an answer to the question *Why did $\hat{f}(x) = \pi$?* In our bank example from Sect. 2, had $\mathcal{E}$'s beliefs been such that she did not expect a loan from the bank, she wouldn't have needed to ask, *why did the bank not give me a loan?*[8]

The conundrum comes from a mismatch between $\mathcal{E}$'s understanding of what $\hat{f}$ was supposed to model (our function f) and $\hat{f}$'s actual predictions. So $\mathcal{E}$, in requesting an explanation of $\hat{f}$'s behavior, might also want an explanation of f itself (see the previous section for a discussion). Either $\mathcal{E}$ is mistaken about the nature of $\hat{f}$, or her grasp of $\hat{f}$ is incomplete..[9] More often than not, $\mathcal{E}$ will have certain preconceptions about $\hat{f}$, and then many if not most of the counterfactuals in $\mathbf{S}_{\hat{x},\pi}$ may be irrelevant to $\mathcal{E}$. A relevant or fair and adequate explanation for $\mathcal{E}$ should provide a set $\mathfrak{C}_{\mathcal{E}}^d$ of appropriate Δ_i with $\|\Delta_i(\hat{x}) - \hat{x}\| \leq d$ showing which of $\mathcal{E}$'s assumptions were faulty or incomplete, thus solving her conundrum.

Suppose that the explainee $\mathcal{E}$ requests an explanation why $\hat{f}(\hat{x}) = \eta$, and that $\hat{x}$ is decomposed into $\langle x_{d_1}, x_{d_2} \rangle$.

CI Suppose $\mathcal{E}$'s conundrum based on incompleteness; i.e., the conundrum arises from the fact that for $\mathcal{E}$ $\hat{f}$ only pays attention to the values of dimensions

---

[7] We note that our discussion and constraint make clear the distinction between f and $\hat{f}$ which is implicit in [15,25].

[8] Of course $\mathcal{E}$ might want to know whether her beliefs matched the bank's reasons for denying her a loan, but that's a different question—and in particular it's not a *why* question.

[9] Perhaps $\mathcal{E}$ is also mistaken about or has an incomplete grasph of f or if not, she is mistaken about how $\hat{f}$ differs from f). But we will not pursue this here.

d_1 in the sense that for her $\hat{f}(\langle x_{d_1}, x_{d_2} \rangle) = \hat{f}(\langle x_{d_1}, x'_{d_2} \rangle)$, for any values x'_{d_2}. Then there is a $\Delta \in \mathfrak{C}_{\mathcal{E}}^d$ such that $\Delta(\langle x_{d_1}, x_{d_2} \rangle) = \langle x_{d_1}, y_{d_2} \rangle$ and $\hat{f}(\Delta(\hat{x})) = \hat{f}(\langle x_{d_1}, y_{d_2} \rangle) = \pi$ while $\hat{f}(\hat{x}) = \hat{f}(\langle x_{d_1}, x_{d_2} \rangle) = \eta$.

CM Suppose $\mathcal{E}$'s conundrum is based on a mistake. Then there is a $\Delta \in \mathfrak{C}_{\mathcal{E}}^d$ such that $\Delta(\langle x_{d_1}, x_{d_2} \rangle) = \langle y_{d_1}, x_{d_2} \rangle$ such that $\hat{f}(\langle y_{d_1}, x_{d_2} \rangle) = \hat{f}(\Delta(\hat{x})) = \pi$. I.e., Δ must resolve $\mathcal{E}$'s conundrum by providing the values for the dimensions d_2 of $\hat{x}$ on which $\mathcal{E}$ is mistaken.

A fair and adequate explanation must not only contain counterfactuals that resolve the explainee's conundrum. It must make clear the biases of the system which may account for 0's incomplete understanding of $\hat{f}$; it must lay bare any prejudicial factors P that affect the explainee and thus in effect all overdetermining factors as in Definition 4. An explainee might reasonably want to know whether such biases resulted in a prediction concerning her. E.g., the explanation in (3) might satisfy CM or CI, but still be misleading. Thus:

CB ∀ prejudicial factors P, there is a $\Delta \in \mathfrak{C}_{\mathcal{E}}^d$ such that $\hat{f}(\Delta(\hat{x})) = \pi$ and $P(\Delta(\hat{x})) = \Delta(\hat{x})$.

In our bank loan example, if the bank is constrained to provide an explanation obeying CB, then it must provide an explanation according to which being white and having $\mathcal{E}$'s salary would have sufficed to get the loan.

Definition 11. *A set of counterfactuals provides a* fair and adequate *explanation of $\hat{f}$ for $\mathcal{E}$ at $\hat{x}$ just in case they together satisfy CM, CI and CB within a certain distance d of $\hat{x}$.*

The counterfactuals in $\mathfrak{C}_{\mathcal{E}}^d$ jointly provide a *fair and adequate* explanation of $\hat{f}$ for $\mathcal{E}$, though individually they may not satisfy all of the constraints. We investigate how hard it is to find an adequate local explanation in the next section.

5   The Algorithmic Complexity of Finding Fair and Adequate Explanations

In this section, we examine the computational complexity of finding a fair and adequate explanation. To find an appropriate explanation, we imagine a game played, say, between the bank and the would-be loan taker $\mathcal{E}$ in our example from Sect. 2, in which $\mathcal{E}$ can ask questions of the bank (or owner/ developer of the algorithm) about the algorithm's decisions. We propose to use a two player game, an *explanation game* to get appropriate explanations for the explainee.

The pragmatic nature of explanations already motivates the use of a game theoretic framework. We have argued fair and adequate explanations must obey pragmatic constraints; and in order to satisfy these in a cooperative game the explainer must understand explainee $\mathcal{E}$'s conundrum and respond so as to resolve it. Providing an explanation is a pragmatic act that takes into account an explainee's cognitive state and the conundrum it engenders for the particular fact that needs explaining. A cooperative explainer will provide an explanation in terms of the type he assigns to $\mathcal{E}$, as the type will encode the relevant portions of $\mathcal{E}$'s cognitive state. On the other hand the explainee will need to interpret the putative explanation in light of her model of the explainer's view of his type. Thus, both explainer and explainee naturally have strategies that exploit information about the other. Signaling games [37] are a well-understood and natural formal framework in which to explore the interactions between explainer and explainee; the game theoretic machinery we develop below can be easily adapted into a signaling game between explainer and explainee where explanations succeed when their strategies coordinate on the same outcome.

Rather than develop signaling games however for coordinating on successful explanations, we look at non-cooperative scenarios where the explainer $\hat{f}$ may attempt to hide a good explanation. For instance, the bank in our running example might have encoded directly or indirectly biases into its loan program that are prejudicial to $\mathcal{E}$, and it might not want to expose these biases. The games below provide a formal account of the difficulty our explainee has in finding a winning strategy in such a setting.

To define an explanation game, we first fix a set of two players $\{\mathcal{E}, \mathcal{A}\}$.

The moves or actions $V_{\mathcal{E}}$ for explainee $\mathcal{E}$ are: playing an ACCEPT move—in which $\mathcal{E}$ accepts a proposed Δ_i if it partially solves her conundrum; playing an N-REQUEST move—i.e. requesting a Δ_j where j differs from all i such that Δ_i has been proposed by $\mathcal{A}$ in prior play; playing a P-REQUEST move—i.e. for some particular i, requesting Δ_i. $\mathcal{E}$ may also play a CHALLENGE move, in which $\mathcal{E}$ claims that a set of features $A_1, \ldots A_n$ of the focal point that entails π in the counterfactual model associated with $\hat{f}$. We distinguish three types of ME explanation games for $\mathcal{E}$ based on the types of moves she is allowed: the *Forcing* ME explanation games, in which $\mathcal{E}$ may play ACCEPT, N-REQUEST, P-REQUEST; the more restrictive *Restriction* ME explanation games, in which $\mathcal{E}$ may only play ACCEPT, N-REQUEST; and finally Challenge ME explanation games in which CHALLENGE moves are allowed.

Adversary $\mathcal{A}$'s moves $V_{\mathcal{A}}$ consists of the following: producing Δ_i and computing $\hat{f}(\Delta_i(\hat{x}))$ in response to N-REQUEST or P-REQUEST by $\mathcal{E}$; if $\mathcal{G}$ is a forcing game, $\mathcal{A}$ must play Δ_i at move m in ρ, if $\mathcal{E}$ has played P-REQUEST Δ_i at $m-1$. In reacting to a N-REQUEST, player $\mathcal{A}$ may offer any new Δ_i; if he is noncooperative, he will offer a new Δ_i that is not relevant to $\mathcal{E}$'s conundrum, unless he has no other choice. On the other hand, $\mathcal{A}$ must react to a CHALLENGE move by $\mathcal{E}$ by playing a Δ_i that either completes or corrects the Challenge assumption. A CHALLENGE demands a cooperative response; and since it can involve any implicitly definable prejudicial factor as in Definition 5, it can also establish CB, as well as remedy CI or CM.

We now specify a win-lose, generic *explanation game*.

Definition 12. *An* Explanation game, $\mathcal{G}$, *concerning a polynomially computable function* $\hat{f} \colon X^n \to Y$, *where* X^n *is a space of boolean valued features for the data and* Y *a set of predictions, is a tuple* $((V_{\mathcal{E}} \cup V_{\mathcal{A}})^*, \mathcal{E}, \mathcal{A}, \hat{f} \colon X^n \to Y, \hat{x}, d, \mathfrak{C}_{\mathcal{E}}^d)$ *where:*

i. $\mathfrak{C}_{\mathcal{E}}^d \subseteq \mathbf{B}_{\hat{x},\pi}^d$ resolves $\mathcal{E}$'s conundrum and obeys CB.

ii. $\hat{x} \in X^n$ is the starting position, d is the antecedently fixed distance parameter.

iii. $\mathcal{A}$, but not $\mathcal{E}$ has access to the behavior of $\hat{f}$ and a fortiori $\mathfrak{C}_{\mathcal{E}}^d$.

iv. $\mathcal{E}$ opens $\mathcal{G}$ with a REQUEST or CHALLENGE move

v. $\mathcal{A}$ responds to $\mathcal{E}$'s requests by playing some $\Delta_i, i \leq d$.

vi. $\mathcal{E}$ may either play ACCEPT, in which case the game ends or again play a REQUEST or CHALLENGE move.

$\mathcal{E}$ wins $\mathcal{G}$ just in case in $\mathcal{G}$ she can determine $\mathfrak{C}_{\mathcal{E}}^d$. The game terminates when (a) 0 has determined $\mathfrak{C}_{\mathcal{E}}^d$ (resolved her conundra) or gives up.

$\mathcal{E}$ always has a winning strategy in an explanation game. The real question is how quickly $\mathcal{E}$ can compute her winning condition. An answer depends on what moves we allow for $\mathcal{E}$ in the Explanation game; we can restrict $\mathcal{E}$ to playing a Restriction explanation game, a Forcing game or a Forcing game with CHALLENGE moves.

Proposition 5. *Suppose $\mathcal{G}$ is a forcing explanation game. Then the computation of $\mathcal{E}$'s winning strategy in $\mathcal{G}$ is Polynomial Local Search complete (PLS) [20, 31]. On the other hand if $\mathcal{G}$ is only a Restriction game, then the worst case complexity for finding her strategy is exponential.*

Proof Sketch. Finding $\mathfrak{C}_{\mathcal{E}}^d$ is a search problem using $\hat{f}$. $\mathfrak{C}_{\mathcal{E}}^d$ is finite with, say, m elements. These elements need not be unique; they just need jointly to solve the conundrum. This search problem is PLS just in case every solution element is polynomially bounded in the size of the input instance, $\hat{f}$ is poly-time, the cost of the solution is poly-time and it is possible to find the neighbors of any solution in poly-time. Let $\hat{x}$ be the input instance. By assumption, $\hat{f}$ is polynomial; and given the bound d, the solutions y for $\hat{f}(y) = \pi$ and $y \in \mathfrak{C}_{\mathcal{E}}^d$ are polynomially bounded in the size of the description of $\hat{x}$. Now, finding a point $y \in \mathfrak{C}_{\mathcal{E}}^d$ that solves at least part of $\mathcal{E}$'s conundrum, as well as finding neighbors of y is poly-time, since $\mathcal{E}$ can use P-REQUEST moves to direct the search. To determine the cost c of finding $\mathfrak{C}_{\mathcal{E}}^d$ for $|\mathfrak{C}_{\mathcal{E}}^d| = m$ in poly-time: we set for $y \in \mathfrak{C}_{\mathcal{E}}^d$ the jth element of $\mathfrak{C}$ computed as $c(y) = m - j$; if $y \notin \mathfrak{C}$, $c(y) = m$. Finding $\mathfrak{C}_{\mathcal{E}}^d$ thus involves determining m local minima and is PLS. In addition, determining $\mathfrak{C}_{\mathcal{E}}^d$ encodes the PLS complete problem FLIP [20]: the solutions y in $\mathcal{G}$ have the same edit distance as the solutions in FLIP, $\hat{f}$ encodes a starting position, and our cost function can be recoded over the values of the Boolean features defining y to encode the cost function of FLIP and the function that compares solutions in FLIP is also needed and constructible in $\mathcal{G}$. So finding $\mathfrak{C}_{\mathcal{E}}^d$ is PLS complete in $\mathcal{G}$ as it encodes FLIP.

The fact that forcing explanation games are PLS complete makes getting an appropriate explanation computationally difficult. Worse, if $\mathcal{G}$ is a Restriction Explanation game, then $\mathcal{A}$ can force $\mathcal{E}$ to enumerate all possible Δ_i within radius d of $\hat{x}$ to find $\mathfrak{C}_{\mathcal{E}}^d$.   □

Proposition 6. *Suppose $\mathcal{G}$ is a Challenge explanation game. Then $\mathcal{E}$ has a winning strategy in $\mathcal{G}$ that is linear time computable.*

Proof Sketch. $\mathcal{A}$ must respond to $\mathcal{E}$'s CHALLENGE moves by correcting or completing $\mathcal{E}$'s proposed list of features. $\mathcal{E}$ can determine $\mathfrak{C}_\mathcal{E}^d$ in a number of moves that is linear in the size of $\mathfrak{C}_\mathcal{E}^d$. □

A Challenge explanation game mimics a coordination game where $\mathcal{A}$ has perfect information about $\mathfrak{C}_\mathcal{E}^d$, because it forces cooperativity and coordination on the part of $\mathcal{A}$. Suppose $\mathcal{E}$ in our bank example claims that her salary should be sufficient for a loan. In response to the challenge, the bank could claim the salary is not sufficient; but that's not true—the salary *is* sufficient *provided* other conditions hold. That is, $\mathcal{E}$'s conundrum is an instance of CI. Because of the constraint on CHALLENGE answers by the opponent, the bank must complete the missing element: *if you were white with a salary of €50K,...* Proposition 6 shows that when investigating an $\hat{f}$ in a challenge game, exploiting a conundrum is a highly efficient strategy.

The flip degree of $\mathbf{S}_{\hat{x},\pi}^d$ and the number of overdetermining factors $O(x,\pi)$ (Definition 4) typically affect the size of $\mathfrak{C}$ and thus the complexity of the conundrum and search for fair and adequate explanations and their logical valid associates. More particularly, when $|O(\pi,\hat{x})| = n$ and the cost of the prediction is as in the proof of Proposition 5, $\mathcal{E}$'s conundrum and the explanations resolving it may require n local minima. When the flip degree of $\mathbf{S}_{\hat{x},\pi}^d$ is m, $\mathcal{E}$ may need to compute m local minima.

To develop practical algorithms for fair and adequate explanations for AI systems, we need to isolate $\mathcal{E}$'s conundrum. This will enable us to exploit the efficiencies of Challenge explanation games. Extending the framework to discover $\mathcal{E}$'s conundrum behind her request for an explanation is something we plan to do using epistemic games from [3] with more developed linguistic moves. In a more restricted setting where Challenge games are not available, our game framework shows that clever search algorithms and heuristics for PLS problems will be essential to providing users with relevant, and provably fair and adequate counterfactual explanations. This is something current techniques like enumeration or finding closest counterparts, which may not be relevant [18,19,22]—do not do.

6   Conclusion

We have shown that counterfactual explanations can deliver partial, but epistemically accessible and adequate explanations. We have also shown that any counterfactual explanation can be extended to a valid deductive one. We have shown that pragmatic factors dramatically affect the complexity of finding adequate explanations, and we introduced Explanation Games, which provided to represent finding fair and adequate counterfactual explanations as a PLS complete search problem. In addition, we explored how the complexity of the set of counterfactuals describing a local neighborhood around the focal point can affect both the complexity of fair and adequate explanations and our evaluation of the learning algorithm as a model.

Our paper fills in part of the gap for finding fair and adequate explanations in a computationally reasonable way. Nevertheless moving from an explanation provided by an explanation game to a proof from a minimal set of sufficient premises as in Proposition 2 is still computationally difficult. In future work we will look at efficient heuristics for this step. In future work, we will alo look at how explanation games help us to formally explore interactive machine learning, in particular "human in the loop" or interactive explainability for machine learning function behavior [2,17]. Such game theoretic investigations may have special relevance in medical domains [16].

Acknowledgement. We thank the ANR PRCI grant SLANT, the ICT 38 EU grant COALA and the 3IA Institute ANITI funded by the ANR-19-PI3A-0004 grant for research support. We alo thank the reviewers for their insightful comments.

References

1. Achinstein, P.: The Nature of Explanation. Oxford University Press, Oxford (1980)
2. Amershi, S., Cakmak, M., Knox, W.B., Kulesza, T.: Power to the people: the role of humans in interactive machine learning. AI Mag. **35**(4), 105–120 (2014)
3. Asher, N., Paul, S.: Strategic conversation under imperfect information: epistemic message exchange Games. Logic, Lang. Inf. **27**(4), 343–385 (2018)
4. Bachoc, F., Gamboa, F., Halford, M., Loubes, J.M., Risser, L.: Entropic variable projection for explainability and intepretability. arXiv preprint arXiv:1810.07924 (2018)
5. Bromberger, S.: An approach to explanation. In: Butler, R. (ed.) Analytical Philsophy, pp. 72–105. Oxford University Press, Oxford (1962)
6. Chang, C.C., Keisler, H.J.: Model theory. Elsevier (1990)
7. De Raedt, L., Dumančić, S., Manhaeve, R., Marra, G.: From statistical relational to neuro-symbolic artificial intelligence. arXiv preprint arXiv:2003.08316 (2020)
8. Doshi-Velez, F., Kim, B.: Towards a rigorous science of interpretable machine learning. arXiv preprint arXiv:1702.08608 (2017)
9. Dube, S.: High dimensional spaces, deep learning and adversarial examples. arXiv preprint arXiv:1801.00634 (2018)
10. Fan, X., Toni, F.: On computing explanations in argumentation. In: Bonet, B., Koenig, S. (eds.) Proceedings of the Twenty-Ninth AAAI Conference on Artificial Intelligence, pp. 1496–1502. AAAI Press (2015)
11. Friedrich, G., Zanker, M.: A taxonomy for generating explanations in recommender systems. AI Mag. **32**(3), 90–98 (2011)
12. Gärdenfors, P., Makinson, D.: Revisions of knowledge systems using epistemic entrenchment. In: Vardi, M.Y. (ed.) Proceedings of the Second Conference on Theoretical Aspects of Reasoning about Knowledge, pp. 83–95. Morgan Kaufmann, San Francisco (1988)
13. Ginsberg, M.L.: Counterfactuals. Artif. Intell. **30**(1), 35–79 (1986)
14. Hempel, C.G.: Aspects of Scientific Explanation. Free Press, New York (1965)
15. Holzinger, A., Carrington, A., Müller, H.: Measuring the quality of explanations: the system causability scale (scs). KI-Künstliche Intelligenz, pp. 1–6 (2020)
16. Holzinger, A., Malle, B., Saranti, A., Pfeifer, B.: Towards multi-modal causability with graph neural networks enabling information fusion for explainable ai. Inf. Fusion **71**, 28–37 (2021)

17. Holzinger, A., Plass, M., Kickmeier-Rust, M., Holzinger, K., Crişan, G.C., Pintea, C.M., Palade, V.: Interactive machine learning: experimental evidence for the human in the algorithmic loop. Appl. Intell. **49**(7), 2401–2414 (2019)
18. Ignatiev, A., Narodytska, N., Asher, N., Marques-Silva, J.: On relating "why?" and "why not?" explanations. In: Proceedings of AI*IA 2020 (2020)
19. Ignatiev, A., Narodytska, N., Marques-Silva, J.: On relating explanations and adversarial examples. In: Advances in Neural Information Processing Systems (2019)
20. Johnson, D.S., Papadimitriou, C.H., Yannakakis, M.: How easy is local search? J. Comput. Syst. Sci. **37**(1), 79–100 (1988)
21. Junker, U.: Preferred explanations and relaxations for over-constrained problems. In: AAAI-2004 (2004)
22. Karimi, A.H., Barthe, G., Balle, B., Valera, I.: Model-agnostic counterfactual explanations for consequential decisions. In: International Conference on Artificial Intelligence and Statistics, pp. 895–905. PMLR (2020)
23. Kurakin, A., Goodfellow, I., Bengio, S.: Adversarial examples in the physical world. arXiv preprint arXiv:1607.02533 (2016)
24. Kusner, M.J., Loftus, J., Russell, C., Silva, R.: Counterfactual fairness. In: Advances in Neural Information Processing Systems, pp. 4066–4076 (2017)
25. Laugel, T., Lesot, M.-J., Marsala, C., Renard, X., Detyniecki, M.: Unjustified classification regions and counterfactual explanations in machine learning. In: Brefeld, U., Fromont, E., Hotho, A., Knobbe, A., Maathuis, M., Robardet, C. (eds.) ECML PKDD 2019. LNCS (LNAI), vol. 11907, pp. 37–54. Springer, Cham (2020). https://doi.org/10.1007/978-3-030-46147-8_3
26. Lewis, D.: Causation. J. Philos. **70**(17), 556–567 (1973)
27. Lundberg, S.M., Lee, S.: A unified approach to interpreting model predictions. In: NIPS, pp. 4765–4774 (2017)
28. Miller, T.: Explanation in artificial intelligence: insights from the social sciences. Artif. Intell. **267**, 1–38 (2019)
29. Molnar, C.: Interpretable machine learning. Lulu. com (2019)
30. Murdoch, W.J., Singh, C., Kumbier, K., Abbasi-Asl, R., Yu, B.: Definitions, methods, and applications in interpretable machine learning. Proc. Natl. Acad. Sci. **116**(44), 22071–22080 (2019)
31. Papadimitriou, C.H., Schäffer, A.A., Yannakakis, M.: On the complexity of local search. In: Proceedings of the Twenty-Second Annual ACM Symposium on Theory of Computing, pp. 438–445 (1990)
32. Pearl, J.: System Z: a natural ordering of defaults with tractable applications to nonmonotonic reasoning. In: Proceedings of the 3rd Conference on Theoretical Aspects of Reasoning about Knowledge (TARK 1990), pp. 121–135 (1990)
33. Peyré, G., et al.: Computational optimal transport: with applications to data science. Found. Trends Mach. Learn. **11**(5–6), 355–607 (2019)
34. Ribeiro, M.T., Singh, S., Guestrin, C.: why should i trust you?: explaining the predictions of any classifier. In: KDD, pp. 1135–1144 (2016)
35. Ribeiro, M.T., Singh, S., Guestrin, C.: Anchors: high-precision model-agnostic explanations. In: AAAI, pp. 1527–1535 (2018)
36. Salzberg, S.: Distance metrics for instance-based learning. In: Ras, Z.W., Zemankova, M. (eds.) ISMIS 1991. LNCS, vol. 542, pp. 399–408. Springer, Heidelberg (1991). https://doi.org/10.1007/3-540-54563-8_103
37. Spence, A.M.: Job market signaling. J. Econ. **87**(3), 355–374 (1973)

38. Wachter, S., Mittelstadt, B., Russell, C.: Counterfactual explanations without opening the black box: automated decisions and the gpdr. Harv. JL Tech. **31**, 841 (2017)
39. Williamson, T.: First-order logics for comparative similarity. Notre Dame J. Formal Logic **29**(4) (1988)
40. Younes, L.: Diffeomorphic learning. arXiv.1806.01240 (2019)

Mining Causal Hypotheses in Categorical Time Series by Iterating on Binary Correlations

Bora I. Kumova⬤ and Dirk Saller(✉)

Baden-Württemberg Cooperative State University, Mosbach, Germany
{bora.kumova,dirk.saller}@mosbach.dhbw.de

Abstract. Machine errors can propagate in a production field and reduce the efficiency of smart manufacturing execution systems. Since every highly automated machine can have many possible status reports, their causalities can only be detected by means of statistical analyses. We present a highly automatable methodology for iteratively analysing machine state time series and for detecting machine error causality hypotheses. First, the categorical status time series of all machines are analysed for binary correlations in two iteration steps using pairwise cross-correlation. Out of all correlations, significantly high correlations are then combined and can be validated for causalities by means of plausibility and semantic criteria. Our experimental results are presented on anonymised real production state time series and a simple representational concept for further causal interpretation is introduced.

Keywords: Time series analysis · Correlation functions · Smart manufacturing execution systems · Production process optimisation · Production analytics · Causal analysis · Failure propagation analysis

1 Introduction

It has broadly been recognised that smart manufacturing increases the efficiency and productivity through data-driven decision making. For that purpose, the actual production state of every machine is optimised on continuously obtained feedback information from the machine via its sensory system. Hence, sensor statuses constitute the basic data for statistical analysis and information extraction in Smart Manufacturing Execution Systems SMES [13]. Therefore, timestamped sensor status messages are collected in production state time series (PSTS) for all machines that produce in a specific production plan. A PSTS is serialised for each machine on timestamps. Although PSTS emerge in real-time while machines operate, batch processing tools are more widely adopted than real-time stream processing framework in SMES [11]. Unlike big data processing in other areas, in SMES all data processing is well structured, hierarchically organised, standardised and data lifecycles well studied [24]. Therefore, time series data acquired through legacy MES machine states are collected by continuously polling machines [14]. In an ideal production world, all production states, production process states or material

© IFIP International Federation for Information Processing 2021
Published by Springer Nature Switzerland AG 2021
A. Holzinger et al. (Eds.): CD-MAKE 2021, LNCS 12844, pp. 99–114, 2021.
https://doi.org/10.1007/978-3-030-84060-0_7

states would be digitally recorded in a relevant and complete manner via modular concepts and multiple sensors in order to gain knowledge (digital twin). This claim is part of Industry 4.0. In particular, the multivariate resolution of the complex states of a single production machine, for example through continuous measurement of temperature, pressure or vibration, even at several points and at comparable times, is an important milestone there. On the other hand, in the real world of many small and medium-sized companies, this is often still a long way off. In many cases on the part of the production companies, the states of each production machine – even if the machine might have more sensor and status data available - are only registered univariately, that is, by only one categorical value at any instant of time, whereby the assumed values, such as "standard operation", "maintenance", "mechanical malfunction", "electrical malfunction", "jam" or "lack of material", have a more complex semantic meaning. The specification of these categorical values is usually company-specific. In any case, polling the univariate or multivariate statuses of all machines synchronously with respect to an elementary time step allows collecting continuously snapshots of the whole production process, containing more or less information about the dynamical production system. In production processes with low rate of status changes relatively to the number of measurement time instants, too many redundant data would be collected by polling this way. The more frequent snapshots are then taken, the more difficult real-time processing data analysis will be feasible. Under such circumstances batch processing the data analysis constitutes an alternative. However, real-time processing SMES is inevitable for controlling Internet of Things (IoT)-based sensory systems. For instance, monitoring sensor statuses for predicting faults based on abnormal events, such as outliers [22].

In our approach, we focus our attention on event-based univariate time series, where the PSTS provide all those instants of time, when the categorical complex status of the machine changes its value. The analysis of event-based big data has become increasingly popular. For instance, for detecting clusters in sensor data streams, modelling state transitions between the clusters as a Markov chain and discovering anomalies [1]. In contrast to time series, the timestamps of a data stream may not monotonously increase, i.e. some older data may occur in the stream after some younger data. Another issue is the quality of sensor data, which may be restricted due to limited sensor precision, failures or malfunctioning. In order to avoid faulty decisions derived from incorrect or misleading sensor data streams and provide some quality level of machine operation, the lifetime of the sensors may continuously be projected, based on their data streams [12].

Within our approach, we apply cross-correlation analysis to pairs of binary coded (univariate) time series, recovered from event-based raw data, in order to extract correlation features. The methodology is validated in providing significant results on real production data and, additionally, equips the correlation-results with algebraic properties of pairwise cross-correlated machines [18]. This additional information is one source of associating causal hypotheses to the correlation results.

The normalized cross-correlation provides a value for any time shift between the two time series, which can be regarded as a probability measure for statistical similarities. Hence, by computing the normalized cross-correlation over a range of time shifts, a sharp maximum of the values indicates a high probability of correlation (with respect to status changes) together with a value for the time shift that corresponds to this correlation. Usually, the algorithm searches within a given time window of the series, that is supposed to contain statistically significant information for a correlation, where the width of the window is determined based on initial evidence for a possible correlation. In case of time series, correlation analysis is the preferred choice for feature extraction and classification. Standard algorithms for the correlation analysis of two time series perform best for noise-free data. However, in most cases real data does come noisy and overlaid with data from multiple logical sources. For such cases the literature is rich on various specialised filters, often for the elimination of both, noise and undesired data from uninterested logical sources. In some cases, filters are designed parametric, such that the focus on interested data may be adjusted, such as filtering outlier data [22], to allow filtering just the relevant data for a specific experiment out of the initial logical sources.

Cross-correlation is frequently employed in the literature for correlation analysis and measurement of univariate linear time series. It has been used for detecting time lags in two series of information in various domains. For instance, in economics, for detecting correlations between advertising and sales [10]. Or in the analysis of neural information, for detecting spatio-temporal correlations between neural discharges. The filter discussed here, allows detecting transitive signal delays within a specific frequency range [20]. Cross-correlation is a popular similarity measure for pixels. It has been reported to be more effectively, when applied on specific features of pixels, such as on brightness values [6]. Cross-correlation itself may be used as a linear filter that separates noise in functional magnetic resonance imaging, by detecting signals that correlate with neural regions of similar activations [7], or for filtering slow components, in order to improve detecting faster components out of binary or continuous time series [16]. However, in the last two approaches noise was already filtered out of the time series, so cross-correlation can more effectively be adapted for filtering undesired correlations.

In case of autocorrelation, a specific pattern of the series is used to detect similar patters inside the same series. Since such individually autocorrelated series do not comprise independent values, cross-correlation may be non-significant and therefore misleading. For avoiding such cases, detecting, revising and verifying them, the reader is referred to related work [5].

The analysis of series with a time warped window allows detecting correlations between the series by means of a distance measure. Dynamic time warp (DTW) was initially introduced in conjunction with dynamic programming for the detection of similar acoustic signals in speech [16]. Since than it has been improved in speed and accuracy [19] and was popular in speech recognition and more recently in data mining.

It has been demonstrated that the success of DTW depends on the warping distance for the given time series and suggested to learn the warping window from the data, for instance with a k-nearest neighbour classifier via cross-validation [9]. Fast learning the warping distance has also been proposed [23]. The warping distance was further used for clustering time-delayed user activities in social media accounts and compared with cross-correlation-based random projection [2].

In manufacturing execution systems (MES) the sensory infrastructure has predominantly real-time constraints and therefore status data is timestamped at machine level in real-time, as soon as a status is read from a sensor.

Common to all correlation measures is the calculation of distance between dependent regressive models found in the data. Although numerous binary distance and correlation measures were introduced in the history of statistical analysis, every application focuses mostly on just one measure, by tuning it to a specific algorithm, in order to optimise the training data. However, knowing the similarities and differences between alternative measures can help in choosing the most appropriate one for the training data [3].

For the verification of the statistical significance of causality of a cross-correlation, a t-test or Granger causality is well discussed in the literature [4, 5, 8, 10, 15]. For causality analysis of cyclic correlations Granger causality estimates simultaneous autoregressive models of two time series, which represent jointly significant lags, ie a highly probable causality. However, causality analysis of time-varying lead-lag structures for detecting structural breaks or regime patterns, requires additional concepts, like the causal path, and large-scaled time series for the verification of a correlation as a causality [21].

The paper is organized as follows. In Sect. 2, the main contribution of this work, namely the extended methodology with respect to [18] is developed. Important issues are (i) the binary classifications with respect to specific, non-standard operating status values and (ii) the net graphs for simply representing the results of the correlation analysis for further causal interpretation. In Sect. 3, some boundary conditions, such as the complexity of the algorithms or the experimental setting is shortly reflected. In Sect. 2.6 we provide shortly the results and give an outlook on further steps.

2 Methodology

The raw data for our methodology consists of exactly one event-based (univariate) time series of possible categorical status changes per machine (for a given time window, such as a day or week).

In [18] it is shown, how to apply cross-correlation to two such event-based time series, namely by (i) interpolating the event-based time series by holding the status value until the next status change event occurs, (ii) fixing the time window, hence, getting the same length of time-status-pairs, and (iii) classifying and converting the multi-categorical status values into binary categorical values "0" for "normal operation" and "1" for "not normal operation". The last step is necessary in order to make the algorithm for cross-correlation applicable to the (non-metrical) categorical status data via the binarized PSTS.

The main contribution of this work is the extension of the methodology in order to recover the multi-categorical information which was lost in classifying by "0" for "standard operation" and "1" for "non-standard operation".

The idea is, to classify "1" for "a single status value out of the set of all status of type *non-standard operation*" and "0" for all other status. Cross-correlation is then applied iteratively. First, we compute cross-correlation to the original classification in order to get a first correlation picture of all machines with respect to the (simplified) on-off-analysis. We therefore say the computation of cross-correlations with the original classification to be *iteration 1*. In the following, we simply say iteration 1 to be the "on-off-classification".

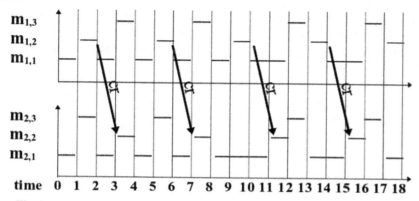

Fig. 1. Correlation cr between single status values $m_{1,1}$ and $m_{2,1}$, $(m_{1,1}, cr, m_{2,1})$.

Then, again, we compute the cross-correlation for all significantly correlated machine pairs of iteration 1 after converting with respect to all other binary classifications, said to be *iteration 2*.

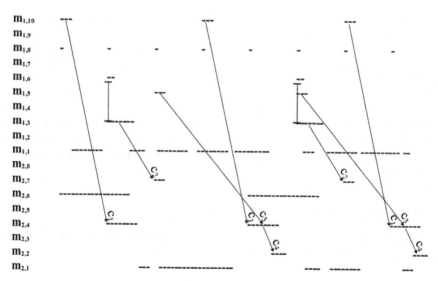

Fig. 2. Sample single correlations between different status values $(m_{1,10}, c_1, m_{2,4})$, $(m_{1,3}, c_2, m_{2,7})$, $(m_{1,5}, c_3, m_{2,4})$ and composed correlations $(m_{1,6} \wedge m_{1,3}, c_2, m_{2,7})$, $(m_{2,4} \wedge c_3, c_4, m_{2,2})$ of machine M_1, M_2.

We want to stress, that the methodology in both iteration steps can easily be automated for the raw data for different companies by fixing a few configuration parameters.

In order to get an impression of what is the meaning of our computation from two PSTS by applying our methodology, we look at some simulated univariate status time series with noise-free statistical correlation. Figure 1 illustrates a single correlation between only one specific status value of each machine. Figure 2 illustrates several pairwise correlations between specific pairs of status values.

Figure 3 gives an impression of the preprocessing steps of the raw data towards an application of cross-correlation.

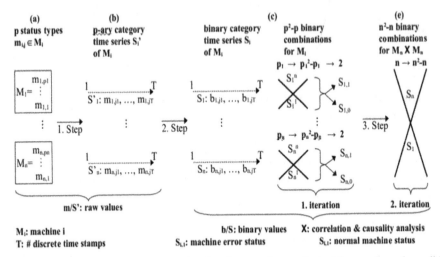

Fig. 3. Preprocessing of status time series for the causality analysis. The number of possible combinations is here considered only in orders of p_i, without considering coefficients. See Sect. 2.4 for more details.

Hence, this work extends the correlation and causality analysis, that we have introduced earlier [18], by an exhaustive search for all possible correlations between all status combinations of machines of a production field and by a selection schema for causality hypotheses.

2.1 Program

Following the description above, the program of our methodology consists of the following two iteration steps:

I.   First iteration:

A.   Input: Raw status data in form of PSTSs of all machines that are used in the execution of a specific production plan in the same time window and with same adapted data length.

B.   Binarisation: Transformation of categorical status values into distinguished binary classification by "0" for "standard operation" and "1" for "non-standard operation".

C.   Correlation analysis: Calculation of all possible correlations between all status time series of all machines with respect to the distinguished binary classification.

1. Causality analysis: Search for significant correlations among all previously calculated correlations and out of them for transitive correlations [18],
2. Output: Highly probable causalities in iteration 1

II. Second iteration:

A. Input: Raw status data in form of PSTSs of selected machines with significant correlations in iteration 1
B. Binarisation: Transformation of categorical sensor values into all binary classification by "1" for "a single status value out of the set of all status of type *not-standard operation*" and "0" for all other status.
C. Correlation analysis: Calculation of all possible correlations between all status time series of selected machines and with respect to all binary classifications.

1. Causality analysis: Search for significant correlations among all previously calculated correlations and out of them for transitive correlations [18]. Validate the semantic meaning of the correlated status values for causal probabilities.
2. Output: Highly probable causalities in iteration 2

We now describe the methodology, that we will apply to our real data below, stepwise by describing the data source, the data model and the application of cross-correlation for the binary converted data.

2.2 Data Sources

The data sources of our methodology lie in production fields that consist of production machines. The raw status data of the machines first need to be normalised for all machines of the production plan.

Every machine M_i may produce a different number p_i of possible status values $m_{i,pi}$ (Fig. 3(a)) in an arbitrary sequence in time T. For any production plan, we know that some inputs of some machines M_j will depend on some outputs of some machines M_i. However, we do not know such relationships exactly a priory. Moreover, we know that multiple such machine relationships may occur in complex transitive dependencies, making the efficiency of the production plan principally dependent on any state m_i of any machine M_i. Such relationships may develop too complex to be detected manually.

As described above, production machines are complex systems each might having sensory sub-systems. However, due to the situation in small and medium enterprises, we refer to the following terms interchangeably: status/error type, status value, sensor message. Hence, for the sake of simplicity, we assume that there is a correspondence between the (multivariate) sensor messages of a machine at any instant of time and the univariate categorical status values of the machine. We could still avoid the notion of sensor in this work, since we still treat machines as black boxes with complex status values. However, even if this is a quite strong simplification, it is helpful for future investigations to keep in mind the working postulates. In a future work, we want to drop this postulate and extend our correlation analysis from univariate categorical (overall) machine status values to multivariate sensor values. We therefore assume that any machine M_i has exactly p_i different categorical status values $\{m_{i,1}, ..., m_{i,pi}\}$, where at least one categorical status value has the semantic meaning of "standard operating".

That is our raw data item. Status messages that indicate lower efficiency, may propagate in the production field and thus reduce the overall efficiency of the production plan. Hence the objective here is to detect such sensor status causalities, in order to optimise the overall efficiency of the plan. The status changes of the machine M_i are serialised to a single time series S_i' over the course of a production plan. For n machines $\{M_1, ..., M_n\}$, n time series $\{S_1', ..., S_n'\}$ of raw sensor data are collected, all having equal number T of discrete values (Fig. 3(b)).

2.3 Data Model

From the above discussed data sources that were collected for a specific production plan, we construct stepwise a suitable data model for our cross-correlation analysis.

For the duration of a production plan, we assume following properties for all n raw/binary time series of iteration 1 and 2, $S'_i/S_i^1/S_i^2{}_h$, respectively, of all machines M_i, $i \in \{1, n\}$.

- Every time series $S'_i = (m_{i,j1}, ..., m_{i,jT})$ consists of T arbitrary status values $m_{i,jk} \in \{m_{i,1}, ..., m_{i,pi}\}, k \in \{1, ..., T\}$
- The execution time T of the production plan determines the total number of elements of a sequence
- Timestamps increase continuously $[1..T]$ in a sequence $S'_i/S_i^1/S_i^2{}_h$ respectively
- In iteration 1, Categorical values $m_{i,jk} \in \{m_{i,1}, ..., m_{i,pi}\}$ are uniquely converted into the binary values $b_{i,jk} \in \{0,1\}$, where $0 =$ standard operation; $1 =$ else[1]
- In iteration 2, Categorical values $m_{i,jk} \in \{m_{i,1}, ..., m_{i,pi}\}$ are converted in all $p_i - 1$ possible ways into binary values $b_{i,jk} \in \{0, 1\}$, where 1 specific non-standard operation, $0 =$ else (see footnote 1)

Hence, a in a test setting of machines $\{M_1, ..., M_n\}$, each machine M_i is associated with a unique categorical sequence S_i' of raw data, a unique induced binary sequence S_i^1 in iteration 1 and $p_i - 1$ induced binary sequences, $S_i^2{}_h$ in iteration 2.

2.4 Correlation Analysis

The objective of our correlation analysis is, to detect correlations between the machines and its status values of any pair of time series in S_i^1 and S_j^1 (iteration 1) or $S_i^2{}_h$ and $S_j^2{}_1$ (iteration 2) out of all n machines $\{M_1, ..., M_n\}$. Consequently, we will find the correlations of the on-off-classification and – for those machine pairs with significant on-off-correlations – the relevant status-specific correlations. Let us assume, for the sake of simplicity, that the number of status values p_i of two machines M_i and M_j is equal, that is $p_i = p_j$.

---

[1] Any other status value of the machine.

Then, there are $p_i^2 - p_i$ possible binary cross-correlations $S_i^2{}_h$ and $S_j^2{}_l$ to be computed between two Machines M_i and M_j (Fig. 3(c), (d)).

It is known from real executions of production plans in production fields, that some error states of machines may cause other error states on the same machine or on other machines. This can be observed in cyclically reoccurring patterns of the time series. By means of our iterations, at the same time we calculate cross-correlations between the status values of different machines and between different status values of the same machine. In this sense, we already take into account auto-correlations.

We use the `xcorr` function of the Python library `matplotlib` for discrete series. The cross-correlation series for $x_n = S_i$ and $y_n = S_j$ of length $n = T$ is given by:

$$R_{xy}(m) = E\{x_{n+m}y_n^*\} = E\{x_n y_{n-m}^*\}$$

where $0 < n < 86400$ s; E: expected value operator.

The above correlation is calculated for each step $m \in \{-W, W\}$ for a given time window size $W < T$, in order to find a significant maximum. Recalling the number of possible binary classifications and related binary time series in iteration 2, we get the following three types of computations and their respective complexities, where we have again set $p = pi = pj$ for all n, for simplicity (Fig. 3(c), (d)).

[1] Intra-machine computations for p status types: $\frac{(p-1)(p-2)}{2}$
[2] Intra-machine computations for n machines: $\frac{n(p-1)(p-2)}{2}$
[3] Inter-machine with intra-machine iterations: $\frac{n(n-1)(p-1)(p-1)}{2}$

Out of all above computed correlations, we finally investigate further the most significant ones in our causality analysis.

2.5 Causality Analysis

Passing from correlation to causality requires context knowledge. So far, we have considered, as the only Data Source, the PSTS of all machines in the considered production field. In fact, as we have seen before that we can recover from those in iteration 2.

(1) the cross-correlation distribution functions for any machine pair and any status-pair (with respect to the time shift),
(2) the respective time lags (for a sufficiently peaked distribution) and
(3) the respective categorical values of the respective status, which provides semantic information

For a causality analysis, first, we filter out from the previously calculated correlations the most significant ones and, secondly, try to find for them further evidence for causality.

In (Fig. 4 and 5) some significant results in iteration 1 and 2 for real production data of two different production lines L1 and L2 are shown. They have following common properties of the cross-correlation data plots: (i) Peak y-value > 0.25 value, (ii) Peak score > 0.25 (iii) Smooth curve: smoothly increasing/decreasing curve before/after the vault.

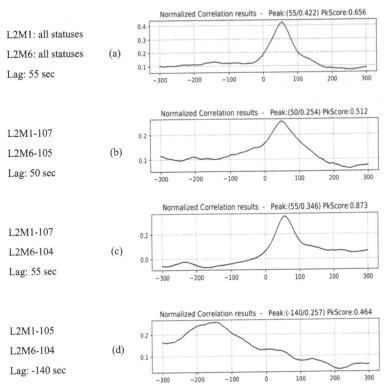

Fig. 4. Significant On-off-correlation (iteration 1) and individual status correlations (iteration 2) with 104: feed shortage, 105: sink jam, 107: side stream jam of machines L2M1, L2M6 on date 15.5.18.

Concerning the automation of detection of such characteristic peaks, we have developed software and involved functions from open libraries. We introduce trigger parameters in order to define and extract significant cross-correlation-results and to aggregate them into characteristic numbers associated with the output data (such as "peak score" of the cross-correlation-function or "frequency" and "balance" of the binary sequences.

After the step of preselection/filtering the cross-correlation results, we can take advantage of further results and semantical information within our methodology.

In a previous paper [18], we have seen, that our methodology makes algebraic properties for more than 2 cross-correlated machines visible, i.e. an addition rule for time lags. Those addition rules can be applied as a kind of plausibility check and provides indications for the topological order of the machines in the production plan.

On the other hand, the results of this paper provide the analyst in iteration 2 with another relevant information concerning correlated machines, namely, the specific status for which the machines yield the correlation and the time lag. The categorical status value, such as "technical failure" or "no material", encodes semantic information and might enable the expert to validate a correlation as to be logically causal or not.

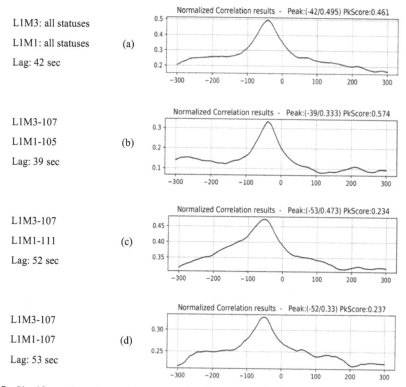

L1M3: all statuses

L1M1: all statuses   (a)

Lag: 42 sec

L1M3-107

L1M1-105   (b)

Lag: 39 sec

L1M3-107

L1M1-111   (c)

Lag: 52 sec

L1M3-107

L1M1-107   (d)

Lag: 53 sec

Fig. 5. Significant On-off-correlation (iteration 1) and individual status correlations (iteration 2) with 104: *feed shortage*, 105: *sink jam*, 107: *side stream jam* 111: *intrinsic disturbance* of machines L1M1, L1M3 on date Dec. 16th 2013.

Hence, the pairwise correlations can be used as starting points for targeted investigations on failure status, as the removal of every such error promises potential increase of the efficiency of the production plan.

2.6 Result Representation

As we have already mentioned, an important aspect of the methodology is that it is highly automatable.

In other words, in our laboratory we start with normalized raw data from a MES System, hence the relevant PSTS are given in a unique data format for any production company (which uses the same MES).

The processing steps, which are described in Subsect. 2.1, are highly supported by prototype software apart from a few manual customizing steps concerning time windows and trigger parameters. In order to provide the expert, that is, the person who has knowledge about the production plan and further relevant semantic knowledge, the analysis results with the possible causal hypotheses in a simple way, we have developed a net-graph-representation.

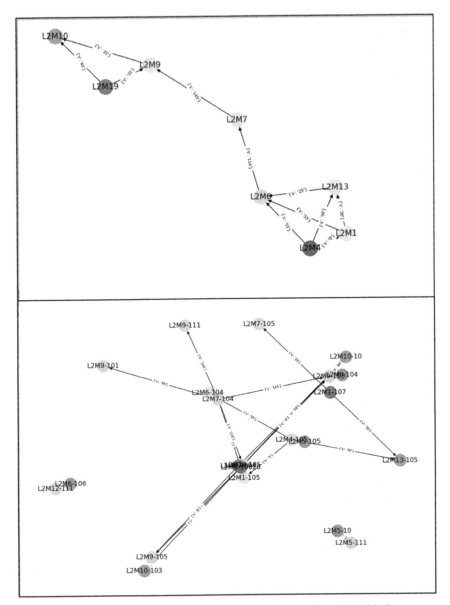

Fig. 6. Net graphs for On-off-correlations (iteration 1, Fig. above) and individual status correlations (iteration 2, Fig. below) with triggered filter (i.e. peak score >0.3) for all machines of
production line L2 on Dec. 16th 2013 with restriction of time lag/sec. to [−150, 150]

Each bubble of the net graph is either a machine (iteration 1) or a machine with a
non-standard status value (iteration 2). The arrow direction takes care of the sign of the
time lag (in the sense of reaction time) and the numbers at the arrows indicate the time
lag value. The colours of the bubbles indicate the position of the bubble in the net graphs.

Red and green bubbles indicate machines or machine status, where arrows only begin or end, respectively.

Figure 6 gives an example of such net-graphs of a production line L2 for iteration 1 (Fig. 6 above) and for iteration 2 (Fig. 6 below). It is not necessary to list all status values here, as we intend a more detailed discussion only for a part of this graph in (Fig. 7). Mit diesem Teil greifen wir das Beispiel aus (Fig. 4) wieder auf. (Fig. 7(a)) illustrates the correlation between machine M1 and M3 in line L2 with the corresponding time lag. At the same time, only two of the iteration 2 status correlations pass through our trigger (Fig. 7(b) and (c)), namely the status correlations M1-107 cr M6-105 and M1-107 cr M6-104. In a sense, this result is plausible, since we find two equal directions of the time lag. In both cases, machine M6 changes its status after M1, in one case, after 50 s., in the other case after 55 s. The third case with an opposite time direction of 140 s. did not pass our filter. On the other hand, one of the remaining cases in iteration 2 is probably dominating (that is, much more frequent than the other) since the overall time lag in iteration 1 corresponds to that time lag in iteration 2. We suppose that the overall time lag in iteration 1 is usually a kind of weighted mean of the time lags in iteration 2. However, such hypotheses need to be validated by experts aware of the production context.

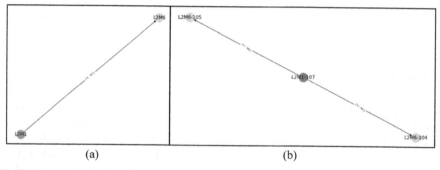

(a)   (b)

Fig. 7. Net graphs for On-off-correlation (iteration 1, (a)) and individual status correlations (iteration 2, (b)) with triggered filter (i.e. peak score > 0.3) for machines M1 and M6 of production line L2 on Dec. 16th 2013 with restriction of time lag/sec. to $[-150, 150]$.

Another interesting example is related with the machines of line L1 in (Fig. 5). The net-graphs are illustrated in (Fig. 8). Both iterations are plausible in the sense, that they reflect the results of (Fig. 5). As in the example above, we observe in iteration 2 that one failure status of M3 induces different failure statuses at M1 after different time lags. Our methodology automatically verifies, that these effects are not due to auto-correlations of M3, that is, that the status 105, 107 and 111 of M1 are not correlated to each other. If this would be the case, our computation of iteration 2 (i.e. type [1] of Sect. 2.4) would have delivered significant correlations for the autocorrelation, which we have verified to fail. As above, we suppose that the 3 different cases "overlap" statistically as possible failure correlations with different time lags., such that the resulting overall time lap is 42 s. The weights of this weighted sum are supposed to depend on the relative frequencies of the

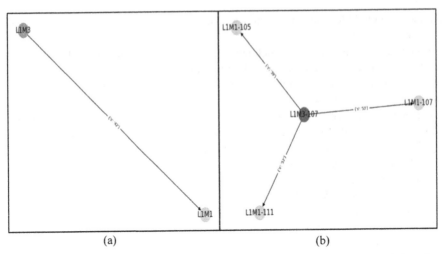

Fig. 8. Net graphs for On-off-correlation (iteration 1, (a)) and individual status correlations (iteration 2, (b)) with triggered filter (i.e. peak score >0.3) for machines M1 and M3 of production line L1 on May 18th 2015.

respective status-correlations. Again, these causal hypotheses need to be validated by experts aware of the production context.

3   Conclusion

In this paper we have developed an extended methodology for computing causality hypotheses in production fields from production status time series, based on cross-correlation analysis and binary classifications of status values. We have also introduced a highly automatable representation method for causal hypotheses based on production net graphs with respect to status values, correlation size parameters and time lags. The results of this work, in particular the results of iteration 2, associate the correlation results with further information that indicates on causality of the status changes of correlated machines. Apart from the algebraic properties concerning the time lags, we can now provide the expert with specific information about the semantic type of correlation, i.e. failure status correlation. Last but not least, the representation method of net graphs provides the expert with a simple information representation for further validation of causal hypotheses. The methodology might be considered as a bridge technology for small and medium production enterprises towards Industry 4.0.

In future work we will improve the efficiency of the iterative algorithm using metaheuristic algorithms, analyse possible transitive causalities between the machines, analyse statistical algebraic properties of cross-correlation and causality analysis as well as further extending our approach to multivariate time series for each machine, i.e. time series that involve different contemporary status values coming from several sensors each production machine.

Acknowledgement. Thanks are due to the former students Martin Stöcker and Tabea von Vulte for useful ideas and implementing supporting software applications. Thanks also to MPDV Microlab GmbH for providing real world production data and domain knowledge.

References

1. Akram, N., et al.: Grand challenge: anomaly detection of manufacturing equipment via high performance RDF data stream processing. In: ACM International Conference on Distributed and Event-based Systems (2017). https://doi.org/10.1145/3093742.3095100
2. Chavoshi, N., Hamooni, H., Mueen, A.: DeBot: Twitter bot detection via warped correlation. In: IEEE International Conference on Data Mining (ICDM) (2016). https://doi.org/10.1109/ICDM.2016.0096
3. Choi, S.S., Cha, S.H., Tappert, C.C.: A survey of binary similarity and distance measures. J. Syst. Cybermat. Inf. (JSCI) **8**, 43–48 (2010)
4. Cryer, J.D., Chan, K.S.: Time Series Analysis. Springer, Heidelberg (2008). https://doi.org/10.1007/978-0-387-75959-3
5. Dean, R.T., Dunsmuir, W.T.M.: Dangers and uses of cross-correlation in analyzing time series in perception, performance, movement, and neuroscience: the importance of constructing transfer function autoregressive models. Behav. Res. Methods **48**(2), 783–802 (2015). https://doi.org/10.3758/s13428-015-0611-2
6. Dietrich, P., Heist, S., Landmann, M., Kühmstedt, P., Notni, G.: BICOS - an algorithm for fast real-time correspondence search for statistical pattern projection-based active stereo sensors. Appl. Sci. **9**, 3330 (2019). https://doi.org/10.3390/app9163330
7. Goutte, C., Toft, P., Rostrup, E., Nielsen, F.Å., Hansen, L.K.: On clustering fMRI time series. Neuroimage **9**, 298–310 (1999). https://doi.org/10.1006/nimg.1998.0391
8. Granger, C.W.J.: Investigating causal relations by econometric models and cross-spectral methods. Econometrica **37**(3) (1969). https://doi.org/10.2307/1912791
9. Gudmundsson, S., Runarsson, T.P.: Sigurdsson: S.: Support vector machines and dynamic time warping for time series. IEEE Xplore (2008). https://doi.org/10.1109/IJCNN.2008.4634188
10. Hanssens, D.: Bivariate time series analysis of the relationship between advertising and sales. Appl. Econ. **12**(3), 329–339 (1982). https://doi.org/10.1080/00036848000000034
11. Ismail, A., Truong, H.-L., Kastner, W.: Manufacturing process data analysis pipelines: a requirements analysis and survey. J. Big Data **6**(1), 1–26 (2019). https://doi.org/10.1186/s40537-018-0162-3
12. Klein, A., Lehner, W.: Representing data quality for streaming and static data. IEEE J. Data Inf. Qual. (2007). https://doi.org/10.1109/ICDEW.2007.4400967
13. Kletti, J. (ed.): MES – Manufacturing Execution System - Moderne Informationstechnologie unterstützt die Wertschöpfung. Springer, Heidelberg (2015). https://doi.org/10.1007/978-3-662-46902-6
14. Lu, Y., Morris, K.C., Frechette, S.: Current standards landscape for smart manufacturing systems. US National Institute of Standards and Technology (2016). https://doi.org/10.6028/NIST.IR.8107
15. Malekpour, S., Sethares, W.A.: Conditional granger causality and partitioned Granger causality: differences and similarities. Biol. Cybern. **109**(6), 627–637 (2015). https://doi.org/10.1007/s00422-015-0665-3
16. Nikolić, D., Mureşan, R.C., Feng, W., Singer, W.: Scaled correlation analysis: a better way to compute a cross-correlogram. Eur. J. Neurosci. **35**, 742–762 (2012). https://doi.org/10.1111/j.1460-9568.2011.07987.x

17. Sakoe, H., Chiba, S.: A dynamic programming approach to continuous speech recognition. In: Seventh International Congress on Acoustics. Akadémiai Kiadó (1971)

18. Saller, D., Kumova, B.I., Hennebold, C.: Detecting causalities in production environments using time lag identification with cross-correlation in production state time series. In: Rutkowski, L., Scherer, R., Korytkowski, M., Pedrycz, W., Tadeusiewicz, R., Zurada, J.M. (eds.) ICAISC 2020. LNCS (LNAI), vol. 12416, pp. 243–252. Springer, Cham (2020). https://doi.org/10.1007/978-3-030-61534-5_22

19. Salvador, S., Chan, P.: FastDTW: toward accurate dynamic time warping in linear time and space. Intell. Data Anal. 11(5), 561–580 (2004)

20. Schneider, G., Havenith, M., Nikolić, D.: Spatiotemporal structure in large neuronal networks detected from cross-correlation. Neural Comput. 18, 2387–2413 (2006). https://doi.org/10.1162/neco.2006.18.10.2387

21. Stübinger, J., Adler, K.: How to identify varying lead–lag effects in time series data: implementation, validation, and application of the generalized causality algorithm. Algorithms 13, 95 (2020). https://doi.org/10.3390/a13040095

22. Syafrudin, M., Alfian, G., Fitriyani, N.F., Rhee, J.: Performance analysis of IoT-based sensor, big data processing, and machine learning model for real-time monitoring system in automotive manufacturing. Sensors 18, 2946 (2018). https://doi.org/10.3390/s18092946

23. Tan, C.W., Herrmann, M., Forestier, G., Webb, G., Petitjean, F.: Efficient search of the best warping window for dynamic time warping. In: SIAM International Conference on Data Mining (2012)

24. Tao, F., Qi, Q., Liu, A., Kusiak, A.: Data-driven smart manufacturing. J. Manuf. Syst. 48, 157–169 (2018). https://doi.org/10.1016/j.jmsy.2018.01.006

Active Finite Reward Automaton Inference and Reinforcement Learning Using Queries and Counterexamples

Zhe Xu[1](✉), Bo Wu[2], Aditya Ojha[2], Daniel Neider[3], and Ufuk Topcu[2]

[1] Arizona State University, Tempe, AZ, USA
xzhe1@asu.edu
[2] University of Texas at Austin, Austin, TX, USA
{adiojha629,utopcu}@utexas.edu
[3] Max Planck Institute for Software Systems, Kaiserslautern, Germany
neider@mpi-sws.org

Abstract. Despite the fact that deep reinforcement learning (RL) has surpassed human-level performances in various tasks, it still has several fundamental challenges. First, most RL methods require intensive data from the exploration of the environment to achieve satisfactory performance. Second, the use of neural networks in RL renders it hard to interpret the internals of the system in a way that humans can understand. To address these two challenges, we propose a framework that enables an RL agent to reason over its exploration process and distill high-level knowledge for effectively guiding its future explorations. Specifically, we propose a novel RL algorithm that learns high-level knowledge in the form of a *finite reward automaton* by using the L* learning algorithm. We prove that in episodic RL, a finite reward automaton can express any non-Markovian bounded reward functions with finitely many reward values and approximate any non-Markovian bounded reward function (with infinitely many reward values) with arbitrary precision. We also provide a lower bound for the episode length such that the proposed RL approach almost surely converges to an optimal policy in the limit. We test this approach on two RL environments with non-Markovian reward functions, choosing a variety of tasks with increasing complexity for each environment. We compare our algorithm with the state-of-the-art RL algorithms for non-Markovian reward functions, such as Joint Inference of Reward machines and Policies for RL (JIRP), Learning Reward Machine (LRM), and Proximal Policy Optimization (PPO2). Our results show that our algorithm converges to an optimal policy faster than other baseline methods.

1 Introduction

Despite the fact that deep reinforcement learning (RL) has surpassed human-level performances in various tasks, it still has several fundamental challenges.

© IFIP International Federation for Information Processing 2021
Published by Springer Nature Switzerland AG 2021
A. Holzinger et al. (Eds.): CD-MAKE 2021, LNCS 12844, pp. 115–135, 2021.
https://doi.org/10.1007/978-3-030-84060-0_8

First, most RL methods require intensive data from the exploration of the environment to achieve satisfactory performance. Second, the use of neural networks in RL renders it hard to interpret the internals of the system in a way that humans can understand [13].

To address these two challenges, we propose a framework that enables an RL agent to reason over its exploration process and distill high-level knowledge for effectively guiding its future explorations. Specifically, we learn high-level knowledge in the form of *finite reward automata*, a type of Mealy machine that encodes non-Markovian reward functions. The finite reward automata can be converted to a deterministic finite state machine, allowing a practitioner to more easily reason about what the agent is learning [14]. Thus, this representation is more interpretable than frameworks that use neural networks.

In comparison with other methods that also learn high-level knowledge during RL, the one proposed in this paper *actively* infers a finite reward automaton from the RL episodes. We prove that in episodic RL, a finite reward automaton can express any non-Markovian bounded reward functions with finitely many reward values and approximate any non-Markovian bounded reward function (with infinitely many reward values) with arbitrary precision. As the learning agent infers this finite reward automaton during RL, it also performs RL (specifically, q-learning) to maximize its obtained rewards based on the inferred finite reward automaton. The inference method is inspired by the L* learning algorithm [4], and modified to the framework of RL. We maintain two q-functions, one for incentivizing the learning agent to answer the *membership queries* during the explorations and the other one for obtaining optimal policies for the inferred finite reward automaton (in order to answer the *equivalence queries*). Furthermore, we prove that the proposed RL approach almost surely converges to an optimal policy in the limit, if the episode length is longer than a theoretical lower bound value.

We implement the proposed approach and three baseline methods (JIRP-SAT [33], LRM-QRM [24], and PPO2 [22]) in the Office world [33] and Minecraft world [3] scenarios. The results show that, at worst the approach converges to an optimal policy 88.8% faster than any of the baselines, and at best the approach converges while the other baselines do not.

1.1 Related Works

Our work is closely related to the use of formal methods in RL, such as RL for finite reward automata [17] and RL with temporal logic specifications [1,2,9,19, 24,27]. The current methods assume that high-level knowledge, in the form of reward machines or temporal logic specifications, is known *a priori*. However, in real-life use cases, such knowledge is implicit and must be inferred from data.

Towards the end of inferring high-level knowledge, several approaches have been proposed [16,20,33,37]. In these methods, the agent jointly learns the high-level knowledge *and* RL-policies *concurrently*. In [37], the inferred high-level knowledge is represented by *temporal logic* formulas and used for RL-based transfer learning. The methods for inferring temporal logic formulas from data can be found in [5, 7, 8, 10, 11, 15, 18, 21, 23, 25, 31, 32, 34–36].

In comparison with temporal logic formulas, the finite reward automata used in this paper are more expressive in representing the high-level structural relationships. Moreover, even if the inferred finite reward automaton is incorrect during the first training loop, the agent is still able to self-correct and learn more complex tasks. In order to learn finite reward automata, the authors in [33] proposed using *passive* inference of finite reward automata and utilizing the inferred finite reward automata to expedite RL. In [16], the authors proposed a method to infer reward machines to represent the memories of Partially observable Markov decision processes (POMDP) and perform RL for the POMDP with the inferred reward machines.

In contrast, our method *actively* infers the finite reward automaton in environments with non-Markovian reward functions. The active inference is facilitated by L* learning. This algorithm assumes the existence of a teacher who can answer the membership and equivalence queries [4, 28–30]. In our approach, an RL engine fulfills the role of the teacher, and the queries are answered through interaction with the environment through the RL engine.

During the submission of this paper, an interesting method was proposed by the authors of [12] which also used L* learning for non-Markovian Rewards. While superficially similar to our work, the two approaches differ in three ways:

(1) The proposed approach in this paper uses finite reward automata, while [12] works with deterministic finite automata (DFAs) and general automata. This is a notable distinction because we prove that finite reward automata can express any non-Markovian bounded reward function with finitely many reward values in episodic RL.
(2) The authors of [12] use only maintain one type of DFA for answering the equivalence queries. The proposed approach in this paper maintains two types of finite reward automata, using one to answer equivalence queries, and the other to answer membership queries. This additional finite reward automaton can incentivize the agent to answer membership queries during the exploration.
(3) We provide a lower bound for the episode length such that the proposed RL approach almost surely converges to an optimal policy in the limit.

2   Finite Reward Automata

In this section we introduce necessary background on reinforcement learning and finite reward automata.

2.1   Markov Decision Processes and Finite Reward Automata

Let $\mathcal{M} = (X, x_{init}, A, p, \mathcal{P}, R, L)$ be a labeled *Markov decision process* (labeled MDP), where the state space X and action set A are finite, $x_{init} \in X$ is a set of initial states, $p : X \times A \times X \rightarrow [0, 1]$ is a probabilistic transition relation, $\mathcal{P}$ is a set of propositional variables (i.e., labels), $R : (X \times A)^+ \times X \to \mathbb{R}$ is a reward function, and $L : X \times A \times X \to 2^{\mathcal{P}}$ is a labeling function.

We define the size of $\mathcal{M}$, denoted as $|\mathcal{M}|$, to be $|X|$ (i.e., the cardinality of the set X). A policy $\pi : X \times A \to [0, 1]$ specifies the probability of taking each action for each state. The *action-value function*, denoted as $q_\pi(x, a)$, is the expected discounted reward if an agent applies policy π after taking action a from state x. A finite sequence $x_0 a_0 \ldots x_k a_k x_{k+1}$ generated by $\mathcal{M}$ under certain policy π is called a *trajectory*, starting from $x_1 = x_{init}$ and satisfies $\sum_{a \in A} \pi(x_k, a) P(x_k, a, x_{k+1}) > 0$ for all $k \geq 1$. Its corresponding *label sequence* is $\ell_0 \ell_1 \ldots \ell_k$ where $L(x_i, a_i, x_{i+1}) = \ell_i$ for each $i \leq k$. Similarly, the corresponding *reward sequence* is $r_1 \ldots r_k$, where $r_i = R(x_0 a_0 \ldots x_i a_i x_{i+1})$, for each $i \leq k$. We call the pair $(\lambda, \rho) := (\ell_1 \ldots \ell_k, r_1 \ldots r_k)$ a *trace*.

Definition 1. *Let $\mathcal{M} = (X, x_{init}, A, p, \mathcal{P}, R, L)$ be a labeled Markov decision process. We define a sequence $(\ell_1, r_1), \ldots, (\ell_k, r_k)$ to be attainable if $k \leq eplength$ and $p(x_i, a_i, x_{i+1}) > 0$ for each $i \in \{0, \ldots, k\}$.*

Definition 2. *A finite reward automaton $\mathcal{A} = (W, w_{init}, 2^{\mathcal{P}}, \mathcal{R}, \delta, \eta)$ consists of a finite, nonempty set W of states, an initial state $w_{init} \in W$, an input alphabet $2^{\mathcal{P}}$, an output alphabet $\mathcal{R}$, a (deterministic) transition function $\delta \colon W \times 2^{\mathcal{P}} \to W$, and an output function $\eta \colon W \times 2^{\mathcal{P}} \to \mathcal{R}$, where $\mathcal{R}$ is a finite set of reward values ($\mathcal{R} \subset \mathbb{R}$). We define the size of $\mathcal{A}$, denoted as $|\mathcal{A}|$, to be $|W|$ (i.e., the cardinality of the set W).*

Remark 1. A finite reward automaton is actually a Mealy machine (Shallit 2008) where the output alphabet is a finite set of values. When the output alphabet is an infinite set of values, it is called a reward machine in [17, 33].

The run of a finite reward automaton $\mathcal{A}$ on a sequence of labels $\ell_1 \ldots \ell_k \in (2^{\mathcal{P}})^*$ is a sequence $w_0(\ell_1, r_1) w_1(\ell_2, r_2) \ldots w_{k-1}(\ell_k, r_k) w_k$ of states and label-reward pairs such that $w_0 = w_{init}$ and for all $i \in \{0, \ldots, k\}$, we have $\delta(w_i, \ell_i) = w_{i+1}$ and $\eta(w_i, \ell_i) = r_i$. We write $\mathcal{A}[\ell_1 \ldots \ell_k] = r_1 \ldots r_k$ to connect the input label sequence to the sequence of rewards produced by the machine $\mathcal{A}$ [33].

Definition 3. *We define that a finite reward automaton $\mathcal{A}$ encodes the reward function R of a labeled MDP $\mathcal{M}$ if for every trajectory $x_0 a_0 \ldots x_k a_k x_{k+1}$ of finite length and the corresponding label sequence $\ell_1 \ldots \ell_k$, the reward sequence equals $\mathcal{A}[\ell_1 \ldots \ell_k]$.*

Definition 4 (Reward Product MDP). *Let $\mathcal{M} = (X, x_{init}, A, p, \mathcal{P}, R, L)$ be a labeled MDP and $\mathcal{A} = (W, w_{init}, 2^{\mathcal{P}}, \mathcal{R}, \delta, \eta)$ a finite reward automaton encoding its reward function. We define the product MDP $\mathcal{M}_{\mathcal{A}} = (X', x'_I, A', p', \mathcal{P}', R', L')$ by*

- $X' = X \times W$;
- $x'_I = (x_{init}, w_{init})$;
- $A' = A$;

- $p'\big((x,w),a,(x',w')\big)$
$$= \begin{cases} p(x,a,x') & \text{if } w' = \delta(w, L(x,a,x')); \\ 0 & \text{otherwise;} \end{cases}$$
- $\mathcal{P}' = \mathcal{P}$;
- $R'\big((x,w),a,(x',w')\big) = \eta\big(w, L(x,a,x')\big)$; and
- $L' = L$.

2.2 Reinforcement Learning with Finite Reward Automata

Q-learning [26] is a form of model-free reinforcement learning (RL). Starting from state x, the system selects an action a, which takes it to state x' and obtains a reward R. The Q-function values will be updated by the following rule:

$$q(x,a) \leftarrow (1-\alpha)q(x,a) + \alpha(R + \gamma \max_a q(x',a)). \tag{1}$$

The q-learning algorithm can be modified to learn an optimal policy when the general reward function is encoded by a finite reward automaton [17]. Starting from state (x,w) in the product space, the system selects an action a, which takes it to state (x',w') and obtains a reward R. The Q-function values will be updated by the following rule.

$$q(x,w,a) \leftarrow (1-\alpha)q(x,w,a) + \alpha(R + \gamma \max_a q(x',w',a)). \tag{2}$$

We consider episodic Q-learning in this paper, and we use *eplength* to denote the episode length.

3 Expressivity of Finite Reward Automata

In this section, we show that any non-Markovian reward function in episodic RL which has finitely many reward values can be encoded by finite reward automata, while any non-Markovian bounded reward function in episodic RL can be approximated by finite reward automata with arbitrary precision.

Theorem 1. *For a labeled MDP* $\mathcal{M} = (X, x_{init}, A, p, \mathcal{P}, R, L)$ *with a non-Markovian reward function* $R : (2^{\mathcal{P}})^{+} \rightarrow \mathcal{R}_f$ *in episodic RL, where* $\mathcal{R}_f$ *is a finite set of values in* $\mathbb{R}$*, there exists at least one finite reward automaton* $\mathcal{A} = (W, w_{init}, 2^{\mathcal{P}}, \mathcal{R}, \delta, \eta)$ *that can encode the reward function* R.

Proof. We use $\mathcal{I}$ to denote the set of trajectories of length at most *eplength* and $\mathcal{T}_i$ to denote the maximal time index for trajectory $i \in \mathcal{I}$. We construct a finite reward automaton $\mathcal{A} = (W, w_{init}, 2^{\mathcal{P}}, \mathcal{R}, \delta, \eta)$,

- $W = w_{init} \cup \{w_{i,t}\}_{i \in \mathcal{I}, 1 < t \leq \mathcal{T}_i}$, $w_{i,1} = w_{init}$, $\forall i \in \mathcal{I}$;
- $\mathcal{R} = \mathcal{R}_f$;
- $\forall t \leq \mathcal{T}_i, \delta\big(w_{i,t}, L(x_{i,t}, a_{i,t}, x_{i,t+1})\big) = w_{i,t+1}$; and
- $\eta\big(w_{i,t}, \ell_i\big) = R(\ell_{i,1}, \ldots, \ell_{i,t})$, where $\ell_{i,t} = L(x_{i,t}, a_{i,t}, x_{i,t+1})$.

Then, it can be easily shown that for every trajectory $x_0 a_0 \ldots x_{i,k} a_{i,k} x_{i,k+1}$ and corresponding label sequence $\ell_1 \ldots \ell_k$ ($k \leq eplength$), the reward sequence equals $\mathcal{A}[\ell_1 \ldots \ell_k]$, i.e., $\mathcal{A}$ encodes the reward function R.

Remark 2. For a labeled MDP $\mathcal{M} = (X, x_{init}, A, p, \mathcal{P}, R, L)$ with a finite horizon, there can only be finitely many reward values, each corresponding to finitely many possible trajectories for the reward functions.

Theorem 2. *For a labeled MDP $\mathcal{M} = (X, x_{init}, A, p, \mathcal{P}, R, L)$ with a non-Markovian bounded reward function $R : (2^{\mathcal{P}})^+ \to \mathbb{R}$ in episodic RL, there exists at least one finite reward automaton $\mathcal{A} = (W, w_{init}, 2^{\mathcal{P}}, \mathcal{R}, \delta, \eta)$ that can approximate the non-Markovian reward function R with arbitrary precision.*

Proof. For a bounded reward function taking values in $[r_{\min}, r_{\max}]$, where $r_{\min}, r_{\max} \in \mathbb{R}$, $r_{\min} \leq r_{\max}$, we construct a finite set $\mathcal{R}_\epsilon = \{r_{\min}, r_{\min} + \epsilon, r_{\min} + 2\epsilon, \ldots, r_{\min} + n_{\max}\epsilon\}$, where $\epsilon \in \mathbb{R}$, $\epsilon > 0$, $n_{\max} = \max\{n \mid r_{\min} + n\epsilon \leq r_{\max}\}$. We use $\mathcal{I}$ to denote the set of trajectories of length at most *eplength* generated from the labeled MDP $\mathcal{M}$ and $\mathcal{T}_i$ to denote the maximal time index for trajectory $i \in \mathcal{I}$. We construct a finite reward automaton $\mathcal{A} = (W, w_{init}, 2^{\mathcal{P}}, \mathcal{R}, \delta, \eta)$, where

- $W = w_{init} \cup \{w_{i,t}\}_{i \in \mathcal{I}, 1 < t \leq \mathcal{T}_i}$, $w_{i,1} = w_{init}$, $\forall i \in \mathcal{I}$;
- $\mathcal{R} = \mathcal{R}_\epsilon$;
- $\forall t \leq \mathcal{T}_i, \delta(w_{i,t}, L(x_{i,t}, a_{i,t}, x_{i,t+1})) = w_{i,t+1}$; and
- $\eta(w_{i,t}, \ell_{i,t}) = \arg \min_{r \in \mathcal{R}_\epsilon} |r - R(\ell_{i,1}, \ldots, \ell_{i,t})|$, where $\ell_{i,t} = L(x_{i,t}, a_{i,t}, x_{i,t+1})$.

Then, it can be easily shown that for any $\epsilon > 0$ and every trajectory $x_0 a_0 \ldots x_{i,k} a_{i,k} x_{i,k+1}$ ($k \leq eplength$), we have $|\mathcal{A}[\ell_{i,1}, \ldots, \ell_{i,t}] - R[\ell_{i,1}, \ldots, \ell_{i,t}]| < \epsilon$, i.e., $\mathcal{A}$ approximates the reward function R with arbitrary precision.

4   Active Finite Reward Automaton Inference and Reinforcement Learning (AFRAI-RL)

In this section, we introduce the Active Finite Reward Automaton Inference and Reinforcement Learning (AFRAI-RL) algorithm. Figure 1 shows the block diagram of the AFRAI-RL approach, and Algorithm 1 shows the procedures of the AFRAI-RL approach. The AFRAI-RL approach consists of an active finite reward automaton inference engine and an RL engine. In the following two subsections, we will introduce the two engines and their interactions for obtaining the optimal RL policy for tasks with non-Markovian rewards.

4.1   Active Finite Reward Automaton Inference Engine

In this subsection, we introduce the active finite reward automaton inference engine which is based on L* learning [4], which is an algorithm that learns a *minimal* deterministic finite automaton (DFA) that accepts an unknown regular

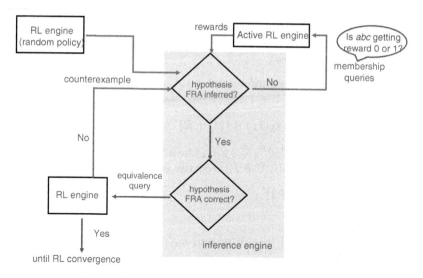

Fig. 1. Block diagram of the AFRAI-RL approach. Initially, a random RL engine is used generate counterexample traces for the inference engine. The inference engine alternately performs two tasks: (a) it creates membership queries for the RL engine if there are counterexamples that are inconsistent with the inferred Finite Reward Automaton (FRA); (b) it generates and sends equivalence queries to the RL-Engine. The active RL Agent will obtain rewards from the environment to answer the membership queries and the equivalence queries.

language $\mathcal{L}$ by interacting with a *teacher*. In the setting of this paper, the role of teacher is fulfilled by the RL engine. For simplicity, we only consider RL tasks with non-Markovian reward functions with finitely many reward values (hence there exists at least one finite reward automaton that can encode the non-Markovian reward function).

We first show that a finite reward automaton can be converted to a DFA.

Definition 5. *A deterministic finite automaton (DFA) is a five-tuple* $\mathfrak{A} = (V, v_I, \Sigma, \delta, F)$ *consisting of a nonempty, finite set v of states, an initial state $v_I \in V$, an input alphabet Σ, a transition function $\delta : V \times \Sigma \rightarrow V$, and a set $F \subseteq V$ of final states. The size of an DFA, denoted by $|\mathfrak{A}|$, is the number $|V|$ of its states.*

A *run* of an DFA $\mathfrak{A} = (V, v_I, \Sigma, \delta, F)$ on an input word $\tau = \tau_1 \ldots \tau_n$ is a sequence $v_0 \ldots v_n$ of states such that $v_0 = v_I$ and $v_i = \delta(v_{ix}, a_i)$ for each $i \in \{1, \ldots, n\}$. A run $v_0 \ldots v_n$ of $\mathfrak{A}$ on a word u is *accepting* if $v_n \in F$, and a word u is *accepted* if there exists an accepting run. The *language* of an DFA $\mathfrak{A}$ is the set $L(\mathfrak{A}) = \{u \in \Sigma^* \mid \mathfrak{A} \text{ accepts } u\}$. As usual, we call two DFAs $\mathfrak{A}_1$ and $\mathfrak{A}_2$ *equivalent* if $L(\mathfrak{A}_1) = L(\mathfrak{A}_2)$.

We show that every finite reward automaton over the input alphabet $2^{\mathcal{P}}$ and output alphabet $\mathcal{R}$ can be translated into an "equivalent" DFA as defined below. This DFA operates over the combined alphabet $2^{\mathcal{P}} \times \mathcal{R}$ and accepts a word $(\ell_0, r_0) \dots (\ell_k, r_k)$ if and only if $\mathcal{A}$ outputs the reward sequence $r_0 \dots r_k$ on reading the label sequence $\ell_0 \dots \ell_k$.

Lemma 1. *Given a finite reward automaton $\mathcal{A} = (W, w_{init}, 2^{\mathcal{P}}, \mathcal{R}, \delta, \eta)$, one can construct a DFA $\mathfrak{A}_{\mathcal{A}}$ with $|\mathcal{A}| + 1$ states such that*

$$\mathcal{L}(\mathfrak{A}_{\mathcal{A}}) = \{(\ell_0, r_0) \dots (\ell_k, r_k) \in (2^{\mathcal{P}} \times \mathcal{R})^* \mid A[\ell_0 \dots \ell_k] = r_0 \dots r_k\}.$$

Proof. Let $\mathcal{A} = (W_{\mathcal{A}}, w_{init}{}^{\mathcal{A}}, 2^{\mathcal{P}}, \mathcal{R}, \delta_{\mathcal{A}}, \eta_{\mathcal{A}})$ be a finite reward automaton. Then, we define a DFA $\mathfrak{A}_{\mathcal{A}} = (V, v_I, \Sigma, \delta, F)$ over the combined alphabet $2^{\mathcal{P}} \times \mathcal{R}$ by

- $V = W_{\mathcal{A}} \cup \{\bot\}$ with $\bot \notin W_{\mathcal{A}}$;
- $v_I = w_{init}{}^{\mathcal{A}}$;
- $\Sigma = 2^{\mathcal{P}} \times \mathcal{R}$;
- $\delta(w, (\ell, r)) = \begin{cases} w' & \text{if } \delta_{\mathcal{A}}(w, \ell) = w' \text{ and } \eta_{\mathcal{A}}(w, \ell) = r; \\ \bot & \text{otherwise}; \end{cases}$
- $F = W_{\mathcal{A}}$.

In this definition, $\bot$ is a new sink state to which $\mathfrak{A}_{\mathcal{A}}$ moves if its input does not correspond to a valid input-output pair produced by $\mathcal{A}$. A straightforward induction over the length of inputs to $\mathfrak{A}_{\mathcal{A}}$ shows that it indeed accepts the desired language. In total, $\mathfrak{A}_{\mathcal{A}}$ has $|\mathcal{A}| + 1$ states.

During the learning process, the inference engine maintains an *observation table* $\mathcal{O} = (S, E, T)$ where $S \subseteq \Sigma^*$ is a set of prefixes ($\Sigma = 2^{\mathcal{P}}$), $E \subseteq \Sigma^*$ is a set of suffixes and $T : (S \cup S \cdot \Sigma) \times E \to \{0, 1\}$. Σ^* denotes finite traces from alphabet set Σ. For $s \in S \cup S \cdot \Sigma$, $e \in E$, if $s \cdot e \in \mathcal{L}$, then $T(s, e) = 1$ and if $s \cdot e \notin \mathcal{L}$ then $T(s, e) = 0$. Membership queries assign the correct value (0 or 1) to $T(s, e)$. For simplicity, we denote $row(s) = (T(s, e_1), \dots, T(s, e_n)) \in \{0, 1\}^n, |E| = n$. The inference engine will always keep the observation table *closed* and *consistent* as defined below.

Definition 6. *An observation table $\mathcal{O} = (S, E, T)$ is closed if for each $t \in S \cdot \Sigma$, we can find some $s \in S$ such that $row(s) = row(t)$.*

Definition 7. *$\mathcal{O}$ is consistent if whenever for $s_1, s_2 \in S$, $row(s_1) = row(s_2)$, then for any $\sigma \in \Sigma$, we have $row(s_1 \sigma) = row(s_2 \sigma)$.*

Remark 3. If an observation table $\mathcal{O} = (S, E, T)$ is closed and consistent, it is possible to construct a DFA $\mathcal{M}(\mathcal{O}) = (Q, \Sigma, \delta, Q_0, F)$ as the acceptor, where

- $Q = \{row(s) | s \in S\}$;
- $q_0 = row(\epsilon)$;
- $\delta(row(s), \sigma) = row(s\sigma), \forall \sigma \in \Sigma$;
- $F = \{row(s) | s \in S, T(s) = 1\}$.

Definition 8. *For a closed and consistent observation table $\mathcal{O} = (S, E, T)$, we define a corresponding hypothesis finite reward automaton $A^h(\mathcal{O}) = (W^h, w_{init}^h, 2^{\mathcal{P}}, \mathcal{R}, \delta^h, \eta^h)$ as follows:*

- $W^h = \{row(s) : s \in S\}$,
- $w_{init}^h = row(\epsilon)$,
- $\delta^h(row(s), \sigma) = row(s\sigma)$,
- $\eta^h(w, \lambda) = 1$, if $T(\lambda) = 1$; and $\eta^h(w, \lambda) = 0$, otherwise.

Algorithm 1. AFRAI-RL

1: Initialize $\mathcal{O} = (S, E, T)$, $sample$, $Nsample$, C, q_{m}, q_{h}
2: **while** there exists counterexample (λ, ρ) **do**
3:    Add (λ, ρ) and its prefixes to S
4:    $ChangeT \leftarrow 0$
5:    **while** $\mathcal{O}$ is neither closed nor consistent $\wedge$ $(ChangeT = 0)$ **do**
6:       $\chi \leftarrow CheckObsTable(\mathcal{O})$
7:       $T, sample, Nsample, q_{\mathrm{m}} \leftarrow MQuery(T, \chi, sample, Nsample, C, q_{\mathrm{m}})$
8:    **end while**
9:    $(\lambda, \rho), \mathcal{O}, sample, Nsample, q_{\mathrm{h}} \leftarrow EQuery(\mathcal{O}, sample, Nsample, q_{\mathrm{h}})$
10: **end while**

Algorithm 1 shows the AFRAI-RL algorithm. Algorithm 1 starts by checking whether there are any counterexample traces that need to be added to the observation table. These traces and their suffixes are added to the observation table (Algorithm 1, Line 3), and the subroutine $CheckObsTable$ (see Algorithm 4) is used to find *membership query traces* (noted as χ in all algorithms).

Then, Algorithm 1 proceeds to answer two types of queries, namely the *membership* query (Algorithm 1, Line 7) and the *equivalence* query (Algorithm 1, Line 9).

Answering Membership Queries: The detailed procedures to answer membership queries are shown in Algorithm 2. The subroutine sets $T(\zeta)$ to 1 if the trace ζ can be accepted by the DFA converted from a finite reward automaton (that can encode an unknown non-Markovian reward function), and sets $T(\zeta)$ to 0 otherwise. We maintain a set $sample$ of accepted traces and use the subroutine $CheckSample$ (see Algorithm 5) to check whether a membership query can already be answered by the set $sample$ (Algorithm 2, Lines 6 to 8). If it can, we provide the answer to $T(\zeta)$ in the observation table $\mathcal{O}$ (Algorithm 2, Line 8); otherwise, we perform RL to answer the membership query (see Algorithm 7 for details). If the membership query trace ζ is *inconsistent* with a trace in RL from the environment (e.g., the membership query trace $\zeta = (\ell_1, 0), (\ell_2, 1), (\ell_3, 1)$, while a trace $(\ell_1, 0), (\ell_2, 0)$ is observed from the environment), then a flag is set (Algorithm 2, Line 14) to stop the loop, $T(\zeta)$ is set to zero immediately (Algorithm 2, Line 25), and move on to the next trace. Otherwise we add the trace to $sample$ (Algorithm 2, Line 16). To boost efficiency, we set a limit on how many episodes we perform to answer a membership query. This limit is $C \in \mathbb{Z}_{>0}$. We answer the membership query as 0 if after C episodes the membership query still cannot be answered (Algorithm 2, Lines 24-25). Such traces are recorded in the set $Nsample$ (Algorithm 2, Line 26). Afterwards, if the trace for the membership

---

Algorithm 2. MQuery

 1: **Input:** $T, \chi, sample, Nsample, C, q_m$
 2: $Query \leftarrow$ membership
 3: **for** each $\zeta \in \chi$ **do**
 4:     Construct a query finite reward automaton $\mathcal{A}^m(\zeta)$
 5:     $counter \leftarrow 0$
 6:     $PrevAnswer \leftarrow check(\zeta, sample)$
 7:     **if** $PrevAnswer \neq Null$ **then**
 8:         $T(\zeta) = PrevAnswer$
 9:     **else**
10:         $Inconsistent \leftarrow 0$
11:         **while** $check(\zeta, sample) = Null \wedge (counter < C) \wedge Inconsistent = 0$ **do**
12:             $\lambda, \rho, q_m \leftarrow$ RL-Engine($Query, \mathcal{A}^m(\zeta), q_m$)
13:             **if** ζ is inconsistent with (λ, ρ) **then**
14:                 $Inconsistent \leftarrow 1$
15:             **else**
16:                 Add (λ, ρ) to $sample$
17:                 $ChangeT, T \leftarrow$ CheckNSample($(\lambda, \rho), Nsample, T$)
18:                 **if** $ChangeT = 1$ **then**
19:                     **return** $T, sample, Nsample, q_m$
20:                 **end if**
21:                 $counter \leftarrow counter + 1$
22:             **end if**
23:         **end while**
24:         **if** $(counter > C) \vee (Inconsistent = 1)$ **then**
25:             $T(\zeta) \leftarrow 0$
26:             Add ζ to $Nsample$
27:         **else**
28:             $T(\zeta) \leftarrow 1$
29:         **end if**
30:     **end if**
31: **end for**
32: **return** $T, sample, Nsample, q_m$

---

query is encountered in the environment, we use the subroutine $CheckNSample$ (see Algorithm 6) to change the original answers in the observation table $\mathcal{O}$ accordingly. This is performed during both membership (Algorithm 2, Line 17) and equivalence queries (Algorithm 3, Line 7). If the answer was changed during a membership query(i.e., $ChangeT = 1$), we exit the Algorithm 2 (Line 19) to generate the additional membership query traces created from changing the table (Algorithm 1, Line 5).

Algorithm 3. EQuery

1: **Input** $\mathcal{O}$, $sample$, $Nsample$, q_h
2: Construct a hypothesis finite reward automaton $\mathcal{A}^\text{h}(\mathcal{O})$
3: $Query \leftarrow$ equivalence
4: **Do**
5:     $\lambda, \rho, q_\text{h} \leftarrow$ RL-Engine($Query$,$\mathcal{A}^\text{h}$, q_h)
6:     Add (λ, ρ) to $sample$
7:     $ChangeT, T \leftarrow$CheckNSample((λ, ρ),$Nsample$,T)
8: **Until** Find counterexample (λ, ρ)
9: **Return** (λ, ρ), $\mathcal{O}$, $sample$, $Nsample$, q_h

---

Answering Equivalence Queries: The detailed procedures to answer equivalence queries are shown in Algorithm 3. We perform RL with the hypothesis finite reward automaton, updating $sample$ along the way (Algorithm 3, Line 6), until a *counterexample* is found (Algorithm 3, Line 8). A counterexample is a trace where the rewards given by environment are different from the rewards given by hypothesis finite reward automaton. Specifically, there are two types of counterexamples. A *positive counterexample* is a trace that is accepted by the DFA converted from the current hypothesis finite reward automaton, but is not accepted by the DFA converted from any finite reward automaton that can encode the unknown non-Markovian reward function. A *negative counterexample* is a trace that is not accepted by the DFA converted from the current hypothesis finite reward automaton, but is accepted by the DFA converted from any finite reward automaton that can encode the unknown non-Markovian reward function. The RL-engine returns counterexamples to the inference engine for another round of inference (Algorithm 3, Line 5 and 9).

Algorithm 4. CheckObsTable

1: **Input:** $\mathcal{O}$
2: **if** $\mathcal{O}$ is not consistent **then**
3:     Find s_1, $s_2 \in S$, $\sigma \in \Sigma$ and $e \in E$ such that $row(s_1) = row(s_2)$ and $T(s_1\sigma e) \neq T(s_2\sigma e)$
4:     add σe to E
5:     $\chi \leftarrow (S \cup S\Sigma)\sigma e$
6: **else if** $\mathcal{O}$ is not closed **then**
7:     Find $s \in S$ and $\sigma \in \Sigma$ such that $\forall s \in S, row(s\sigma) \neq row(s)$
8:     add $s\sigma$ to S
9:     $\chi \leftarrow (s\sigma \cup s\sigma\Sigma)E$
10: **end if**
11: Return χ

Algorithm 4 generates *membership query traces* based on whether or not the observation table $\mathcal{O}$ is closed or consistent. If the table is not consistent, then we add σe to E and each $\zeta \in (S \cup S\Sigma)\sigma e$ forms a *membership query trace*. If the

---

Algorithm 5. CheckSample
___
1: **Input:** ζ, *sample*
2: **for** each trace (λ, ρ) in *sample* **do**
3:     **if** ζ is prefix of (λ, ρ) **then**
4:         Return 1
5:     **end if**
6:     **if** ζ is inconsistent with (λ, ρ) **then**
7:         Return 0
8:     **end if**
9: **end for**
10: Return *Null*

___

table is not closed, then we add add $s\sigma$ to S and each $\zeta \in (s\sigma \cup s\sigma\Sigma)E$ forms a *membership query trace*.

Algorithm 5 shows the subroutine *CheckSample*. It returns 1 or 0 if the membership query for ζ has already been answered, and *Null* otherwise. For each trace (λ, ρ) in *sample*, if a membership query trace ζ is prefix of (λ, ρ), then ζ must be accepted by the DFA converted from the finite reward automaton that encodes the unknown non-Markovian reward function; hence *CheckSample* returns 1. If ζ is *inconsistent* with a trace in *sample*, then ζ cannot be accepted by the DFA mentioned above and *CheckSample* returns 0.

Algorithm 6 shows the subroutine *CheckNSample*. Each trace ζ' in *Nsample* is checked to see if it is a prefix of (λ, ρ) (the recent answer from the RL-Engine). If the trace is a prefix, then its answer in the observation table is changed (Line 4). A flag, *ChangeT*, is set so that the observation table is rechecked for being closed and consistent in Algorithm 1.

By answering the membership and equivalence queries, L^* algorithm is guaranteed to converge to the minimum DFA accepting the unknown regular language $\mathcal{L}$ using $O(|\Sigma|n^2 + n\log c)$ membership queries and at most $n - 1$ equivalence queries, where n denotes the number of states in the final DFA and c is the length of the longest counterexample from the RL engine when answering equivalence queries [4].

Algorithm 6. CheckNSample
___
1: **Input:** (λ, ρ), *Nsample*, T
2: **for** each $\zeta' \in Nsample$ **do**
3:     **if** ζ' is a prefix of (λ, ρ) **then**
4:         $ChangeT = 1$, $T(\zeta') = 1$
5:     **end if**
6: **end for**
7: Return *ChangeT*, T

___

4.2   Active Reinforcement Learning Engine

In this subsection, we introduce the active reinforcement learning engine. We first define a *query finite reward automaton* corresponding to a membership query trace $\zeta = (\ell_1, r_1), \ldots, (\ell_k, r_k)$ as follows.

Definition 9. *For a membership query trace* $\zeta = (\ell_1, r_1), \ldots, (\ell_k, r_k)$, *we define a corresponding query finite reward automaton* $\mathcal{A}^m(\zeta) = (W^m, w_{init}^m, 2^\mathcal{P}, \mathcal{R}, \delta^m, \eta^m)$ *as follows:*

- $W^m = \{w_0^m, w_1^m, \ldots, w_k^m\}$,
- $w_{init}^m = w_0^m$,
- *for any* $i \in [0, k-1]$, $\delta^m(w_i^m, \ell_{i+1}) = w_{i+1}^m$, $\delta^m(w_i^m, \ell) = w_i^m$ *for any* $\ell \neq \ell_i$, *and*
- *for any* $w^m \in W^m$ *and any* $\ell \in 2^\mathcal{P}$, $\eta^m(w^m, \ell) = 1$, *if* $\delta^m(w^m, \ell) \neq w^m$; *and* $\eta^m(w^m, \ell) = 0$, *otherwise.*

Intuitively, the query finite reward automaton corresponding to a membership query trace $\zeta = (\ell_1, r_1), \ldots, (\ell_k, r_k)$ is a finite reward automaton that outputs a reward of one every time a new label ℓ_i ($i \in [1, k]$) is achieved (and the state of the finite reward automaton moves from w_i^m to w_{i+1}^m). Therefore, in performing RL with the query finite reward automaton, the rewards obtained from the query finite reward automaton serve as incentives to encounter the label sequence in the membership query trace and hence to answer the membership query.

In the RL engine, we maintain two different types of q-functions: *query q-functions* for answering membership queries, denoted as q_m; and *hypothesis q-functions* for maximizing the cumulative rewards from the environment (also answering equivalence queries), denoted as q_h.

We update the query q-functions as follows:

$$q_m(x, w^m, a) \leftarrow (1 - \alpha)q_m(x, w^m, a) \\ + \alpha(r_m + \gamma \max_a q_m(x', w'^m, a)), \tag{3}$$

Similarly, we update the hypothesis q-functions as follows:

$$q_h(x, w^h, a) \leftarrow (1 - \alpha)q_h(x, w^h, a) \\ + \alpha(r_h + \gamma \max_a q_h(x', w'^h, a)), \tag{4}$$

Algorithm 7 shows the procedure to run each RL episode to answer an membership query or equivalence query. Algorithm 7 first initializes the initial state x of the MDP, the initial state of the (query or hypothesis) finite state automaton w and the trace (as the empty trace). Specifically, w is initialized as w_{init}^m if it is in the membership query phase ($\mathcal{A} = \mathcal{A}_m$), and initialized as w_{init}^h if it is in the equivalence query phase ($\mathcal{A} = \mathcal{A}_h$) (Line 6). Algorithm 8 is used to run one step through the environment (see next paragraph for details). We feed the query q-function and finite reward automaton into Algorithm 8 if a membership query

---

Algorithm 7. RL-Engine

1: **Hyperparameters:** learning rate α, discount factor γ, episode length *eplength*
2: **Input:** Variable *Query*, a reward automaton $\mathcal{A}$, q-function q
3: $x \leftarrow InitialState(); (\lambda, \rho) \leftarrow []$
4: $w \leftarrow w_{init}$
5: **for** $0 \leq t < eplength$ **do**
6:     $x', w', q \leftarrow Step(Query, \mathcal{A}, q, x, w)$
7:     append $(L(x, a, x'), r)$ to (λ, ρ)
8:     $x \leftarrow x', w \leftarrow w', t \leftarrow t+1$
9: **end for**
10: **return** (λ, ρ, q)

Algorithm 8. Step

1: **Input:** Variable *Query*, a finite reward automaton $\mathcal{A}$, a q-function q, an MDP state x, and an FRA state w
2: $a = GetEpsilonGreedyAction(q, w, x)$
3: $x' = ExecuteAction(x, a)$
4: **if** *Query* = membership **then**
5:     $r \leftarrow \eta(w, L(x, a, x'))$;
6: **else**
7:     Observe r from the environment;
8: **end if**
9: $w' = \delta(w, L(x, a, x'))$
10: $q(x, w, a) \leftarrow (1 - \alpha)q(x, w, a) + \alpha(r + \gamma \max_a q(x', w', a))$
11: **for** $\hat{w} \in W \setminus \{w\}$ **do**
12:     $\hat{w}' = \delta(\hat{w}, L(x, a, x'))$
13:     $\hat{r} = \eta(\hat{w}, L(x, a, x'))$
14:     $q(x, \hat{w}, a) \leftarrow (1 - \alpha)q(x, \hat{w}, a) + \alpha(\hat{r} + \gamma \max_a q(x', \hat{w}', a))$
15: **end for**
16: **return** x', w', q

---

is being asked, and the hypothesis q-function and automaton otherwise. Algorithm 7 returns the trace, the query q-function and the hypothesis q-function (Line 10).

Algorithm 8 runs one step through the environment, updating the q-function for either a membership or equivalence query. First, at state x an action a is selected according to the q-function using the epsilon-greedy approach and executed to reach a new state x' (Line 2). Then rewards are collected based on the type of query being asked. If a membership query is being asked, then the automaton's reward function is used; otherwise the reward is observed from the environment. The next mealy state for the automaton is calculated (Line 9) and the q-function supplied is updated (Line 10).

Lemma 2. *Let $\mathcal{M}$ be a labeled MDP and $\mathcal{A}$ the finite reward automaton encoding the reward function of $\mathcal{M}$. Then, AFRAI with eplength $\geq 2^{|\mathcal{M}|+1}(|\mathcal{A}|+1)-1$*

almost surely learns a finite reward automaton in the limit that is equivalent to $\mathcal{A}$ on all attainable label sequences.

Proof. Given a *eplength* $= 2^{|\mathcal{M}|+1}(|\mathcal{A}|+1) - 1$, we are almost sure to experience every possible attainable trace from the environment. The proof of this is similar to the proof provided for Lemma 2 in [33] (Proof is located in Appendix C of [33]). Given every attainable trace, AFRAI-RL will answer all membership and equivalence queries correctly with probability 1 in the limit, because the observation table $\mathcal{O}$ can be changed as the Algorithm 1 runs (Algorithm 1, Lines 7 and 9). Furthermore, the author in [4] shows that if all membership and equivalence queries can be answered, a DFA can be formed from the observation table (as in Remark 3). This DFA will be the smallest DFA that can accept the language defined in Lemma 1. In this RL context, this language will match the language of the DFA converted from the finite reward automaton; i.e., the language will encode the finite reward automaton that is equivalent to $\mathcal{A}$. Therefore, the observation table encodes the finite reward automaton that is equivalent to $\mathcal{A}$.

With Lemma 2, we can proceed to prove Theorem 3.

Theorem 3. *Let $\mathcal{M}$ be a labeled MDP and $\mathcal{A}$ the reward machine encoding of the reward function of $\mathcal{M}$. Then, AFRAI-RL with eplength $\geq 2^{|\mathcal{M}|+1} \cdot (|\mathcal{A}|+1) - 1$ almost surely converges to an optimal policy in the limit.*

Proof. Lemma 2 shows that, eventually, the reward machine learned by AFRAI-RL, will be equivalent to $\mathcal{A}$ on all attainable label sequences. Let $\mathcal{H}$ be the reward machine learned by AFRAI-RL and $\mathcal{M}_H$ be the product MDP.

Thus an optimal policy for $\mathcal{M}_H$ will also be optimal for $\mathcal{M}$. When running episodes of QRM (Algorithm 7) under the reward machine $\mathcal{H}$, an update of a q-function connected to a state of $\mathcal{H}$ corresponds to updating the q function for $\mathcal{M}_H$. Since *eplength* $\geq |M|$, the fact that QRM uses the epsilon-greedy strategy and that updates are done in parallel for all states of $\mathcal{H}$ implies that every state-action pair of the $\mathcal{M}_H$ will be seen infinitely often. Hence, the convergence of q-learning for $\mathcal{M}_H$ to an optimal policy is guaranteed by [26]. Therefore, as the number of episodes goes to infinity, with *eplength* $\geq 2^{|\mathcal{M}|+1} \cdot (|\mathcal{A}|+1) - 1$, AFRAI-RL converges towards an optimal policy.

5   Case Studies

In this section, we apply the proposed approach to the office world scenario adapted from [17] and the craft world scenario from [3]. We perform the following four different methods:

– AFRAI-RL: We use the libalf [6] implementation of active automata learning as the algorithm to infer finite reward automata.

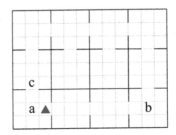

Fig. 2. The map in the office world scenario.

- JIRP-SAT: We use the libalf [6] implementation of SAT-solving (see Sect. 3.2 of [33]) as the algorithm to infer finite reward automata.
- LRM-QRM: We use the QRM implementation from [17], adapted to test the agent at the end of each episode.
- PPO2: We use the Stable Baselines implementation of PPO2 [22]. The state space is a history of the past states the agent has been in. This was added because PPO2 doesn't have any way of remembering it's previous states.

In [33], the authors have shown that JIRP-SAT and JIRP-RPNI outperform q-learning in augmented state space (QAS), hierarchical reinforcement learning (HRL), and deep reinforcement learning with double q-learning (DDQN) in three case studies. Therefore, if we can show that AFRAI-RL outperforms JIRP-SAT and JIRP-RPNI, then we can deduce that AFRAI-RL outperforms QAS, HRL and DDQN as well.

5.1 Office World Scenario

We consider the office world scenario in the 9×12 grid-world. Figure 2 shows the map in the office world scenario. We use the triangle to denote the initial position of the agent. The agent has four possible actions at each time step: move north, move south, move east and move west. After each action, the robot may slip to each of the two adjacent cells with probability of 0.05. In Algorithm 1, we set $C = 500$.

We consider the following three tasks:

Task 1: first go to a, then go to b and a in this order, and finally return to c. Episode Length was set to 200, and total training time was set to 1,000,000.
Task 2: first go to b, then go to c and a in this order, then repeat the sequence. Episode length was set to 800, and total training time was set to 2,000,000.
Task 3: first go to c, then go to b and a in this order, then go to b and c in this order and return to a. Episode length was set to 800, and total training time was set to 6,000,000.

Figure 3 shows the attained rewards of 10 independent simulation runs for each task, averaged every 10 training steps. For task 1, it can be seen that, on average, the proposed AFRAI-RL approach converges to an optimal policy in about 0.2 million training steps, while JIRP-SAT, LRM-QRM, and PPO2 do not converge to an optimal policy. For task 2, on average the proposed AFRAI-RL approach converges to an optimal policy in about 1.8 million training steps, while JIRP-SAT, LRM-QRM, and PPO2 do not converge to an optimal policy. For task 3, on average the proposed AFRAI-RL approach converges to an optimal policy in about 4.0 million training steps, while JIRP-SAT converges to an optimal policy in about 4.5 million training steps and LRM-QRM, and PPO2 do not converge to an optimal policy.

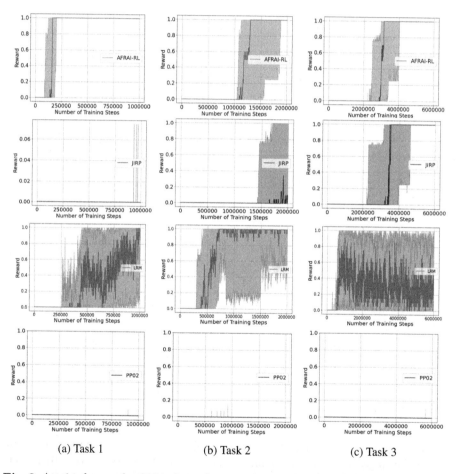

(a) Task 1   (b) Task 2   (c) Task 3

Fig. 3. Attained rewards of 10 independent simulation runs of the office world scenario, averaged for every 10 training steps in AFRAI-RL (First row), JIRP-SAT (Second Row), LRM-QRM (Third Row), and PPO2 (Fourth Row): (a) Task 1; (b) Task 2; (c) Task 3.

5.2 Minecraft world scenario

We consider the Minecraft world scenario in the 21×21 grid-world [3]. The four actions and the slip rates are the same as in the office world scenario. We train on two tasks: making a hammer and spear. In Algorithm 1, we set $C = 500$.

We consider the following two tasks, noting the symbol used for each object in parenthesis:

Task 1: Build a Hammer. Agent must collect string (b), stone (e), iron (f), stone (e) (in this order) and travel to the workbench (c) to make the hammer. Episode length was set to 400, and total training time was set to 400,000.

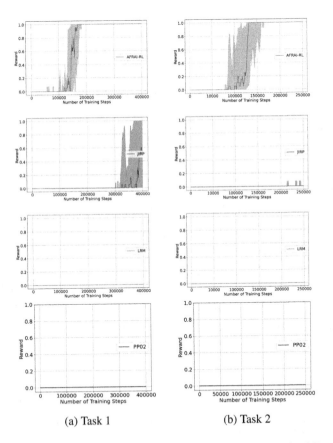

(a) Task 1   (b) Task 2

Fig. 4. Attained rewards of 10 independent simulation runs of the Minecraft world scenario, averaged for every 10 training steps in AFRAI-RL (First row), JIRP-SAT (Second Row), LRM-QRM (Third Row) and PPO2 (Fourth Row): (a) Task 1; (b) Task 2.

Task 2: Build a Spear. The agent must collect string, stone, wood (a), string and travel to the workbench to make the spear. Episode length was set to 400, and total training time was set to 250,000. Figure 4 shows the attained rewards of 10 independent simulation runs for each task, averaged every 10 training steps. For task 1, it can be seen that, on average, the proposed AFRAI-RL approach converges to an optimal policy in about 190,000 training steps, while JIRP-SAT, LRM-QRM, and PPO2 do not converge to an optimal policy. For task 2, on average the proposed AFRAI-RL approach converges to an optimal policy in about 170,000 million training steps, while JIRP-SAT, LRM-QRM, and PPO2 do not converge to an optimal policy.

6 Conclusions

We propose an active reinforcement learning approach that infers finite reward automata during the RL process through interaction with the environment. The algorithm can actively guide the RL towards finding answers to the queries needed for inferring the finite reward automata, thus making the finite reward automaton learning process more efficient. The case studies show that this algorithm is more efficient than recent baseline algorithms for the environments used.

This work has potential in multiple future directions. First, we assume that the same state-action sequences lead to the same reward values in this paper (i.e., the method depends on the correct labeling of the trajectories). We will investigate the scenarios where incorrect labels can occur and the same state-action sequences may lead to different reward values. Second, current active RL work usually adopts a model-based framework and search for states that are less visited, thus improving the sample efficiency of RL. Our proposed approach adopts a model-free framework and actively recovers the reward structure, which turns out to be effective in the non-Markov RL scenarios. To make connections with the other work in active RL, we will investigate model-based RL with active finite reward automaton inference. Finally, as we consider RL for a single agent in this paper, we will extend the work to multi-agent settings for collaborative or non-collaborative games.

Acknowledgment. This material is based upon work supported by the Defense Advanced Research Projects Agency (DARPA) under Contract No. HR001120C0032, ARL W911NF2020132, ARL ACC-APG-RTP W911NF, NSF 1646522, and Deutsche Forschungsgemeinschaft (DFG, German Research Foundation) - grant no. 434592664. Any opinions, findings and conclusions or recommendations expressed in this material are those of the author(s) and do not necessarily reflect the views of DARPA.

References

1. Aksaray, D., Jones, A., Kong, Z., Schwager, M., Belta, C.: Q-learning for robust satisfaction of signal temporal logic specifications. In: IEEE CDC 2016, December 2016, pp. 6565–6570 (2016)

2. Alshiekh, M., Bloem, R., Ehlers, R., Könighofer, B., Niekum, S., Topcu, U.: Safe reinforcement learning via shielding. In: AAAI 2018 (2018)

3. Andreas, J., Klein, D., Levine, S.: Modular multitask reinforcement learning with policy sketches. In: Proceedings of the 34th International Conference on Machine Learning-Volume 70. JMLR.org, 2017, pp. 166–175 (2017)

4. Angluin, D.: Learning regular sets from queries and counterexamples. Inf. Comput. **75**(2), 87–106 (1987)

5. Baharisangari, N., Gaglione, J.R., Neider, D., Topcu, U., Xu, Z.: "Uncertainty-aware signal temporal logic inference" (2021). https://arxiv.org/abs/2105.11545

6. Bollig, B., Katoen, J.-P., Kern, C., Leucker, M., Neider, D., Piegdon, D.R.: The automata learning framework. In: Touili, T., Cook, B., Jackson, P. (eds.) CAV 2010. LNCS, vol. 6174, pp. 360–364. Springer, Heidelberg (2010). https://doi.org/10.1007/978-3-642-14295-6_32

7. Bombara, G., Vasile, C.I., Penedo, F., Yasuoka, H., Belta, C.: A decision tree approach to data classification using signal temporal logic. In: Proceedings of the HSCC 2016, pp. 1–10 (2016)

8. Cai, M., Hasanbeig, M., Xiao, S., Abate, A., Kan, Z.: Modular deep reinforcement learning for continuous motion planning with temporal logic (2021)

9. Fu, J., Topcu, U.: Probably approximately correct MDP learning and control with temporal logic constraints. Robotics: Science and Systems. abs/1404.7073 (2014)

10. Furelos-Blanco, D., Law, M., Russo, A., Broda, K., Jonsson, A.: Induction of sub-goal automata for reinforcement learning. In: Proceedings of the AAAI Conference on Artificial Intelligence, vol. 34, no. 04, pp. 3890–3897, April 2020. https://ojs.aaai.org/index.php/AAAI/article/view/5802

11. Gaglione, J.R., Neider, D., Roy, R., Topcu, U., Xu, Z.: Learning linear temporal properties from noisy data: a maxsat-based approach. In: ATVA 2021, Gold Coast, Australia, 18–22 October. Lecture Notes in Computer Science. Springer (2021)

12. Gaon, M., Brafman, R.: Reinforcement learning with non-markovian rewards. In: Proceedings of the AAAI Conference on Artificial Intelligence, vol. 34, no. 04, pp. 3980–3987, April 2020. https://ojs.aaai.org/index.php/AAAI/article/view/5814

13. Holzinger, A., Malle, B., Saranti, A., Pfeifer, B.: Towards multi-modal causability with graph neural networks enabling information fusion for explainable AI. Inf. Fus. **71**, 28–37 (2021). https://www.sciencedirect.com/science/article/pii/S1566253521000142

14. Hopcroft, J.E., Motwani, R., Ullman, J.D.: Introduction to Automata Theory, Languages, and Computation, 3rd edn. Addison-Wesley Longman Publishing Co., Inc., Boston, MA, USA (2006)

15. Hoxha, B., Dokhanchi, A., Fainekos, G.: Mining parametric temporal logic properties in model-based design for cyber-physical systems. Int. J. Softw. Tools Technol. Transfer **20**(1), 79–93 (2017). https://doi.org/10.1007/s10009-017-0447-4

16. Toro Icarte, R., Waldie, E., Klassen, T., Valenzano, R., Castro, M., McIlraith, S.: Learning reward machines for partially observable reinforcement learning. In: NeurIPS 2019 (2019)

17. Icarte, R.T., Klassen, T.Q., Valenzano, R.A., McIlraith, S.A.: Using reward machines for high-level task specification and decomposition in reinforcement learning. In: ICML 2018, Stockholmsmässan, Stockholm, Sweden, 10–15 July 2018, pp. 2112–2121 (2018)

18. Kong, Z., Jones, A., Belta, C.: Temporal logics for learning and detection of anomalous behavior. IEEE TAC **62**(3), 1210–1222 (2017)

19. Li, X., Vasile, C.-I., Belta, C.: Reinforcement learning with temporal logic rewards. In: Proceedings of the IROS 2017, September 2017, pp. 3834–3839 (2017)

20. Neider, D., Gaglione, J.R., Gavran, I., Topcu, U., Wu, B., Xu, Z.: Advice-guided reinforcement learning in a non-markovian environment. In: AAAI 2021 (2021)
21. Neider, D., Gavran, I.: Learning linear temporal properties. In: Formal Methods in Computer Aided Design (FMCAD) 2018, pp. 1–10 (2018)
22. Schulman, J., Wolski, F., Dhariwal, P., Radford, A., Klimov, O.: "Proximal policy optimization algorithms" (2017)
23. Shah, A., Kamath, P., Shah, J.A., Li, S.: Bayesian inference of temporal task specifications from demonstrations. In: Bengio, S., Wallach, H., Larochelle, H., Grauman, K., Cesa-Bianchi, N., Garnett, R. (eds.) NeurIPS, Curran Associates Inc., 2018, pp. 3808–3817 (2018). http://papers.nips.cc/paper/7637-bayesian-inference-of-temporal-task-specifications-from-demonstrations.pdf
24. Toro Icarte, R., Klassen, T.Q., Valenzano, R., McIlraith, S.A.: Teaching multiple tasks to an RL agent using LTL. In: AAMAS 2018, Richland, SC, 2018, pp. 452–461 (2018)
25. Vazquez-Chanlatte, M., Jha, S., Tiwari, A., Ho, M.K., Seshia, S.A.: Learning task specifications from demonstrations. In: Proceedings of the NeurIPS 2018, pp. 5372–5382 (2018)
26. Watkins, C.J., Dayan, P.: Q-learning. Mach. Learn. **8**(3), 279–292 (1992). https://doi.org/10.1007/BF00992698
27. Wen, M., Papusha, I., Topcu, U.: Learning from demonstrations with high-level side information. In: Proceedings of the IJCAI 2017, pp. 3055–3061 (2017)
28. Wu, B., Lin, H.: Counterexample-guided permissive supervisor synthesis for probabilistic systems through learning. In: American Control Conference (ACC) 2015, pp. 2894–2899. IEEE (2015)
29. Wu, B., Zhang, X., Lin, H.: Permissive supervisor synthesis for markov decision processes through learning. IEEE Trans. Autom. Control **64**(8), 3332–3338 (2018)
30. Zhang, X., Wu, B., Lin, H.: Supervisor synthesis of pomdp based on automata learning. Automatica (2021 to appear). https://arxiv.org/abs/1703.08262
31. Xu, Z., Birtwistle, M., Belta, C., Julius, A.: A temporal logic inference approach for model discrimination. IEEE Life Sci. Lett. **2**(3), 19–22 (2016)
32. Xu, Z., Belta, C., Julius, A.: Temporal logic inference with prior information: an application to robot arm movements. In: Proceedings of the Analysis and Design of Hybrid Systems, vol. 48, no. 27, Atlanta, GA, USA, October 2015, pp. 141–146 (2015)
33. Xu, Z., et al.: Joint inference of reward machines and policies for reinforcement learning. In: Proceedings of the International Conference on Automated Planning and Scheduling, vol. 30, pp. 590–598 (2020)
34. Xu, Z., Julius, A.: Census signal temporal logic inference for multiagent group behavior analysis. IEEE Trans. Autom. Sci. Eng. **15**(1), 264–277 (2018)
35. Xu, Z., Ornik, M., Julius, A.A., Topcu, U.: Information-guided temporal logic inference with prior knowledge. In: Proceedings of the 2019 American control conference (ACC), pp. 1891–1897. IEEE (2019). https://arxiv.org/abs/1811.08846
36. Xu, Z., Saha, S., Hu, B., Mishra, S., Julius, A.: Advisory temporal logic inference and controller design for semiautonomous robots. IEEE Trans. Autom. Sci. Eng. **16**, 1–19 (2018)
37. Xu, Z., Topcu, U.: Transfer of temporal logic formulas in reinforcement learning. In: IJCAI-19. International Joint Conferences on Artificial Intelligence Organization, July 2019, pp. 4010–4018 (2019). https://doi.org/10.24963/ijcai.2019/557

Rice Seed Image-to-Image Translation Using Generative Adversarial Networks to Improve Weedy Rice Image Classification

Atthakorn Petchsod[(✉)] and Tanasai Sucontphunt

National Institute of Development Administration, Bangkok, Thailand
atthakorn.pet@stu.nida.ac.th, tanasai@as.nida.ac.th

Abstract. Rice is a staple food for more than half of the world's population. Furthermore, rice is the main export crop of Thailand which produces 21% world's market share. However, weedy rice is a major counterproductive plant that reduces rice productivity by more than 80% in Thailand. Previous research attempted to develop image classification models to recognize types of rice using images captured in closed environments, which is not practical for farmers with typical mobile phone cameras. This research develops a specific Generative Adversarial Network (GAN) architecture to translate an input image from a typical mobile phone cameras into the closed environment setting. Our GAN architecture can translate mobile phone images and achieves 90.06% weedy rice recognition accuracy, as compared to 58.10% without the translation.

Keywords: Image-to-image translation · Computer vision · Generative Adversarial Networks

1 Introduction

Rice is a daily life food supplies consumed by more than half of the world's population [19], especially in Thailand. In 2019, Thailand's rice fields reached 27,147,673 acres, accounting for 46% of the total agricultural land in the country [16], and exported 7,583,662 tons of rice, worth more than 130 billion Baht [23]. Thailand is the world's second largest exporter of rice, with a 21% market share; hence, rice is considered as an important economic crop for Thailand [21].

Since 2001, farmers in Thailand have faced severe weedy or wild rice spread. The first outbreak occurred around the Kanchanaburi area and expanded throughout the central and northern regions of the country. There are various species of weedy rice but they have similar undesirable characteristics, that is, they are defective and tend to fall before the harvest. This characteristic of weedy rice damages productivity and reduces the price of cultivated rice by more than 80%. Traditionally, Thailand's Rice Department has suggested 4 steps to prevent the growth of weedy rice. First, select cultivated rice seeds without

© IFIP International Federation for Information Processing 2021
Published by Springer Nature Switzerland AG 2021
A. Holzinger et al. (Eds.): CD-MAKE 2021, LNCS 12844, pp. 137–151, 2021.
https://doi.org/10.1007/978-3-030-84060-0_9

contaminated weedy rice. Second, always clean agricultural machinery before planting. Third, select compost, organic fertilizers, and bio-fertilizers produced without any contaminated weedy rice. Finally, use a net to fish out any weedy rice that may be floating with water [6]. Thus, distinguishing weedy rice from cultivated rice seeds is a crucial step.

Over the past few years, several studies have attempted to develop models to help farmers classify rice grains by applying image processing and computer vision [1,2].

Because cultivated and weedy rice seeds are very similar in details and characteristics that are difficult to distinguish even by humans, as shown in Fig. 1 to construct an efficient rice image classification model, the training images must contain as much details as possible. They are therefore typically captured with a high resolution camera in a closed environment.

Since most previous research has developed classification models using rice seed images captured in closed environments. This research aims to generalize the classification model by using image-to-image translation in a GAN architecture model drawing on rice seed images in various environments captured using different settings, in order to make the application more practical. In other words, the proposed GAN architecture is used to translate rice seed images captured under any settings into the same settings as used in the classification model.

2   Related Work

2.1   Rice Classifications

Many researchers have carried out rice classification using computer vision techniques. Zhao-yan et al. [12] and Chathurika et al. [20] used color thresholding to segment rice seeds then feature extraction and neural networks to classify them. Kuo et al. [22] used Multifocus image fusion and feature extraction to generate the images and classify them. Kittinun et al. [1] used Mask R-CNN [9] to segment and classify rice seeds in images. However, although there is much work on rice classification, none focuses on weedy rice classification with mobile cameras.

2.2   Generative Adversarial Networks

Generative adversarial networks (GANs) [8] have succeeded in various kind of applications in the image field, such as image generation (FCGAN [8], DCGAN [17]) or image super-resolution (SRGAN [11]). What makes various kinds of GANs successful is the idea of adversarial loss, which builds a game where a generator and discriminator race each other; the generator tries to fool the discriminator by generating fake images that looks as much as possible like real ones, while the discriminator tries to distinguish between real images and the fake ones created by the generator. This loss serves to make the generated images indistinguishable from the real ones.

2.3   Conditional Adversarial Networks

In general, GANs are unconditional: however, they can be extended and become a conditional model by adding some information y, such as class labels or images, into the generator and discriminator [14]. Conditional GANs (CGAN) [14] make use of conditional adversarial networks by adding class numbers as labels for a CGAN trained on the MNIST dataset, creating a model that can generate digit images with the same number (class) as specified. Another interesting work is the Auxiliary Classifier GAN (ACGAN) [15], which extends GANs by adding a class to the generator and a discriminator head to classify the generated images.

2.4   Image-to-Image Translation

Image-to-image translation is a problem of transforming input images from one domain into output images in the style of another domain using several techniques, including GANs such as pix2pix [10], which uses a conditional adversarial network, or BicycleGAN [25], which combines cVAE-GAN [3] and cLR-GAN [7]. Although pix2pix and BicycleGAN are excellent for solving image-to-image translation problems, they are limited to paired image-to-image translation.

2.5   Unpaired Image-to-Image Translation

Unpaired image-to-image translation is an image-to-image translation problem in which the training data are unpaired. The goal is slightly different from mapping input images x from domain X to output images y in the style of domain Y into relating two domains X and Y. Several techniques have been developed to tackle this kind of problem, such as Rosales et al.'s Unsupervised Image Translation [18], which proposed a Bayesian framework with a patch-based Markov random field, or the GAN-based, state-of-the-art unpaired image-to-image translation method CycleGAN [24] which uses the idea of cycle consistency to constrain the networks.

3   Dataset

In this work, a mobile image dataset is constructed from 10 species of paddy rice seed captured by 6 models of camera. The 10 species are divided into 2 categories, cultivated and weedy rice. Cultivated rice consists of 6 species: Chai Nat 2, Khao Dawk Mali 105, Leuang Pratew 123, Pathum Thani 1, Phitsanulok 2 and RD43. Weedy rice consists of 4 species: Short (Ded Tia), Striped (Lai), Tail (Hang) and Tall (Ded Sung). The 6 models of camera are the Canon 77D, iPhone 6s, iPhone 7, iPhone 7 Plus, iPhone 11, and Redmi Note 5. The camera models are divided into two groups: the Canon 77D camera, which is a digital single-lens reflex camera, is DSLR and the others are Mobile. The images captured by the DSLR camera were taken in a closed environment that controls the background, lighting, and distance between rice and lens. After that, the images were masked and segmented using traditional image processing techniques. The images captured by Mobile cameras were taken in less controlled settings by various models of mobile cameras, with different lighting, and varying distances between rice and lens. The images were then masked and segmented manually. The number of images for each species and camera model are shown in Table 1 and sample images are shown in Fig. 1.

Table 1. Number of images for each species and camera models in the full dataset

	Canon 77D	iPhone 6s	iPhone 7	iPhone 7 Plus	iPhone 11	Redmi Note 5	Total
Cultivated Rice							
Chai nat 2	45,318	1,729	1,808	2,349	1,465	1,583	54,252
Khao dawk mali 105	46,750	1,479	1,829	1,758	1,718	1,688	55,222
Leuang pratew 123	38,419	1,257	1,859	2,453	1,633	1,586	47,207
Pathum thani 1	38,809	1,289	1,954	1,885	1,571	1,300	46,808
Phitsanulok 2	37,873	1,888	2,111	2,064	1,587	1,334	46,857
RD43	30,115	1,314	1,665	1,739	1,658	1,749	38,240
Weedy rice							
Short	11,453	1,325	1,564	1,799	1,364	1,246	18,751
Striped	28,109	1,517	2,002	1,720	1,214	1,483	36,045
Tail	11,656	1,017	961	1,294	961	996	16,885
Tall	42,312	1,261	1,651	1,683	1,514	1,497	49,918
Total	330,814	14,076	17,404	18,744	14,685	14,462	410,185

The entire dataset is divided into 3 subsets: training, validation and test set. Rice seed images captured by the iPhone 7 Plus camera are assigned to the test set. The others are split into training and validation sets with a ratio of 80:20.

4   Approach

4.1   Main Components

Generator. For all generators in this research, the architecture was adopted from CycleGAN [24]. The network consists of a convolutional layer, two strided convolutional layers, six residual blocks, and two fractionally strided convolutional layers, as shown in Fig. 2.

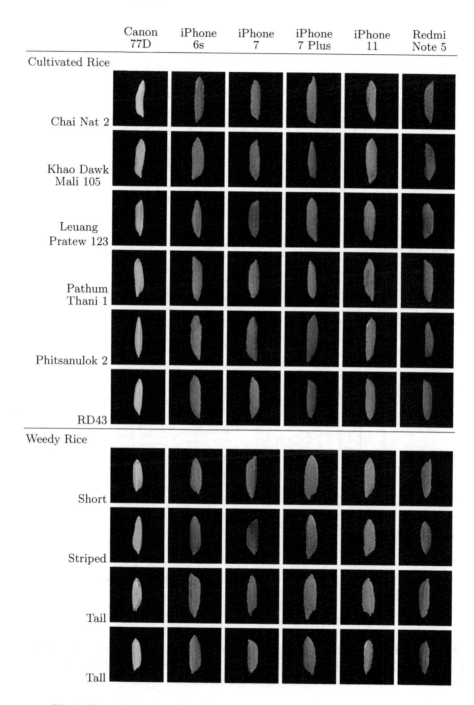

Fig. 1. Samples image of paddy rice for each species and camera model

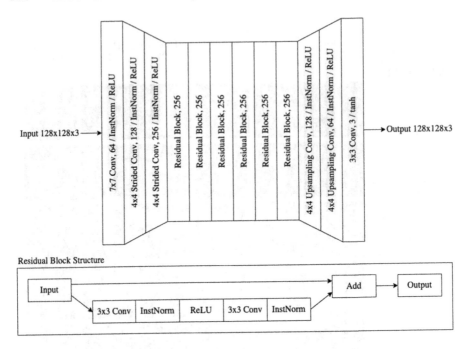

Fig. 2. Architecture of generators used in this research.

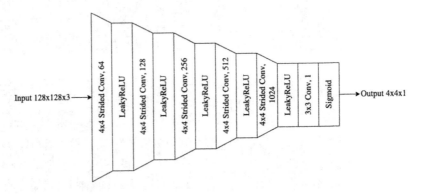

Fig. 3. Architecture of discriminators used in this research.

Discriminator. For the discriminator, the architecture was also adopted from CycleGAN [24]. The discriminator network consists of a convolutional layer and five strided convolutional layers, as shown in Fig. 3.

Classifier. The classifier architecture is nearly identical to the discriminator but a convolutional layer without padding was added to form the output layer for the classification task, as shown in Fig. 4.

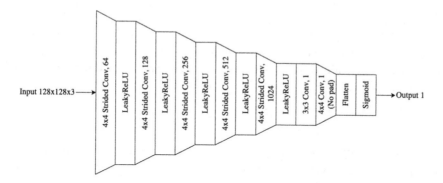

Fig. 4. Architecture of classifiers used in this research.

4.2 Objective Formulation

Let $k \in K = \{$iPhone 6s camera, iPhone 7 camera, iPhone 11 camera, Redmi Note 5 camera$\}$, Our goal is to learn a many-to-one mapping function, from rice seed images captured by mobile cameras (X) to rice seed images captured by a DSLR camera (Y) given samples $\{x_{ki}\}_{i=1}^{N}$, where $x_{ki} \in X_k \subset X$, and $\{y_j\}_{j=1}^{N}$, where $y_i \in Y$. Let the data distribution be denoted as $x_k \sim p_{data}(x_k)$ and $y \sim p_{data}(y)$. Our model consists of 5 mappings of the forms $F_k : Y \rightarrow X_k$, and a sixth mapping, $G : X_k \rightarrow Y$, an adversarial discriminator D_Y, whose objective is to distinguish between y and $G(x_k)$, and a classifier C, whose objective is to classify the species of Y.

The objective consists of four loss types: adversarial loss to match the distribution of the generated images into the target domain, cycle consistency loss to make the translation invertible given a specific camera model k, identity loss to constrain the translation to only perform on $x \in X$, and classification loss to make the generated images the correct seed class.

Adversarial Loss. Adversarial loss is applied to the mapping function $G : X \rightarrow Y$ and its corresponding discriminator D_Y. This loss leads the generator G to translate images from $x \in X$ into $G(x)$, which is distributed identically to Y. The objective is shown in Eq. 1.

$$L_{GAN}(G, D_Y) = \mathbb{E}_{y \sim p_{data}(y)}[log(D_Y(y))]$$
$$+ \mathbb{E}_{x \sim p_{data}(x)}[log(1 - D_Y(G(x)))] \tag{1}$$

Cycle Consistency Loss. Adversarial loss performs well on the leading generator to translate images from X to Y. However, given a sufficiently large amount of training data, the generator G can map the set of input images x onto an image $y \in Y$ and the generator will successfully *generate* an image distributed identically to Y but fail to *translate* the image. Hence, we include cycle consistency loss in our objective. This adds an additional constraint to reduce the plausible mapping set from X to Y. Cycle consistency means that for each image $x \in X$, the image translation cycle should be able to recall x, i.e., $F(G(x)) \approx x$. Since the model is not trained directly with X, but rather with five mappings of the form $X_k \subset X$, our objective is expressed as Eq. 2.

$$L_{cycle}(G, F_k) = \mathbb{E}_{x_k \sim p_{data}(x_k)}[\|F_k(G(x_k)) - x_k\|_1] \qquad (2)$$

Identity Loss. Since the main objective is to train a generator that can translate image $x \in X$ into $G(x) \in Y$, therefore, putting $y \in Y$ into G should returned y, since $y \notin X$. Thus, the objective can be expressed as Eq. 3.

$$L_{identity}(G) = \mathbb{E}_{x \sim p_{data}(x)}[\|G(y) - y\|_1] \qquad (3)$$

Classification Loss. The idea of classification loss is adapted from InfoGAN[4], which adds a classifier to the traditional GAN's architecture, and ACGAN[15], which adds a classifier head to the traditional GAN discriminator. The classification loss is shown in Eq. 4.

$$L_{classification}(G, C) = \mathbb{E}_{x \sim p_{data}(x)}[log(C(G(x)))] \qquad (4)$$

Full Objective. The full objective is a weighted sum of adversarial loss, cycle consistency loss, identity loss and classification loss, as shown in Eq. 5.

$$L(G, F_k, D_Y) = L_{GAN}(G, D_Y) + \lambda_1 \sum_{k \in K} L_{cycle}(G, F_k) + \lambda_2 L_{identity}(G)$$
$$+\lambda_3 L_{classification}(G, C) \qquad (5)$$

4.3 Model Architecture

This work proposes four model architectures with minor variations, as shown in Fig. 5. The models will be called M.1, M.2, M.3, and M.4, as abbreviations of Model 1, Model 2, Model 3, and Model 4, respectively.

Model 1. The first model is the base for the other three. It consists of 5 inputs, denoted as $y \in Y$ and four $x_k \in X$. y passes through the generator G and takes the identity y', and each x_k also passes through G to generate a fake image $G(x_k)$. Next, each $G(x_k)$ is passed into a generator F_k for each k to forms a cycle for x_k, i.e., $x'_k = F_k(G(x_k))$. Losses are calculated by passing $G(x_k)$ into

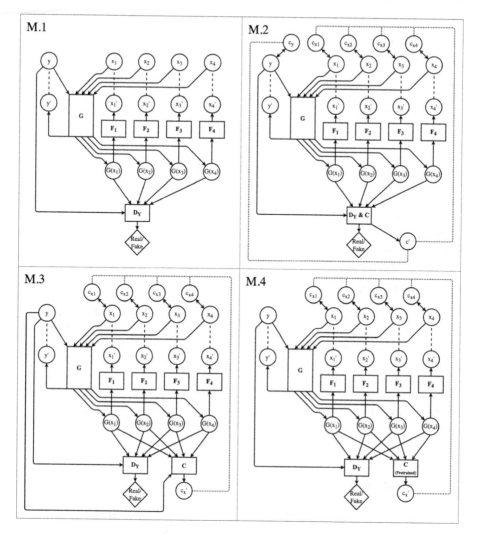

Fig. 5. Architecture of all models used in this research.

the discriminator D_Y as an adversarial loss, y' is compared to y as an identity loss, and each x'_k is compared with x_k as a cycle loss. The losses are combined to form the objective of this model, as expressed in Eq. 6.

$$L(G, F_k, D_Y) = L_{GAN}(G, D_Y) + \lambda_1 \sum_{k \in K} L_{cycle}(G, F_k) + \lambda_2 L_{identity}(G) \quad (6)$$

Model 2. The second model extends the first by adding a classifier head that classifies a rice image according to whether it is weedy rice. The idea of adding a classifier head to a discriminator is inspired by [15]. Thus, there is a *classification* loss added to the objective, in addition to the Model 1 loss, as expressed in Eq. 5.

Model 3. The third model adds an additional classifier, which determines whether the rice image is weedy rice, into the GAN loop. Adding an additional classifier into GAN has been done before in InfoGAN [4]. Therefore, the objective is the same as for Model 2.

Model 4. The last model has the same architecture as Model 3, except the additional classifier is pretrained with DSLR images. A pretrained model has been used to help calculate loss before, by SRGAN[11], which uses a pretrained VGG16 network to calculate loss. Thus, the pretrained classifier has helped calculate the loss instead of training a classifier within the GAN loop. The objective is also as expressed in Eq. 5.

4.4   Training Details

For all experiments, the model is trained using *epoch* = 20, and batch size = 1, on an Adam optimizer with learning rate = 0.0001 for the first 10 epochs and then linearly decreasing by 0.00001 for the next 10. The linearly decay learning rate is the same as for CycleGAN [24]. For the objectives' λ, λ_1 is set the same way as in CycleGAN [24] i.e. $\lambda_1 = 10$; however, λ_2 is also set to 10 because it is the main objective, while λ_3 is set to 0.01. The least-squares loss technique [13] is also applied to the models, hence the L_{GAN} objective while training G is $\mathbb{E}_{x \sim p_{data}(x)}[(D_Y(G(x)) - 1)^2]$, and while training D_Y, it is $\mathbb{E}_{y \sim p_{data}(y)}[(D_Y(y) - 1)^2] + \mathbb{E}_{x \sim p_{data}(x)}[D_Y(G(x))^2]$.

5   Evaluations

To evaluate the results, a benchmark model is trained with the same architecture as the proposed classifier. Moreover, this benchmark model is used as a pretrained classifier that provides loss for Model 4. The evaluation is carried out by passing x and y into the generator G and then comparing the translated images created using this model to the benchmark score.

The validation process is done in two parts according to the validation dataset, DSLR or Mobile. The validation of the DSLR dataset measures how much Generator G decreases weedy rice classification performance. For Mobile dataset, the validation measures how much Generator G increases weeedy rice classification performance. The final decision on which model is better is achieved by picking the model that performs best on test dataset, which is the unseen mobile camera model.

5.1 Benchmarks

The quantitative results of the benchmark model on the validation sets (both DSLR and Mobile subset) and test set are shown in Table 2. The benchmark model performs very well on the DSLR validation set as it should, since it was trained on that set. However, the performance on the Mobile validation set and test set are poor, for the same reasons as described.

Table 2. Weedy rice classifier model benchmarks

Dataset	Subset	Rice			Weed			Accuracy
		Precision	Recall	F1-Score	Precision	Recall	F1-Score	
Validation	DSLR	99.57%	99.52%	99.54%	98.77%	98.91%	98.84%	99.34%
	Mobile	89.78%	32.06%	47.25%	43.16%	93.39%	59.03%	53.88%
Test		88.49%	41.25%	56.27%	44.79%	89.89%	59.79%	58.10%

5.2 Validation Set

The quantitative results for all models on the DSLR and Mobile validation sets are shown in Table 3. All model performances on the Mobile validation set are greatly increased, especially for models with a classifier in the architecture, which increase to more than 90% accuracy. However, the performance on the DSLR validation set slightly decreases from the benchmark in all cases except model 2, the accuracy of which decreases by more than 10%.

For the qualitative results, the translated images are shown in Fig. 6. The first 3 models translate and generate images that look like rice images captured by the DSLR camera. But for the fourth model, the results look like rice seeds with a more crimson color and red borders around them.

Table 3. Weedy rice classifier model validation results

Model	Subset	Rice			Weed			Accuracy	Macro F1
		Precision	Recall	F1	Precision	Recall	F1		
M.1	DSLR	94.19%	99.35%	96.70%	98.08%	84.46%	90.76%	95.14%	93.73%
	Mobile	81.30%	95.74%	87.93%	88.62%	60.13%	71.65%	83.07%	79.79%
M.2	DSLR	88.04%	99.82%	93.56%	99.30%	65.59%	79.00%	90.14%	86.28%
	Mobile	95.98%	97.57%	96.77%	95.46%	92.61%	94.01%	95.80%	95.39%
M.3	DSLR	95.61%	99.71%	97.62%	99.17%	88.39%	93.47%	96.51%	95.54%
	Mobile	92.18%	99.58%	95.74%	99.10%	84.70%	91.34%	94.28%	93.54%
M.4	DSLR	99.49%	99.5%	99.50%	98.74%	98.70%	98.72%	99.28%	99.11%
	Mobile	98.02%	96.85%	97.43%	94.42%	96.45%	95.42%	96.71%	96.43%

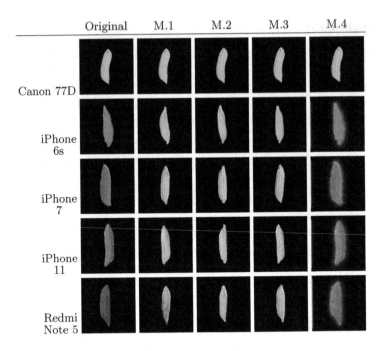

Fig. 6. Samples of validation image translated by each model's G

5.3   Test Set

The results on the test set in this section are the most important part. The quantitative results are shown in Table 4. The table indicates that the second model performs best on the test set, with the highest values for both accuracy and macro-f1 score. Moreover, the qualitative results shown in Fig. 7 also reveal that the realistic images generated by the second model from all rice species in the dataset, as compared to the first and third models, also perform well on most species. The rice seed images translated by the fourth model are again more crimson with red borders around them, as with the validation set.

Table 4. Weedy rice classifier model test results

Model	Rice			Weed			Accuracy	Macro F1
	Precision	Recall	F1	Precision	Recall	F1		
M.1	83.18%	89.18%	86.08%	76.39%	66.01%	70.82%	81.15%	78.45%
M.2	96.14%	88.33%	92.07%	80.92%	93.32%	86.68%	90.06%	89.38%
M.3	85.71%	97.71%	91.32%	94.12%	69.29%	79.82%	87.86%	85.57%
M.4	95.51%	85.33%	90.13%	76.97%	92.44%	84.00%	87.79%	87.07%

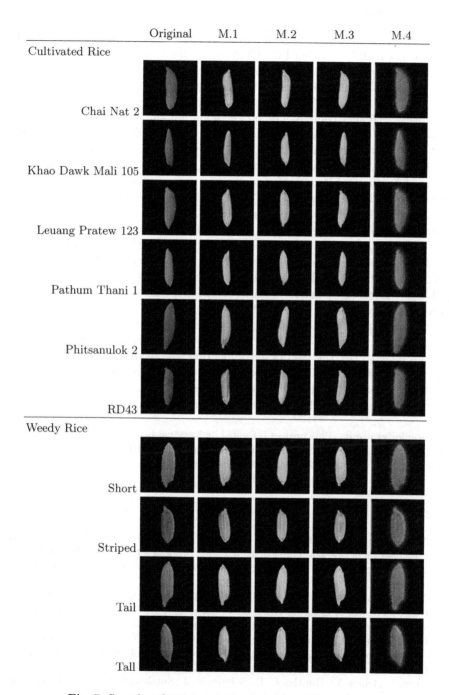

Fig. 7. Samples of test image translated by each model's G

6   Conclusions

In this work, four architectures adapted from CycleGAN [24], with additional components mainly inspired by ACGAN [15], InfoGAN [4], and SRGAN [11], are proposed to tackle the problem of translating rice seed images captured by various models of mobile camera into images similar to those captured by a specific DSLR camera. The quantitative results for all models, with any kind of rice species classifier components, on all validation and test sets are above 80%. However, qualitative results indicate that the fourth model cannot translate rice seed images into DSLR-like images since all results are more crimson than usual and have red borders around them, which is not characteristic of rice seed images captured by the DSLR camera. For the two reasons discussed above, the first and the fourth models are not ideal to tackle the problem. Both the other models have good qualitative results on the validation and test sets. The second model performs worse than the third on the DSLR validation set; this means it is necessary to sacrifice more on DSLR to archieve this level of performance on the mobile validation set. However, the quantitative results of the second model on the test set outperform those for the third model by 2.20% in terms of accuracy and 3.81% in terms of macro-f1 score. Therefore, the second model is the best since it performs best on the objective.

One problem of all the architectures proposed is that some of the translated images have different orientations or sometimes different appearances than the original images, i.e., the results looks like different seeds than the originals. A suggestion to solve this problem would be to add more constraints on image masks using the idea proposed by SemGAN [5], to force a similarity between the original image masks and the translated image masks.

Acknowledgement. This project is funded by National Research Council of Thailand on Research and Development for Rice Seed and Weedy Rice Classifcation Models Detecting Impurity using 2D Image Analytics and Machine Learing on Website and Mobile Applications for Laypersons project.

References

1. Aukkapinyo, K., Sawangwong, S., Pooyoi, P., Kusakunniran, W.: Localization and classification of rice-grain images using region proposals-based convolutional neural network. Int. J. Autom. Comput. **17**, 1–14 (2019). https://doi.org/10.1007/s11633-019-1207-6
2. Aznan, A.A., Rukunudin, I.H., Shakaff, A.Y.M., Ruslan, R., Zakaria, A., Saad, F.S.A.: The use of machine vision technique to classify cultivated rice seed variety and weedy rice seed variants for the seed industry. Int. Food Res. J. **23**, S31 (2016)
3. Bao, J., Chen, D., Wen, F., Li, H., Hua, G.: CVAE-GAN: fine-grained image generation through asymmetric training. In: Proceedings of the IEEE International Conference on Computer Vision, pp. 2745–2754 (2017)
4. Chen, X., Duan, Y., Houthooft, R., Schulman, J., Sutskever, I., Abbeel, P.: Infogan: interpretable representation learning by information maximizing generative adversarial nets. In: Advances in Neural Information Processing Systems, pp. 2172–2180 (2016)

5. Cherian, A., Sullivan, A.: Sem-GAN: semantically-consistent image-to-image translation. CoRR abs/1807.04409 arxiv:1807.04409 (2018)
6. Division of Rice Research and Development, Rice Department Weed in rice fields (2009). http://www.ricethailand.go.th/Rkb/weed/index.php-file=content.php&id=43.htm
7. Donahue, J., Krahenbuhl, P., Darrell, T.: Adversarial feature learning. arXiv preprint arXiv:1605.09782 (2016)
8. Goodfellow, I.J., et al.: Generative adversarial networks (2014). arXiv:1406.2661
9. He, K., Gkioxari, G., Dollár, P., Girshick, R.: Mask R-CNN. In: Proceedings of the IEEE International Conference on Computer Vision, pp. 2961–2969 (2017)
10. Isola, P., Zhu, J.-Y., Zhou, T., Efros, A.A.: Image-to-image translation with conditional adversarial networks. In: The IEEE Conference on Computer Vision and Pattern Recognition (CVPR), July 2017
11. Ledig, C., et al.: Photo-realistic single image super-resolution using a generative adversarial network. In: Proceedings of the IEEE Conference on Computer Vision and Pattern Recognition, pp. 4681–4690 (2017)
12. Liu, Z.-Y., Cheng, F., Ying, Y.-B., et al.: Identification of rice seed varieties using neural network. J. Zhejiang Univ. Sci. **6B**(11), 1095–1100 (2005)
13. Mao, X., Li, Q., Xie, H., et al.: Least squares generative adversarial networks (2017). arXiv:1611.04076v3
14. Mirza, M., Osindero, S.: Conditional generative adversarial nets. arXiv preprint arXiv:1411.1784 (2014)
15. Odena, A., Olah, C., Shlens, J.: Conditional image synthesis with auxiliary classifier GANs. arXiv preprint arXiv:1610.09585 (2016)
16. Office of Agricultural Economics (2019) Agricultural Statistics of Thailand (2018)
17. Radford, A., Metz, L., Chintala, S.: Unsupervised representation learning with deep convolutional generative adversarial networks. In: Proceeding of International Conference on Learning Representations (2016)
18. Rosales, R., Achan, K., Frey, B.J.: Unsupervised image translation. In: ICCV, pp. 472–478 (2003)
19. Saneei, P., Larijani, B., Esmaillzadeh, A.: Rice consumption, incidence of chronic diseases and risk of mortality: meta-analysis of cohort studies. Public Health Nutr. **20**, 233–244 (2017)
20. Silva, C.S., Sonnadara, U.: Classification of rice grains using neural networks. In: Proceedings of Technical Sessions, Sri Lanka, September 2013, pp 9–14 (2013)
21. Sowcharoensuk, C.: Rice Industry: Business/Industrial trend 2019–2021, PP.1–10 (2019)
22. Kuo, T.Y., Chung, C.L., Chen, S.Y., Lin, H.A., Kuo, Y.F.: Identifying rice grains using image analysis and sparse-representation-based classification. Comput. Electron. Agric. **127**, 716–725 (2016)
23. Thai Rice Exporters Association (2020) Rice Export Quantity and Value (2019). http://www.thairiceexporters.or.th/statistic_2019.html
24. Zhu, J.Y., Park, T., Isola, P., Efros, A.A.: Unpaired image-to-image translation using cycle-consistent adversarial networks. In: IEEE International Conference on Computer Vision, pp. 2242–2251 (2017)
25. Zhu, J.Y.: Toward multimodal image-to-image translation. In: Proceedings of the 31st International Conference on Neural Information Processing Systems, NIPS 2017. Curran Associates Inc., Red Hook, NY, USA, pp. 465–476 (2017)

Plischke, M., Sullivan, A. Scott... [illegible faded text]
Journal Wells, the [1967] [illegible] [illegible]

Redfield, et Bose... [illegible] and [illegible]... Part Department, W. A. [illegible]... [illegible]... 1970... [illegible] date... [illegible] 1986... [illegible]
[1988]... 29-41

[illegible faded lines]

Reliable AI Through SVDD and Rule Extraction

Alberto Carlevaro[1,2]($\boxtimes$) and Maurizio Mongelli[2]($\boxtimes$)

[1] Department of Electrical,
Electronics and Telecommunications Engineering and Naval Architecture (DITEN),
University of Genoa, Genoa, Italy
`alberto.carlevaro@edu.unige.it`
[2] Institute of Electronics, Computer and Telecommunication Engineering,
National Research Council of Italy, Genoa, Italy
{`alberto.carlevaro,maurizio.mongelli`}`@ieiit.cnr.it`

Abstract. The proposed paper addresses how Support Vector Data Description (SVDD) can be used to detect safety regions with zero statistical error. It provides a detailed methodology for the applicability of SVDD in real-life applications, such as Vehicle Platooning, by addressing common machine learning problems such as parameter tuning and handling large data sets. Also, intelligible analytics for knowledge extraction with rules is presented: it is targeted to understand safety regions of system parameters. Results are shown by feeding data through simulation to the train of different rule extraction mechanisms.

Keywords: SVDD · Safety regions · Explainable AI

1 Introduction

The study proposed in the paper follows the recent trend dedicated to identifying and handling assurance under uncertainties in AI systems [23]. It falls in the category of improving reliability of prediction confidence. The topic remains a significant challenge in machine learning, as learning algorithms proliferate into difficult real-world pattern recognition applications. The intrinsic statistical error introduced by any machine learning algorithm may lead to criticism by safety engineers. The topic has recieved a great interest from industry [25], in particular in the automotive [27] and avionics [7] sectors. In this perspective, the conformal predictions framework [5] studies methodologies to associate reliable measures of confidence with pattern recognition settings including classification, regression, and clustering. The proposed approach follows this direction, by identifying methods to circumvent data-driven safety envelopes with statistical zero errors. We show how this assurance may limit considerably the size of the safety envelope (e.g., providing collision avoidance by drastically reducing speed of vehicles) and focus on how to find a good balance between the assurance and the safety space.

© IFIP International Federation for Information Processing 2021
Published by Springer Nature Switzerland AG 2021
A. Holzinger et al. (Eds.): CD-MAKE 2021, LNCS 12844, pp. 153–171, 2021.
https://doi.org/10.1007/978-3-030-84060-0_10

We concentrated our work on specific machine learning methods, the Support Vector Data Description, which by (its) definition is particularly suitable to define safety envelops (see Sect. 2). To it we have added intelligible models for knowledge extraction with rules: intelligibility means that the model is easily understandable, e.g. when it is expressed by Boolean rules. Decision trees (DTs) are typically used towards this aim. The comprehension of neural network models (and of the largest part of the other ML techniques) reveals to be a hard task (see, e.g. Sect. 4 of [14]). Together with DT, we use logic learning machine (LLM), which may show more versatility in rule generation and classification precision.

Our work takes a step forward in these areas due to

- safety regions are tuned on the basis of the radius of the SVDD hypersphere
- simple rule extraction method from SVDD compared with LLM and DT

The article is organized as follows: first, a detailed introduction of SVDD and Negative SVDD is introduced, also focusing on how to choose the best model parameters (Sect. 2.2) and how to handle large datasets (Sect. 2.3). Then Sect. 3 is devoted to rule extraction: LLM and DT are presented and how to extract intelligible rules from SVDD is explained. Finally, an application example is proposed in Sect. 4.

2   Support Vector Data Description

Characterizing a data set in a complete and exhaustive way is an essential preliminary step for any action you want to perform on it. Having a good description of a data set means being able to easily understand if a new observation can contribute to the information brought by the rest of the data or be totally irrelevant. The task of the data domain description is precisely to identify a region, a border, in which to enclose a certain type of information in the most precise possible way, i.e. not adding misinformation or empty spaces. This idea is realized mathematically by a circumference (a sphere, a hypersphere depending on the size of the data space) that encloses as many points with as little area (volume) as possible. Indeed, SVDD can be used also to perform a classification of a specific class of target objects, i.e. it is possible to identify a region (a closed boundary) in which objects which should be rejected are not allowed.

This section is organized as follows: SVDD is introduced as in [28], focusing first on the normal description and then on the description with negative examples [29]. Then we will focus on two proposed algorithms for solving two problems involving SVDD: fast training of large data sets [6] and autonomous detection of SVDD parameters [31]. Finally, the last subsection is devoted to two original methods for finding zero False Negative Rate (FNR) regions with SVDD.

2.1   Theory

Let $\{\mathbf{x}_i\}, i = 1, \ldots, N$ with $\mathbf{x}_i \in \mathbb{R}^d$, $d >= 1$, be a training set for which we want to obtain a description. We want to find a sphere (a hypersphere) of radius R and center $\mathbf{a}$ with minimum volume, containing all (or most of) the data objects.

Normal Data Description. For finding the decision boundary which captures the normal instances and at the same time keeps the hypersphere's volume at minimum, it is necessary to solve the following optimization problem [29]:

$$\min_{R,\mathbf{a}} F(R,\mathbf{a}) = R^2 \text{ s.t. } ||\mathbf{x}_i - \mathbf{a}||^2 \le R^2 \;\; \forall i \tag{1}$$

But to allow the possibility of outliers in the training set, analogously to what happens for the soft-margin SVMs [1], slack variables $\xi_i \ge 0$ are introduced and the minimization problem changes into [29]:

$$\min_{R,\mathbf{a},\xi_i} F(R,\mathbf{a},\xi_i) = R^2 + C\sum_i \xi_i \tag{2}$$

$$\text{s.t.} \begin{cases} ||\mathbf{x}_i - \mathbf{a}||^2 \le R^2 + \xi_i, \\ \xi_i \ge 0 \end{cases} \quad i = 1,\dots,N \tag{3}$$

where the parameter C controls the influence of the slack variables and thereby the trade-off between the volume and the errors.

The optimisation problem is solved by incorporating the constraints (3) into Eq. (2) using the method of Lagrange for positive inequality constraints [12]:

$$
\begin{aligned}
L(R,\mathbf{a},\alpha_i,\gamma_i,\xi_i) = R^2 &+ C\sum_i \xi_i \\
&- \sum_i \alpha_i \left[R^2 + \xi_i - \left(||\mathbf{x_i}||^2 - 2\mathbf{a}\cdot\mathbf{x}_i + ||\mathbf{a}||^2 \right) \right] - \sum_i \gamma_i \xi_i
\end{aligned}
\tag{4}
$$

with the Lagrange multipliers $\alpha_i \le 0$ and $\gamma_i \le 0$. According to [28], L should be minimized with respect to $R, \mathbf{a}, \xi_i$ and maximized with respect to α_i and γ_i.

Setting partial derivatives of $R, \mathbf{a}$ and ξ_i to zero gives the constraints [10]:

$$\frac{\partial L}{\partial R} = 0 : \sum_i \alpha_i = 1, \quad \frac{\partial L}{\partial \mathbf{a}} = 0 : \mathbf{a} = \sum_i \alpha_i \mathbf{x}_i \tag{5}$$

$$\frac{\partial L}{\partial \xi_i} = 0 : C - \alpha_i - \gamma_i = 0 \Rightarrow 0 \le \alpha_i \le C \tag{6}$$

and then, substituting (5) into (4) gives the dual problem of (2) and (3):

$$\max_{\alpha_i} L = \sum_i \alpha_i (\mathbf{x}_i \cdot \mathbf{x}_i) - \sum_{i,j} \alpha_i \alpha_j (\mathbf{x}_i \cdot \mathbf{x}_j) \tag{7}$$

$$\text{s.t} \begin{cases} \sum_i \alpha_i = 1, \\ 0 \le \alpha_i \le C, \quad i = 1,\dots,N \end{cases} \tag{8}$$

Maximimizing (7) under (8) allows to determine all α_i and then the parameters $\mathbf{a}$ and ξ_i can be deduced.

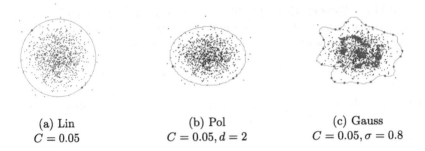

(a) Lin
$C = 0.05$

(b) Pol
$C = 0.05, d = 2$

(c) Gauss
$C = 0.05, \sigma = 0.8$

Fig. 1. SVDD with (a) linear kernel $K(\mathbf{x}_i, \mathbf{x}_j) = \mathbf{x}_i \cdot \mathbf{x}_j$, (b) polynomial kernel $K(\mathbf{x}_i, \mathbf{x}_j) = (1 + \mathbf{x}_i \cdot \mathbf{x}_j)^d$, (c) gaussian kernel $K(\mathbf{x}_i, \mathbf{x}_j) = \exp(-\frac{||\mathbf{x}_i - \mathbf{x}_j||}{2\sigma^2})$ and the respective parameters. In red are plotted the SV (with $\alpha_i < C$) of the description. (Color figure online)

A training object $\mathbf{x}_i$ and its corresponding α_i satisfy one of the following conditions [28, 29]:

$$||\mathbf{x}_i - \mathbf{a}||^2 < R^2 \Rightarrow \alpha_i = 0 \tag{9}$$

$$||\mathbf{x}_i - \mathbf{a}||^2 = R^2 \Rightarrow 0 < \alpha_i < C \tag{10}$$

$$||\mathbf{x}_i - \mathbf{a}||^2 > R^2 \Rightarrow \alpha_i = C \tag{11}$$

Since $\mathbf{a}$ is a linear combination of the objects with α_i as coefficients, only $\alpha_i > 0$ are needed in the description: this object will therefore be called the *support vectors* of the description (SV). So by definition, R^2 is the distance from the center of the sphere to (any of the support vectors on) the boundary, i.e. objects with $0 < \alpha_i < C$. Therefore

$$R^2 = ||\mathbf{x}_k - \mathbf{a}||^2$$
$$= \underbrace{(\mathbf{x}_k \cdot \mathbf{x}_k) - 2\sum_i \alpha_i(\mathbf{x}_k \cdot \mathbf{x}_i) + \sum_{i,j} \alpha_i\alpha_j(\mathbf{x}_i\mathbf{x}_j)}_{T_{\mathbf{a}}(\mathbf{x}_k)} \tag{12}$$

for any $\mathbf{x}_k \in SV_{<C}$, the set of the support vectors which have $\alpha_k < C$.

To test a new object $\mathbf{z}$ it is necessary to calculate its distance $T_{\mathbf{a}}(\mathbf{z})$ from the center of the sphere and compare it with R^2

$$\operatorname{sgn}(R^2 - T_{\mathbf{a}}(\mathbf{z})) = \begin{cases} +1 & \text{if } \mathbf{z} \text{ is inside the sphere} \\ -1 & \text{if } \mathbf{z} \text{ is outside the sphere} \end{cases} \tag{13}$$

As it is common in machine learning theory [32], the method can be made more flexible [28, 29] by replacing all the inner products $(\mathbf{x}_i \cdot \mathbf{x}_j)$ with a kernel function $K(\mathbf{x}_i, \mathbf{x}_j)$ satisfying Mercer's theorem. The data are mapped into a higher dimensional space via a feature map and there the previous spherically classification is computed. The polynomial kernel and the gaussian kernel are discussed in [28, 29].

An example description by SVDD with different kernel functions for a 2 dimensional gaussian data set is shown in Fig. 1. The 1000 data are generated by a gaussian distribution with mean $[0, 0]$ and variance 1. Figures are handmade drawn using Matlab and the description bound is shown by a 2D contour plot.

Negative Examples Data Description. When two (or more) classes of data are available and it is necessary to identify a specific one among the others, SVDD can be trained to recognize objects that should be included in the description from those that should be rejected. This task of SVDD can be very useful in real-world applications where, for example, a safety region must be determined (see Sect. 4).

In the following the target objects are enumerated by indices i, j and the negative examples by l, m. We assume that target objects are labeled $y_i = 1$ and outlier objects are labeled $y_l = -1$.

In the same way as before, we want to solve this optimization problem:

$$\min_{R, \mathbf{a}, \xi_i, \xi_l} F(R, \mathbf{a}, \xi_i, \xi_l) = R^2 + C_1 \sum_i \xi_i + C_2 \sum_l \xi_l \tag{14}$$

$$\text{s.t} \begin{cases} ||\mathbf{x}_i - \mathbf{a}||^2 \leq R^2 + \xi_i, \\ ||\mathbf{x}_l - \mathbf{a}||^2 \geq R^2 - \xi_l, \\ \xi_i \geq 0, \quad \xi_l \geq 0 \quad \forall i, l \end{cases} \tag{15}$$

The constraints are again incorporated in Eq. (14) and the Lagrange multipliers $\alpha_i, \alpha_l, \gamma_i, \gamma_l$ are introduced [29]:

$$L(R, \mathbf{a}, \xi_i, \xi_l, \alpha_i, \alpha_l, \gamma_i, \gamma_l) = R^2 + C_1 \sum_i \xi_i + C_2 \sum_l \xi_l - \sum_i \gamma_i \xi_i - \sum_l \gamma_l \xi_l$$
$$- \sum_i \alpha_i [R^2 + \xi_i - (\mathbf{x}_i - \mathbf{a})^2] - \sum_l \alpha_l [(\mathbf{x}_l - \mathbf{a})^2 - R^2 + \xi_l] \tag{16}$$

with $\alpha_i \geq 0, \alpha_l \geq 0, \gamma_i \geq 0, \gamma_l \geq 0$.

Setting the partial derivatives of L with respect to $R, \mathbf{a}, \xi_i$ and ξ_l to zero gives new constraints [29]:

$$\sum_i \alpha_i - \sum_l \alpha_l = 1, \ \mathbf{a} = \sum_i \alpha_i \mathbf{x}_i - \sum_l \alpha_l \mathbf{x}_l \tag{17}$$

$$0 \leq \alpha_i \leq C_1, 0 \leq \alpha_l \leq C_2 \ \forall i, l \tag{18}$$

and substituting (17) in equation (16) we obtain similarly to before the dual problem of (14) and (15):

Fig. 2. Negative SVDD applied to a two-spirals shaped data set [20]. It is interesting to note that for changing the target objects it is only necessary to flip the labels. The asterisked points are the SV on the edge, depending on the respective class.

$$\max_{\alpha_i, \alpha_l} L = \sum_i \alpha_i (\mathbf{x}_i \cdot \mathbf{x}_i) - \sum_l \alpha_l (\mathbf{x}_l \cdot \mathbf{x}_l) - \sum_{i,j} \alpha_i \alpha_j (\mathbf{x}_i \cdot \mathbf{x}_j)$$
$$+ 2 \sum_{l,j} \alpha_l \alpha_j (\mathbf{x}_l \cdot \mathbf{x}_j) - \sum_{l,m} \alpha_l \alpha_m (\mathbf{x}_l \cdot \mathbf{x}_m) \tag{19}$$

$$\text{s.t} \begin{cases} \sum_i \alpha_i - \sum_l \alpha_l = 1 \\ 0 \le \alpha_i \le C_1 \ \forall i \\ 0 \le \alpha_l \le C_2 \ \forall l \end{cases} \tag{20}$$

Again, solving the previous optimization problem allows to determine α_i and α_l and then we can classify all the data set objects according to the respective Lagrange coefficient:

$$||\mathbf{x}_i - \mathbf{a}||^2 < R^2 \Rightarrow \alpha_i = 0 \ ; \ ||\mathbf{x}_l - \mathbf{a}||^2 < R^2 \Rightarrow \alpha_l = C_2 \tag{21}$$
$$||\mathbf{x}_i - \mathbf{a}||^2 = R^2 \Rightarrow 0 < \alpha_i < C_1 \ ; \ ||\mathbf{x}_l - \mathbf{a}||^2 = R^2 \Rightarrow 0 < \alpha_l < C_2 \tag{22}$$
$$||\mathbf{x}_i - \mathbf{a}||^2 > R^2 \Rightarrow \alpha_i = C_1 \ ; \ ||\mathbf{x}_l - \mathbf{a}||^2 > R^2 \Rightarrow \alpha_l = 0 \tag{23}$$

Similarly, we test a new point $\mathbf{z}$ based on its distance from the center

$$||\mathbf{z} - \mathbf{a}||^2 = (\mathbf{z} \cdot \mathbf{z}) - 2 \left(\sum_i \alpha_i (\mathbf{z} \cdot \mathbf{x}_i) - \sum_l \alpha_l (\mathbf{z} \cdot \mathbf{x}_l) \right)$$
$$+ \sum_{i,j} \alpha_i \alpha_j (\mathbf{x}_i \cdot \mathbf{x}_j) - 2 \sum_{l,j} \alpha_l \alpha_j (\mathbf{x}_l \cdot \mathbf{x}_j) + \sum_{l,m} \alpha_l \alpha_m (\mathbf{x}_l \cdot \mathbf{x}_m) := T_{\mathbf{a}}(\mathbf{z}) \tag{24}$$

and we evaluate it compared to the radius squared

$$\text{sgn}(R^2 - T_{\mathbf{a}}(\mathbf{z})) = \begin{cases} +1 \ \text{if } \mathbf{z} \text{ is inside the sphere} \\ -1 \ \text{if } \mathbf{z} \text{ is outside the sphere} \end{cases} \tag{25}$$

where the radius is calculated as the distance of any SV on the edge ($0 < \alpha_i < C_1, 0 < \alpha_l < C_2$) from the center $\mathbf{a}$

$$R^2 = T_{\mathbf{a}}(\mathbf{x}_k) \text{ for any } \mathbf{x}_k \in SV_{<C_1, <C_2} \tag{26}$$

Similarly to before, it is possible to replace all the inner products $(\mathbf{x}_i \cdot \mathbf{x}_j)$ with a kernel function $K(\mathbf{x}_i, \mathbf{x}_j)$ [28, 29, 32] to obtain a more flexible description.

An example of Negative-SVDD is performed in Fig. 2: gaussian kernel with $\sigma = 3$ is used and the parameters $C1$ and $C2$ are both set to 0.25.

2.2 Autonomous Detection of SVDD Parameters with RBF Kernel

Like most machine learning models, SVDD is massively influenced by the choice of model parameters. It is necessary to find the best trade-off between error and covering by choosing suitable C_1 and C_2 and the best kernel parameter σ that avoids overfitting or underfitting issues.

For this work we will focus on the RBF kernel since it is well known that it is the kernel function that performs well in application methods [28].

The method used to find the best model parameters is inspired by the work presented in [31] in which it is proposed an autonomous detection of the normal SVDD parameters based only in the training set, since in normal SVDD it is not possible to use cross-validation because only true positives and false negatives can occur during the training. In our work instead we joined some techniques in [31] with cross-validation method for finding the best C_1, C_2 and σ parameters for negative SVDD.

The regularisation parameters C_1, C_2 are lower bounded by $1/N_1$ and $1/N_2$ respectively, where N_1 is the number of target objects and N_2 the number of negative examples ($N_1 + N_2 = N$) [28, 29, 31]. When in one class of training objects set no errors are expected we can set $C_i = 1$ ($i = 1, 2$), indicating that all objects of the target class of training set should be accepted ($C_1 = 1$) and all outliers should be rejected ($C_2 = 1$). So the value range for C_1 and C_2 is

$$\frac{1}{N_1} \leq C_1 \leq 1, \quad \frac{1}{N_2} \leq C_2 \leq 1, \tag{27}$$

The second parameter to be optimised is the kernel width σ. For high values of σ the shape of SVDD becomes spherical with the risk of underfitting, while for small values of σ too much objects become support vectors and the model is prone to overfitting.

The search for the best parameters is performed by constructing a grid with C_1, C_2 and σ, on which holdout cross-validation is performed. The optimization criterion is chosen according to [31], selecting the parameters such that the respective misclassification error e and radius R minimize

$$\lambda = \sqrt{e^2 + |1 - R|^2} \tag{28}$$

for each triple C_1, C_2 and σ in the grid. The idea behind (28) is that minimizing the misclassification error means reducing the number of support vectors [28, 29] (and so reducing overfitting) while constraining the radius to be close to 1 means choosing small σ [31] (and so reducing underfitting). Then the balance between these two terms seems the best criterion for finding the best parameters (see Fig. 3).

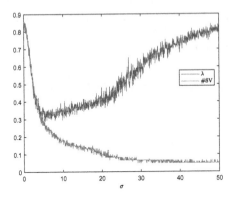

Fig. 3. For too small or too high values of σ the optimization criterion λ (our metric for the 'best error') is high. Also keep in mind the behavior of the SV, which is very similar to the one described in [28,29].

2.3 Fast Training SVDD

The curse of dimensionality is a problem that affects many optimization and machine learning problems, and SVDD is not saved. To overcome this problem, a method based on iterative training of only SV is proposed by [6].

The method iteratively samples from the training data set with the objective of updating a set of support vectors called as the master set of support vectors (SV^*). During each iteration, the method updates SV^* and corresponding threshold R^2 value and center $\mathbf{a}$. As the threshold value R^2 increases, the volume enclosed by the SV^* increases. The method stops iterating and provides a solution when the threshold value R^2 and the center $\mathbf{a}$ converge. At convergence, the members of the master set of support vectors SV^* characterize the description of the training data set.

2.4 Zero FNR Regions with SVDD

Safety regions research is a well-known task for machine learning [13–15] and the main focus is to avoid false negatives, i.e., including in the safe region unsafe points. In this section, two methods for the research of zero FNR regions are proposed: the first one is based simply on the reduction of the SVDD radius until only safe points are enclosed in the SVDD shape, the second one instead performs successive iterations of the SVDD on the safe region until there are no more negative points.

(a) FNR=0.517   (b) FNR=0.095

Fig. 4. Application of Algorithm 1 on a data set of 400 points sampled from a gaussian with mean $[1, 1]$ and variance 1, 200 target objects and 200 negative examples. The algorithm converged in 12 iterations.

Radius Reduction. Since also in the transformed space via feature mapping the shape of SVDD is a sphere, it is reasonable to think that reducing the volume of the sphere the number of negative points misclassifed should reduce. We implemented this simple procedure in Matlab and we tested it on several datasets (see Fig. 4):

Algorithm 1 RadiusReduction
Data set $\mathcal{X} \times \mathcal{Y}$ is divided in training set $\mathcal{X}_{tr} \times \mathcal{Y}_{tr}$ and test set $\mathcal{X}_{ts} \times \mathcal{Y}_{ts}$

SVDD-**cross-validation** on $\mathcal{X}_{tr} \times \mathcal{Y}_{tr}$
$[\mathbf{a}, R^2] = \text{SVDD}(\mathcal{X}_{tr}, \mathcal{Y}_{tr}, C_1, C_2, \text{param})$
maxiter $= 1000$;
$i = 1$;
while(i¡maxiter)
  $R^2 = R^2 - 10\text{e-}5*R^2$;
  Test SVDD on $\mathcal{X}_{ts} \times \mathcal{Y}_{ts}$
   if(FNR$< \varepsilon$)
   return $[\mathbf{a}, R^2]$;
   end
  $i = i + 1$;
end

SVDD Zero FNR Iterative Procedure. Here we present another algorithm for finding zero FNR regions with SVDD. The idea is simply to perform successive SVDDs on the safe regions found with a preliminary SVDD to avoid the presence of unsafe points. Again, we achieve convergence when we reach a fixed number of iterations or when the condition on FNR is satisfied.

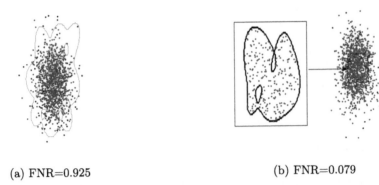

(a) FNR=0.925   (b) FNR=0.079

Fig. 5. Application of Algorithm 2 on a data set of 2000 target objects sampled from a gaussian with mean $[1,1]$ and variance 4 and 100 negative examples sampled from a gaussian with mean $[1,1]$ and variance 5. (a) is the first iteration of the algorithm and (b) is the convergence at the 97th iteration.

Algorithm 2 ZeroFNRSVDD
Data set $\mathcal{X} \times \mathcal{Y}$ is divided in training set $\mathcal{X}_{tr} \times \mathcal{Y}_{tr}$ and test set $\mathcal{X}_{ts} \times \mathcal{Y}_{ts}$

SVDD-**cross-validation** on $\mathcal{X}_{tr} \times \mathcal{Y}_{tr}$
$[\mathbf{a}, R^2] = \mathrm{SVDD}(\mathcal{X}_{tr}, \mathcal{Y}_{tr}, C_1, C_2, \text{param})$
Test SVDD on $\mathcal{X}_{ts} \times \mathcal{Y}_{ts}$
maxiter $= 1000$;
$i = 1$;
while($i < $ maxiter)
    $\mathcal{X}_{tr_i} = $ "safety"$(\mathcal{X}_{ts})$;
    SVDD-**cross-validation** on $\mathcal{X}_{tr_i} \times \mathcal{Y}_{tr_i}$
    $[\mathbf{a}_i, \quad R_i^2] = \mathrm{SVDD}(\mathcal{X}_{tr_i}, \mathcal{Y}_{tr_i}, C_1, C_2, \text{param})$ **Test** SVDD on $\mathcal{X}_{ts} \times \mathcal{Y}_{ts}$
      if(FNR$< \varepsilon$)
      return $[\mathbf{a}_i, R_i^2]$;
      end
    $i = i + 1$;
end

We performed this algorithm in Matlab and tested using data from [19]. In Fig. 5 is reported an example with a 2 dimensional gaussian data set.

3   Rules Extraction

We now consider how to make the SVDD explainable in order to explicit the inherent logic and use the extracted rules for further safety envelope tuning as in [14].

Let us suppose to have an information vector $\mathbf{I}$ and to have to solve a classification problem depending on two classes $\omega = 0$ or 1. Let $\aleph = \{(\mathbf{I}^k, \omega^k), k = 1, \ldots, \beth\}$ be a data set corresponding to the collection of events representing a dynamical system evolution (ω) under different system settings $(\mathbf{I}(\cdot))$.

The classification problem consists of finding the best boundary function $f(\mathbf{I}(\cdot), \cdot)$ separating the $\mathbf{I}^k$ points in $\aleph$ according to the two classes $\omega = 0$ or $\omega = 1$. For the case of SVDD the best boundary f is simply the shape of the hypersphere. Although the shape of the hypersphere is well intelligible (it is enough to have a center and a radius to describe it), it is still interesting to have a rule-based shape to describe it.

3.1   Logic Learning Machine

The derivation of $f(\mathbf{I}(\cdot), \cdot)$)in a rule-based shape is made by DT and LLM (the analysis was performed through the Rulex software suite, developed and distributed by Rulex Inc. (http://www.rulex.ai/)). They are both based on a set of intelligible rules of the type **if** (*premise*) **then** (*consequence*), where (*premise*) is a logical product (AND, $\wedge$) of conditions and (*consequence*) provides a class assignment for the output. In the present study, the two classes correspond to the presence or the absence of anomalous patterns. LLM rules are obtained through a three-step process. In the first phase (*discretisation and latticisation*) each variable is transformed into a string of binary data in a proper Boolean lattice, using the inverse only-one code binarisation. All strings are eventually concatenated in one unique large string per each sample. In the second phase (*shadow clustering*) a set of binary values, called *implicants*, are generated, which allow the identification of groups of points associated with a specific class. (An implicant is defined as a binary string in a Boolean lattice that uniquely determines a group of points associated with a given class. It is straightforward to derive from an implicant an intelligible rule having in its premise a logical product of threshold conditions based on cut-offs obtained during the discretisation step. The optimal placement of these cut-offs is, therefore, an important phase to extract the highest information gain before clustering [2].) During the third phase (*rule generation*) all implicants are transformed into a collection of simple conditions and eventually combined in a set of intelligible rules. The interested reader on shadow clustering and algorithms for efficient rule generation is referred to [16] and references therein.

3.2   Rules Extraction from SVDD

As far as SVDD is concerned, the derivation of intelligible rules is made in this way: after that a SVDD is computed and tested, a new data set of observations is provided and the classification via SVDD is made. The new dynamical system obtained is then exported in Rulex and a LLM algorithm with zero error or a DT algorithm is executed over the data, obtaining then the set of intelligible rules. Algorithm 3 summarizes the procedure:

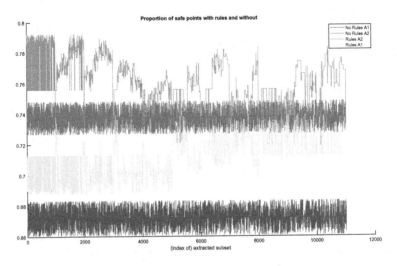

Fig. 6. Using intelligible rules (LLM) the proportion of safe points increases but FNR increases too. (Color figure online)

Algorithm 3 IRulesSVDD
Apply Algorithm 1 or Algorithm 2 on
$\mathcal{X} \times \mathcal{Y}$ data set

---

generate randomly a new data set $\mathcal{X}_{new}$
as a *copy* of $\mathcal{X}$
Classify $\mathcal{X}_{new}$ in $\mathcal{Y}_{new}$ with $[\mathbf{a}, R^2]$ from
Algorithm1/Algorithm2
apply **LLM/DT** algorithm
find an explained safety region $\mathcal{R}$
return $\mathcal{R}$

For example, for the case of vehicle platooning (see Sect. 4) the first three rules for *covering* (i.e. how many points are covered by rule r) of SVDD (Algorithm 2) using LLM are

if $((N < 7) \land (F_0 > -8 \land F_0 <= -3) \land (v(0) > 12 \land v(0) <= 29))$ **then safe**
if $(d(0) > 4.102334 \land d(0) <= 8.993453) \land (v(0) > 12 \land v(0) <= 23))$ **then safe**
if $((N < 6) \land (PER > 0.000827 \land PER <= 0.465396)$ **then safe**

As in [14] we applied these rules with the goal of maximizing the number of safe points while keeping FNR at zero. This is possible by performing rule tuning as in [14] but SVDD allows for much more flexibility.

Figure 6 shows the relationship between the prediction of safety regions of the two proposed algorithms with LLM-based rules (green and yellow) and with only the shape of hypersphere (orange and blue). The vehicle platooning data set is used and 11×10^3 have been done (see Sect. 4). We can say that the behaviour of the prediction is quite similar for both the methodology but when intelligible rules are used there is an increase of FNR (not too high, just up to 1%-5% more).

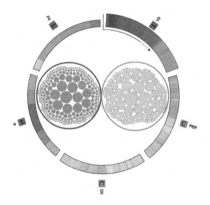

Fig. 7. Rule viewer

Furthemore, Fig. 7 shows, as an example, a summary of the rules extracted with LLM from SVDD, Algorithm 2, in the case of vehicle platooning (see Sect. 4.1). Each circle represents a rule and the larger this is the more the respective rule covers a larger number of points. In this example the classification is done in two classes, green and red, and in the outer crown the input features are shown. The high number of rules is an indication of the complexity of the system: with a two-dimensional example we could say that a large number of rectangles (rules) is needed to best approximate the complicated shape of the SVDD. We will discuss these concepts in more detail in Sect. 4, dedicated to applications.

4   Applications: The Vehicle Platooning Example

Finally in this section we investigate how the SVDD works in a real safety classification problem. We focus on an automotive example of cyber-physical system [21]: the vehicle platooning [22].

4.1   Vehicle Platooning

Vehicle Platooning (VP) is taken as a reference here as being representative of one of the most challenging CPS (Cyber Physical Systems) of the automotive sector [21]. The main goal in VP is finding the best trade-off between performance (i.e. maximising speed and minimising vehicles reciprocal distance) and safety (i.e. avoiding collision) [11]. Most of the literature on this topic focuses on advanced control schemes while abstracting the communication medium. Delay of communication is typically considered as fixed or described through probabilistic models. This allows the analytical derivation of stability models under some hypotheses of the dynamical system [18], but it may be unreliable under realistic conditions. Two branches are evident from the literature in this respect: the derivation of simple models of the delay bound that guarantees safety (see, e.g. Section IV.C of [34]) and extensive simulation with visualisation of safety regions under subsets of parameters when addressing realistic communication [9,26], and realistic vehicles [24].

The following scenario is considered. Given the platoon at a steady state of speed and reciprocal distance of the vehicles, a braking is applied by the leader of the platoon [24,34]. The behaviour of the dynamical system is investigated with respect to the following metrics. Safety is referred to a collision between adjacent vehicles (in the study, it is actually registered when the reciprocal distance between vehicles achieves a lower bound (e.g. 2 m)). For both safety and driving comfort, string stability (SS) is also important. It means that speed and acceleration fluctuations should be attenuated downstream the string of vehicles.

The dynamic of the system is generated by the following differential equations [34]:

$$\begin{cases} \dot{v} = \frac{1}{m_i}(F_i - (a_i + b_i \cdot v_i^2)); \\ \dot{d}_i = v_{i-1} - v_i \end{cases} \tag{29}$$

where v_i is the speed of vehicle i, m_i the mass of the vehicle i, d_i the distance of vehicle i from the previous one $i-1$, a_i is the tyre/road rolling distance, b_i the aerodynamic drag and the control law F_i.

The behaviour of the dynamical system is synthesised by the following vector of features:

$$\mathbf{I} = [N, \boldsymbol{\iota}(0), F_0, \mathbf{m}, \mathbf{q}, \mathbf{p}] \tag{30}$$

$N+1$ being the number of vehicles in the platoon (subscript $i = 0$ defines the index of the leader), $\boldsymbol{\iota} = [\mathbf{d}, \mathbf{v}, \mathbf{a}]$ are the vectors of reciprocal distance, speed, and acceleration of the vehicles, respectively ($\boldsymbol{\iota}(0)$ denotes that the quantities

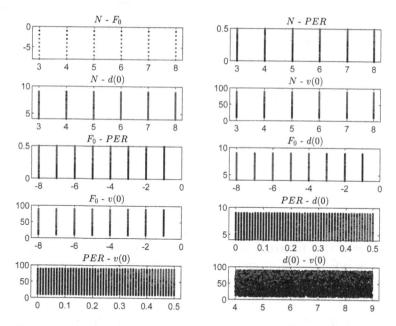

Fig. 8. Scatter plots of the quantities of the platooning dynamical system as in [13–15]. In blue non-collision points are plotted, in red collision ones. (Color figure online)

sampled at time $t = 0$, after which a braking force is applied by the leader [24]. Simulations are set in order to manage possible transient periods and achieve a steady state of ι before applying the braking.), $\mathbf{m}$ are the vectors of weights of the vehicles, F_0 is the braking force applied by the leader, $\mathbf{q}$ is the vector of quality measures of the communication medium, fixed delay and packet error rate (PER) are considered in the study, $\mathbf{p}$ is the vector of tuning parameters of the control scheme.

The Plexe simulator [24,34] is used to register $\beth = 15 \times 10^3$ observations and then we reduced them under the following ranges:

$N \in [3,8]$, $F_0 \in [-8,-1] \times 10^3 N$ (from now on, the notation ($\times 10^3$) is omitted when referring to thresholds applied to F_0), PER $\in [0,0.5]$, $d(0) \in [4,9]$ m, $v(0) \in [10,90]$ Km/h. With these choice the size of the sample has been reduced to 7567 samples (see Fig. 8).

Our goal is to determine the largest region of parameters with no false negatives (i.e. prediction of no collision, but a collision in reality). To do this, we applied the two algorithms proposed in Sect. 2.4 to the 7567 size sample above (a Fast-SVDD is used, see Sect. 2.3) using RBF kernel with $C_1 = 1$, $C_2 = 1$ (indicating that all objects of the specific class of training set should be accepted, $C_1 = 1$, and all the objects of the negative class should be rejected, $C_2 = 1$) and σ determined with cross-validation. The results are shown in Table 1, where FNR is the usual False Negative Rate, % safe is the percentage of safe points, #iter the number of algorithm iterations, #time (s) the time in second for the convergence, R^2 the squared hypersphere's radius, #SV the number of determined support vectors. The last column holds the precision on the negative class $\frac{TN}{TN+FN}$.

Table 1. Results on VP data set.

	FNR	% safe	# iter	# time (s)	R^2	#SV	$prec_{\omega_{-1}}$
Alg 1	0.0993	79.34	15	192.12	0.9220	139	0.9935
Alg 2	0.0556	70.06	42	310.31	0.8158	61	0.9938

Then we tested the performances of the algorithms in different extractions of 10^3 subsets with different sizes from 8% to 50% of the total points available for test (12×10^3); 11×10^3 trials in total. We compared them with other methodologies as in [14] (see Fig. 9) and so a rules extraction has been requested (see Sect. 3). The other methodologies used for the comparison are briefly discussed below, for more details see [14].

- **manual calibration:** rules set by hand. For example, by inspecting Fig. 8, it is intuitive to identify the following safety region: $(N \leq 5) \wedge (PER < 0.23)$. More accurate inspection can bring to more accurate rules.
- **LLM** and **DT** are tuned according to [14] (Sect. 4.4). The procedure can be briefly summarize in this way: (1) manually inspect of the most relevant regions for safety. (2) LLM/DT is trained with zero error when developing the rules. (3) Progressively extraction of unsafe points from the original data set until only safe points are obtained.
- **LLM violation** rules are generated analogously to the previous LLM rules but a slack coefficient $\delta_{(\cdot)}$ is inserted depending on a sensitive analysis performed over the features (see [14], Sect. 4.4.1). For example **if** $((PER \leq 0.325 \cdot \delta_{PER}) \wedge (N \leq 7 \cdot \delta_N) \wedge (d(0) > 4.2385))$ **then safe**, where $\delta_{PER} = 0.325 * 0.1$ and $\delta_N = 6/7$ (i.e. one car is eliminated from the platoon to be safe).

The analysis shows that SVDD performs the best safety region in the chosen ranges of parameters: up to 70% of safe points with almost zero FNR for Algorithm 1 and up to 80% for Algorithm 2. The comparison with the other methods shows as the rules extracted from SVDD are the better ones, but due to the complex form of SVDD boundary function an higher number of them is required: 674 rules for Algorithm 1 and 771 for Algorithm 2. The rules are applied all together in logical OR ($\vee$). To deepen the analysis we used also DT for the SVDD rules (150 in logical OR) extraction (Algorithm 2) and we obtained a lower number of safe points (only up to 60%) obtaining however a number of FNR very close to zero (0.003%).

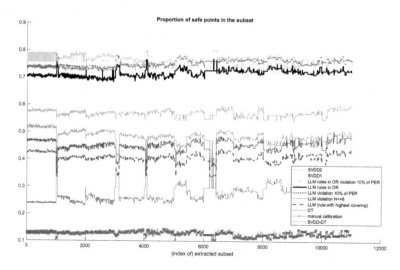

Fig. 9. Performance of safe points with and without rules.

5   Conclusions and Future Works

The study shows how SVDD can be a very useful method for identifying safety regions, even in complex applications such as VP.

This paper also provides a detailed methodology on how to deal with application problems in machine learning, such as parameter tuning and handling large data sets. In addition, a more thorough explanation on negative SVDD has been performed. Thus, the proposed approach could be applied for a wide range of applications.

Its novelty, when compared for example with [15], is that it is possible to manage the shape of the safety region by varying the radius of SVDD or by removing unsafe points using multiple iterations of SVDD in the region. Then, we have shown that rule extraction works better after identifying the safety region through SVDD.

Furthermore, the notion of "safety region" discussed in our paper is an important topic that has been widely studied in machine learning and learning theory in various forms. It seems that research topics such as conformal prediction, selective prediction [33] and three-way decision making [4] are widely related to the notion of safety region. Our goal for the future is to investigate how SVDD can be a useful tool to apply to other types of safety research methodology, such as those previously mentioned. Also, there is a large body of work related to rule extraction from SVM [3,17,35] (which are strongly related with SVDD) which could be interesting to investigate for performing a method for direct rule extraction from SVDD. Our rule extraction method differs from the others in that it is not based directly on the SVDD machine learning model but relies on other rule extraction methods (LLM and DT), after classifying the points through the SVDD. It is our intention to try to define a totally autonomous method of rule extraction from the SVDD (based on those already mentioned for the SVM), and then verify whether the results will actually be better than those obtained now.

References

1. Abe, S.: Support Vector Machines for Pattern Classification. Advances in Pattern Recognition, 2nd edn. Springer, London (2010). https://doi.org/10.1007/978-1-84996-098-4
2. Boros, E., Hammer, P.L., Ibaraki, T., et al.: An implementation of logical analysis of data. IEEE Trans. Knowl. Data Eng. **12**(2), 292–306 (2000)
3. Barakat, N., Bradley, A.P.: Rule extraction from support vector machines: a review. Neurocomputing **74**(1–3), 178–190 (2010). https://doi.org/10.1016/j.neucom.2010.02.016. ISSN 0925–2312
4. Campagner, A., Cabitza, F., Ciucci, D.: Three-way decision for handling uncertainty in machine learning: a narrative review. In: International Joint Conference on Rough Sets (2020)
5. Balasubramanian, V.N., Ho, S.S., Vovk, V.: Conformal Prediction for Reliable Machine Learning, 1st edn. Morgan Kaufmann Elsevier, Waltham (2014). ISBN 9780123985378
6. Chaudhuri, A., et al.: Sampling method for fast training of support vector data description. arXiv e-prints, 2016arXiv160605382C (2006)
7. European Union Aviation Safety Angency: Concepts of Design Assurance for Neural Networks CoDANN. 2020 mar, EASA AI Task Force. Daedalean, AG. https://www.easa.europa.eu/sites/default/files/dfu/EASA-DDLN-Concepts-of-Design-Assurance-for-Neural-Networks-CoDANN.pdf
8. Fisch, D., Hofmann, A., Sick, B.: On the versatility of radial basis function neural networks: a case study in the field of intrusion detection. Inf. Sci. **180**(12), 2421–2439 (2010). http://www.sciencedirect.com/science/article/pii/S0020025510001015
9. Ge, J.I., Orosz, G.: Dynamics of connected vehicle systems with delayed acceleration feedback. Transp. Res. C Emerg. Technol. **46**, 46–64 (2014). Cited By 90
10. Huang, G., Chen, H., Zhou, Z., Yin, F., Guo, K.: Two-class support vector data description. Pattern Recogn. **44**, 320–329 (2011)
11. Jia, D., Lu, K., Wang, J., et al.: A survey on platoon-based vehicular cyber-physical systems. IEEE Commun. Surv. Tutor. **18**(1), 263–284 (2016)
12. Jones, C.A.: Lecture notes: Math2640 introduction to optimisation 4. University of Leeds, School of Mathematics, Technical report (2005)
13. Mongelli, M., Muselli, M., Scorzoni, A., Ferrari, E.: Accelerating PRISM validation of vehicle platooning through machine learning, pp. 452–456 (2019). https://doi.org/10.1109/ICSRS48664.2019.8987672
14. Mongelli, M., Muselli, M., Ferrari, E., Fermi, A.: Performance validation of vehicle platooning via intelligible analytics. IET Cyber-Phys. Syst.: Theory Appl. **4**, 120–127 (2018). https://doi.org/10.1049/iet-cps.2018.5055
15. Fermi, A., Mongelli, M., Muselli, M., Ferrari, E.: Identification of safety regions in vehicle platooning via machine learning. In: 2018 14th IEEE International Workshop on Factory Communication Systems (WFCS), Imperia, Italy, pp. 1–4 (2018). https://doi.org/10.1109/WFCS.2018.8402372
16. Muselli, M., Ferrari, E.: Coupling logical analysis of data and shadow clustering for partially defined positive Boolean function reconstruction. IEEE Trans. Knowl. Data Eng. **23**(1), 37–50 (2011)

17. Nunez, H., Angulo, C., Català, A.: Rule-based learning systems for support vector machines. Neural Process. Lett. **24**, 1–18 (2006)
18. Oncu, S., van de Wouw, N., Nijmeijer, H.: Cooperative adaptive cruise control: tradeoffs between control and network specifications. In: 2011 14th International IEEE Conference on Intelligent Transportation Systems (ITSC), Washington, DC, USA, pp. 2051–2056 (2011)
19. KEEL: Website: KEEL (Knowledge Extraction based on Evolutionary Learning), November 2012. http://sci2s.ugr.es/keel/datasets.php
20. Kools, J.: 6 functions for generating artificial datasets. https://www.mathworks.com/matlabcentral/fileexchange/41459-6-functions-for-generating-artificial-datasets. MATLAB Central File Exchange. Accessed 4 Apr 2021
21. Pop, P., Scholle, D., Hansson, H., et al.: The safecopecsel project: safe cooperating cyber-physical systems using wireless communication. In: 2016 Euromicro Conference on Digital System Design (DSD), Limassol, Cyprus, pp. 532–538 (2016)
22. Pop, P., Scholle, D., Sljivo, I., et al.: Safe cooperating cyber-physical systems using wireless communication. Microprocess. Microsyst. **53**, 42–50 (2017)
23. Czarnecki, K., Salay, R.: Towards a framework to manage perceptual uncertainty for safe automated driving. In: Gallina, B., Skavhaug, A., Schoitsch, E., Bitsch, F. (eds.) SAFECOMP 2018. LNCS, vol. 11094, pp. 439–445. Springer, Cham (2018). https://doi.org/10.1007/978-3-319-99229-7_37
24. Santini, S., Salvi, A., Valente, A.S., et al.: A consensus-based approach for platooning with intervehicular communications and its validation in realistic scenarios. IEEE Trans. Veh. Technol. **66**(3), 1985–1999 (2017)
25. Standardization in the area of Artificial Intelligence, ISO/IEC. Creation date 2017, Washington, DC 20036, USA (2017). https://www.iso.org/committee/6794475.html
26. Segata, M., Cigno, R.L.: Automatic emergency braking: realistic analysis of car dynamics and network performance. IEEE Trans. Veh. Technol. **62**(9), 4150–4161 (2013)
27. Road vehicles Safety of the intended functionality PD ISO PAS 21448:2019. International Organization for Standardization, Geneva, CH
28. Tax, D.M.J., Duin, R.P.W.: Support vector domain description. Pattern Recogn. Lett. **20**, 1191–1199 (1999)
29. Tax, D.M.J., Duin, R.P.W.: Support vector data description. Mach. Learn. **54**, 45–66 (2004)
30. Tax, D.M.: One-class classification, concept-learning in the absence of counter-examples. Ph.D. dissertation, Delft University of Technology (2001)
31. Theissler, A., Dear, I.: Autonomously determining the parameters for SVDD with RBF kernel from a one-class training set. In: Conference: WASET International Conference on Machine Intelligence, Stockholm (2013)
32. Vapnik, V.: The Nature of Statistical Learning Theory. Springer, New York (1995). https://doi.org/10.1007/978-1-4757-2440-0
33. Wiener, Y., El-Yaniv, R.: Agnostic pointwise-competitive selective classification. J. Artif. Intell. Res. **52**, 171–201 (2015)
34. Xu, L., Wang, L.Y., Yin, G., et al.: Communication information structures and contents for enhanced safety of highway vehicle platoons. IEEE Trans. Veh. Technol. **63**(9), 4206–4220 (2014)
35. Zhu, P., Hu, Q.: Rule extraction from support vector machines based on consistent region covering reduction. Knowl.-Based Syst. **42**, 1–8 (2012)

Airbnb Price Prediction Using Machine Learning and Sentiment Analysis

Pouya Rezazadeh Kalehbasti[(⊠)], Liubov Nikolenko, and Hoormazd Rezaei

Stanford University, Stanford, CA 94305, USA
pouyar@stanford.edu, {liubov,hoormazd}@alumni.stanford.edu

Abstract. Pricing a property and evaluating the proposed price for a property are challenges that, respectively, owners and customers of Airbnb rentals face on a daily basis. This paper aims to create a model for predicting the price of an Airbnb listing using property specifications, owner information, and customer reviews for the listing. Owners and customers can use the resulting model to estimate the expected value of an Airbnb listing. Linear regression, tree-based models, K-means Clustering, Support Vector Regression (SVR), and neural networks are trained and tuned on a dataset of Airbnb listings from New York city, and the resulting models are compared in terms of Mean Squared Error, Mean Absolute Error, and R^2 score. Sentiment analysis is used to extract features from the customer reviews which help enhance the performance of the selected predictive models. Feature importance analysis is also used to select the most representative features for predicting the price of the listings. Experimentation shows that SVR model can achieve an R^2 score of 69% and a MSE of 0.147 (defined on ln(price)) on the test set, outperforming the other models considered in the paper. [Link to the repository: github.com/PouyaREZ/AirBnbPricePrediction].

Keywords: AirBNB · Rental property pricing · Machine learning · Sentiment analysis

1 Introduction

Pricing a rental property on Airbnb is a challenging task for the owner as it determines the number of customers for the place. On the other hand, customers have to evaluate an offered price with minimal knowledge of an optimal value for the property. This paper aims to develop a reliable price prediction model using machine learning, deep learning, and natural language processing techniques to aid both the property owners and the customers with price evaluation given minimal available information about the property. Features of the rentals, owner characteristics, and the customer reviews will comprise the predictors, and a range of methods from linear regression to tree-based models, support-vector regression (SVR), K-means Clustering (KMC), and neural networks (NNs) will be used for creating the prediction model.

© IFIP International Federation for Information Processing 2021
Published by Springer Nature Switzerland AG 2021
A. Holzinger et al. (Eds.): CD-MAKE 2021, LNCS 12844, pp. 173–184, 2021.
https://doi.org/10.1007/978-3-030-84060-0_11

2   Related Work

Parts of the existing literature on property pricing focus on non-shared property purchase or rental price predictions. Previously, Yu and Wu [28] tried to implement a real estate price prediction using feature importance analysis along with linear regression, SVR, and Random Forest regression. They also attempted to classify the prices into 7 classes using Naive Bayes, Logistic Regression, SVC and Random Forest. They declared a best RMSE of 0.53 for their SVR model and a classification accuracy of 69% for their SVC model with PCA. In another paper, Ma et al. [18] have applied Linear Regression, Regression Tree, Random Forest Regression and Gradient Boosting Regression Trees to analyzing warehouse rental prices in Beijing. They concluded that the tree regression model was the best-performing model with an RMSE of 1.05 CNY/m^2-day.

Another class of studies, which are more related to this paper, inspect the rental prices in hotels and sharing economy. Wang and Nicolau [25] have studied price determinants of sharing economy by analyzing Airbnb listings using ordinary least squares and quantile regression analysis. In a similar study, Masiero et al. [19] use quantile regression model to analyze the relation between travel traits and holiday homes as well as hotel prices. In a simpler work, Yang et al. [27] applied linear regression to study the relationship between market accessibility and hotel prices in Caribbean. They also included the user ratings and hotel classes as contributing factors in their study. Li et al. [15] also studied a clustering method called Multi-Scale Affinity Propagation and applied Linear Regression to the obtained clusters in an effort to create a price prediction model for Airbnb in different cities. They took the distance of the property to the city landmarks as the clustering feature. Papers by Chiny et al. [5], Zhou and Tong [29], Trang et al. [24], Kokasih and Paramita [14], and Ma et al. [17] are also noteworthy recent publications around rental prices in lodging industry.

This research has tried to improve and add to the experimented methods from the literature by focusing on a variety of feature selection techniques, implementing Neural Networks, and leveraging the customer reviews through sentiment analysis. The last two contributions are novel undertakings in rental price prediction as they were not observed in the existing body of literature prior to the date the preprint of this paper [12] was written (i.e. 2019).

3   Dataset

The public Airbnb dataset for New York City [1] was used as the main data source for this study. The dataset included 50,221 entries, each with 96 features. Figure 1 shows the geographic distribution of the listing prices in this dataset.

For the initial prepossessing, the authors inspected each feature of the dataset to (i) remove features with frequent and irreparable missing fields or set the missing values to zero where appropriate, (ii) convert some features into floats (e.g. by removing the dollar sign in prices), (iii) change boolean features to binaries, (iv) remove irrelevant or uninformative features, e.g. host picture url,

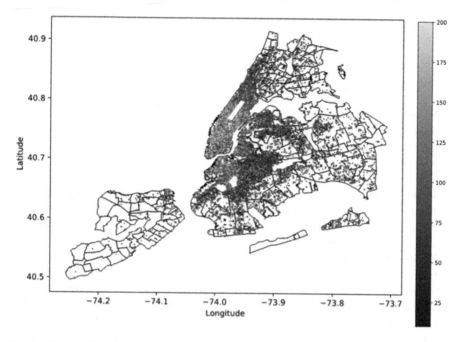

Fig. 1. Geographic spread of price labels (with filtered outliers) across New York City in USD per day

constant-valued fields or duplicate features, and (v) convert the 10 categorical features in the final set, e.g. 'neighborhood name' and 'cancellation policy,' into "one-hot vectors." In addition, the features were normalized and the labels were converted into logarithm of the prices to mitigate the impact of the outliers in the dataset. The data was split into three sets; namely, train set (comprising 90% of the original data), validation set, and test set (both comprising 5% of original data). Since the dataset was relatively large, 10% of the data was deemed sufficient for the accumulated testing and validation sets. The following explains the sentiment analysis conducted on the reviews and the steps taken for selecting the most important features among the available set of features.

3.1 Sentiment Analysis on the Reviews

Given the importance of customer reviews on the pricing of an Airbnb listing, and in order to increase the accuracy of the predictive model, the reviews for each listing were analyzed using TextBlob [16] sentiment analysis library and the results were added to the set of features. This method assigns a score between −1 (very negative sentiment) and 1 (very positive sentiment) to each analyzed text. For every listed property, each review was analyzed using this method and the scores were averaged across all the reviews of that listing. The final score for each listing was included as a new feature in the model. Future work can delve more into other approaches for mining the opinions of the customers [23].

3.2   Feature Selection

After data preprocessing, the feature vector contained 764 elements. Feeding this excessive set of features to the models resulted in a high variance of error. Consequently, using the training set, several feature selection techniques were used to find the features with the most predictive values to both reduce the model variances and reduce the computation time. Based on prior experience of the authors with housing price estimation, the first tried method was manual selection of features to create a baseline for evaluating the other feature selection processes.

The second selection method was tuning the coefficient of linear regression model with Lasso Regularization trained on the train split. Based on this analysis, the model with the best performance over validation split was selected. The resulting set consisted of 78 features with non-zero values, i.e. 90% less than the number of original features.

Finally, lowest p-values of regular linear regression model trained on train split were used to choose the third set of features. An upper limit of 100 features was imposed on the selection procedure. The final set was comprised of 22 features for which linear regression model performed the best on the validation split. As an example to demonstrate the results of the feature selection techniques, Appendix A lists the set of features resulting from this p-value analysis.

The performance of manually selected features as well as p-value and Lasso feature selection schemes were compared using the R^2 score of the linear regression models trained on the validation set. All models outperformed the baseline model, which used the whole feature set, and the second method, Lasso regularization, yielded the highest R^2 score. Figure 2 shows the best R^2 scores obtained using the set of features identified with each feature selection method.

4   Methods

Linear Regression using the entire set of features as model inputs was taken as the baseline model for evaluating the performance of the other methods. After selecting a set of features using Lasso feature selection, several machine learning models were considered in order to find the optimal one. All of the models except neural networks were implemented using Scikit-learn library [9]. The neural network model was implemented with the help of Keras library [8]. The implemented models are introduced in what follows.

4.1   Ridge Regression

Linear Regression with L_2 regularization adds a penalizing term to the squared error cost function in order to help the algorithm converge for linearly separable data and reduce overfitting. Therefore, Ridge Regression minimizes $J(\theta) = ||y - X\theta||_2^2 + \alpha||\theta||_2^2$ with respect to θ, where X is a design matrix and α is a hyperparameter. Since the baseline models were observed to have high variance, Ridge Regression seemed to be an appropriate choice to solve the issue.

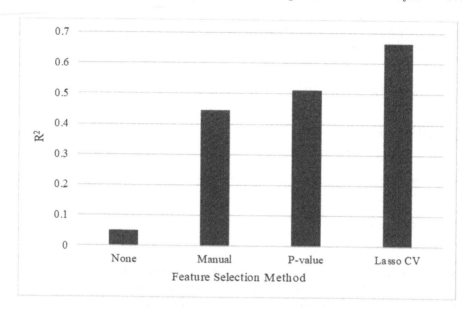

Fig. 2. Best feasible R^2 scores with each selection methods

4.2   K-means Clustering with Ridge Regression

In order to capture the non-linearity of the data, the training examples were split into different clusters using k-means clustering on the features and the Ridge Regression was run on each of the individual clusters. The data clusters were identified using Algorithm 1 given m points and k clusters. The algorithm would converge when the Frobenius norm of the difference between the cluster centers from two consecutive iterations became lower than $10^{(-4)}$ [22].

4.3   Support Vector Regression

In order to model the non-linear relationship between the covariates, the authors employed support vector regression with RBF kernel to identify a linear boundary in a high-dimensional feature space. Using the implementation based on Chang and Lin [3], the algorithm provides a solution for the following optimization problem:

$$\min_{w,b,\xi,\xi^*} \frac{1}{2}||w||^2 + C\sum_{i=1}^{m}\xi_i + C\sum_{i=1}^{m}\xi_i^*, \text{subject to} \tag{1}$$

$$w^T\phi(x^{(i)}) + b - y^{(i)} \leq \epsilon + \xi_i, \tag{2}$$

$$y^{(i)} - w^T\phi(x^{(i)}) - b \leq \epsilon + \xi_i^*, \tag{3}$$

$$\xi_i, \xi_i^* \geq 0, i = 1, ..., m \tag{4}$$

Algorithm 1. K-means Clustering

Initialize cluster centroids $\mu_i, ..., \mu_k$ randomly

repeat

  Assgin each point, $x^{(i)}$, to a cluster, $c^{(i)}$, such that:
  $c^{(i)} = \arg\min_j ||x^{(i)} - \mu_j||_2^2$

  Update each centroid such that:
  $\mu_j = \frac{\sum_{i=1}^m 1\{c^{(i)}=j\}x^{(i)}}{\sum_{i=1}^m 1\{c^{(i)}=j\}}$

  Calculate the loss function for the assignments:
  $J(c,\mu) = \sum_{i=1}^m ||x^{(i)} - \mu_{c^{(i)}}||_2^2$

until convergence

---

where $C > 0, \epsilon > 0$ are given parameters. This problem can be converted into a dual problem that does not involve $\phi(x)$, but involves $K(x, z) = \phi(x)\phi(z)$ instead. Since we are using RBF kernel, $K(x, z)$ was taken as

$$K(x, z) = \exp\left(\frac{||x - z||^2}{2\sigma^2}\right) \qquad (5)$$

4.4   Neural Network

Neural network was used to build a model that combined the input features into high level predictors. The architecture of the optimized network had 3 fully-connected layers: 20 neurons in the first hidden layer with relu activation function, 5 neurons in the second hidden layer with relu activation function, and 1 output neuron with a linear activation function.

4.5   Gradient Boosting Tree Ensemble

Since the relationship between the feature vector and price is non-linear, regression tree seemed like a proper model for this problem. Regression trees split the data points into regions according to the following formula

$$\max_{j,t} L(R_p) - (L(R_1) - L(R_2)) \qquad (6)$$

where j is the feature the dataset is split on, t is the threshold of the split, R_p is the parent region and R_1 and R_2 are the child regions. Squared error is used as the loss function.

Since standalone regression trees have low predictive accuracies individually, gradient boost tree ensemble was used to increase the models' performance. The idea behind a gradient boost is to improve on a previous iteration of the model by correcting its predictions using another model based on the negative gradient of the loss. The algorithm for the gradient boosting is the following [10]:

Algorithm 2. Gradient Boosting

Initialize F_0 to be a constant model
for m = 1,..., number of iterations **do**
  for all training examples $(x^{(i)}, y^{(i)})$ **do**
    Squared error $R(y^{(i)}, F_{m-1}(x^{(i)})) = -\frac{\partial \text{Loss}}{\partial F_{m-1}(x^{(i)})} = y^{(i)} - F_{m-1}(x^{(i)})$
  end for
  Train regression model h_m on $(x^{(i)}, R(y^{(i)}, F_{m-1}(x^{(i)})))$, for all training examples
  $F_m(x) = F_{m-1}(x) + \alpha h_m(x)$, where α is the learning rate
end for
return F_m

5 Experiments and Discussion

Mean absolute error (MAE), mean squared error (MSE) and R^2 score were used to evaluate the trained models. Training (39,980 examples) and validation (4,998 examples) splits were used to choose the best-performing models within each category. The test set, containing 4,998 examples, was used to provide an unbiased estimate of error, with the final models trained on both train and validation splits. Table 1 contains the performance metrics for the final models[1]; namely, linear regression, Ridge regression, Gradient Boosting, K-Means Clustering with Ridge Regression, SVR, and Neural Network.

Table 1. Performance metrics of the trained models

Model name	Train split			Test split		
	MAE	MSE	R^2 Score	MAE	MSE	R^2 Score
Linear reg. (baseline)	0.2744	0.1480	0.690	96895.82	2.4E13	−5.1E13
Ridge reg.	0.2813	0.15461	0.6765	0.2936	0.1613	0.6601
Gradient boost	0.2492	0.1376	0.7121	0.3282	0.1963	0.5864
K-means + Ridge reg.	0.2717	0.1438	0.6992	0.2850	0.1543	0.6748
SVR	0.2132	0.1067	0.7768	0.2761	0.1471	0.6901
Neural net	0.2602	0.1316	0.7246	0.2881	0.1570	0.6692

Table 1 shows that the models had relatively similar R^2 scores. This indicates that the Lasso feature importance analysis has majorly contributed to the performance of the models by reducing the variance, such that all the different models using the selected features have led to close R^2 scores. Even after the feature selection, the resulting input vector was relatively large, and this caused the models to overfit. This explains why Gradient Boost - a tree-based model

---

[1] Optimized models can be found at github.com/PouyaREZ/AirBnbPricePrediction.

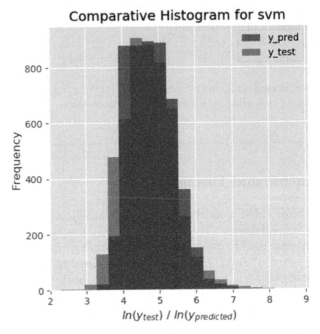

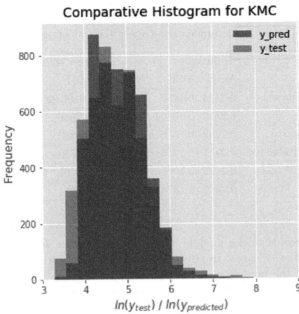

Fig. 3. Comparative histograms of predicted and actual prices for the top 3 models: SVR, KMC, and NN

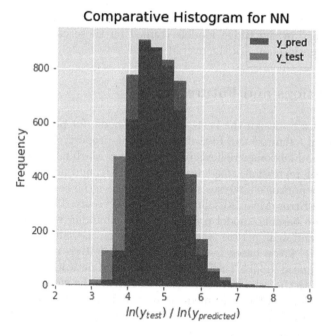

Fig. 4. Comparative histograms of predicted and actual prices for the top 3 models: SVR, KMC, and NN (cont'd)

prone to high variance - has performed worse than the rest of the models while it did not perform the worst on the training set.

Figures 3 and 4 show the comparative histograms of the ln of the test labels against the ln of the predicted labels for the top three models in terms of R^2 score on the test set, i.e., SVR (denoted as "svm" in the figure), K-Means + Ridge Regression (denoted as "KMC" in the figure), and Neural Network (denoted as "NN" in the figure), respectively. These figures show that, compared to the other two models, SVR has produced a more similar data distribution to the test set. Further, despite considering a larger number of features in the feature vector, SVR with RBF kernel yielded the best performing model with the least MAE and MSE and the highest R^2 score on both the train and the test set (Table 1). RBF feature mapping has been able to better model the prices of the apartments which have a non-linear relationship with the apartment features. Since regularization is considered in the SVR optimization problem, parameter tuning has ensured that the model would not overfit (Table 1 and Figs. 3 and 4).

Finally, Table 1 shows that the top three models, i.e., Ridge regression, neural network, and K-means + Ridge regression, had similar R^2 scores even though the last two models were more complex than Ridge regression. The large number of unknown parameters in the neural network model compared to the small size of the training set has probably caused the neural network to overfit the data and to underperform. K-means + Ridge regression model has probably faced a similar issue: too few training examples in some of the clusters has resulted in high variance in the Ridge models trained on those clusters, and this has damaged the performance of the overall model.

6   Conclusions and Future Work

This paper tries to design the best-performing model for predicting the Airbnb prices based on a limited set of features including property specifications, owner information, and customer reviews on the listings. Machine learning techniques including linear regression, tree-based models, SVR, and neural networks along with feature importance analyses are used to achieve the best results in terms of Mean Squared Error, Mean Absolute Error, and R^2 score. The initial experimentation with the baseline model proved that the abundance of features leads to high variance and weak performance of the model on the validation set compared to the train set. Lasso-based feature importance analysis reduced the variance and using advanced models such as SVR and neural networks resulted in higher R^2 score for both the validation and test sets. Among the models tested, Support Vector Regression (SVR) performed the best and produced an R^2 score of 69% and a MSE of 0.147 (defined on ln(price)) on the test set. This level of accuracy is a promising outcome given the heterogeneity of the dataset and the involved hidden factors and interactive terms, including the personal characteristics of the owners, which were impossible to consider.

Future work can (i) study other feature selection schemes such as Random Forest feature importance and correlation-based feature selection [7], (ii) further experiment with neural network architectures and use different machine learning models (e.g., [2,4,6,11,20]) (iii) use specialized hardware to boost the machine learning and deep learning models already used (e.g., [13,21,26]), and (iv) get more training examples from other hospitality services such as VRBO to boost the performance of K-means clustering with Ridge Regression model, in particular. Also, the sentiment analysis can be improved in future studies by, e.g., weighing the more recent reviews more than the dated ones, and including in the training features other metrics in addition to the average sentiment score for each listing.

A   Appendix

List of features selected using p-value importance method (the last 6 feature names are those of one-hot vectors):

'longitude', 'accommodates', 'bathrooms', 'bedrooms', 'beds', 'security_deposit', 'cleaning_fee', 'guests_included', 'Cable_TV', 'Dryer', 'Washer', 'Family/kid_friendly', 'Gym', 'Elevator', 'Entire home/apt', 'Private room', 'Brooklyn', 'Manhattan', 'Brooklyn.1', 'New York', 'Chelsea', 'Midtown'.

References

1. AirBNB public dataset. http://insideairbnb.com/get-the-data.html. Accessed 01 Dec 2018
2. Aggarwal, K., Kirchmeyer, M., Yadav, P., Keerthi, S.S., Gallinari, P.: Conditional generative adversarial networks for regression. arXiv:190512868 Cs Stat. (10) (2019)
3. Chang, C.C., Lin, C.J.: LibSVM: a library for support vector machines. ACM Trans. Intell. Syst. Technol. (TIST) **2**(3), 27 (2011)
4. Chen, T., Guestrin, C.: XGBoost: a scalable tree boosting system. In: Proceedings of the 22nd ACM SIGKDD International Conference on Knowledge Discovery and Data Mining, pp. 785–794 (2016)
5. Chiny, M., Bencharef, O., Hadi, M.Y., Chihab, Y.: A client-centric evaluation system to evaluate guest's satisfaction on AirBNB using machine learning and NLP. Appl. Comput. Intell. Soft Comput. **2021** (2021)
6. Forghani, M., et al.: Application of deep learning to large scale riverine flow velocity estimation. Stoch. Env. Res. Risk Assess. **35**(5), 1069–1088 (2021). https://doi.org/10.1007/s00477-021-01988-0
7. Hall, M.A., Smith, L.A.: Practical feature subset selection for machine learning (1998)
8. Keras: The Python Deep Learning Library. https://keras.io/
9. scikit-learn Machine Learning in Python. https://scikit-learn.org/stable/
10. Johansson, R.: An intuitive explanation of gradient boosting. http://www.cse.chalmers.se/richajo/dit865/files/gb_explainer.pdf
11. Kalehbasti, P.R., Lepech, M.D., Pandher, S.S.: Augmenting high-dimensional nonlinear optimization with conditional GANs. arXiv preprint arXiv:2103.04748 (2021)
12. Kalehbasti, P.R., Nikolenko, L., Rezaei, H.: AirBNB price prediction using machine learning and sentiment analysis. arXiv preprint arXiv:1907.12665 (2019)
13. Kalehbasti, P.R., Ushijima-Mwesigwa, H., Mandal, A., Ghosh, I.: Ising-based Louvain method: clustering large graphs with specialized hardware. arXiv preprint arXiv:2012.11391 (2020)
14. Kokasih, M.F., Paramita, A.S.: Property rental price prediction using the extreme gradient boosting algorithm. IJIIS: Int. J. Informat. Inf. Syst. **3**(2), 54–59 (2020)
15. Li, Y., Pan, Q., Yang, T., Guo, L.: Reasonable price recommendation on AirBNB using multi-scale clustering. In: 2016 35th Chinese Control Conference (CCC), pp. 7038–7041. IEEE (2016)
16. Loria, S., et al.: Textblob: simplified text processing. Secondary TextBlob: Simplified Text Processing (2014)
17. Ma, C., Liu, Z., Cao, Z., Song, W., Zhang, J., Zeng, W.: Cost-sensitive deep forest for price prediction. Pattern Recogn. **107**, 107499 (2020)
18. Ma, Y., Zhang, Z., Ihler, A., Pan, B.: Estimating warehouse rental price using machine learning techniques. Int. J. Comput. Commun. Control **13**(2) (2018)
19. Masiero, L., Nicolau, J.L., Law, R.: A demand-driven analysis of tourist accommodation price: a quantile regression of room bookings. Int. J. Hosp. Manag. **50**, 1–8 (2015)
20. Mnih, V., et al.: Playing Atari with deep reinforcement learning. arXiv preprint arXiv:1312.5602 (2013)
21. Oh, K.S., Jung, K.: GPU implementation of neural networks. Pattern Recogn. **37**(6), 1311–1314 (2004)

22. Pedregosa, F., et al.: Scikit-learn: machine learning in Python. J. Mach. Learn. Res. **12**, 2825–2830 (2011)

23. Petz, G., Karpowicz, M., Fürschuß, H., Auinger, A., Stříteský, V., Holzinger, A.: Opinion mining on the web 2.0-characteristics of user generated content and their impacts. In: Holzinger, A., Pasi, G. (eds.) HCI-KDD 2013. LNCS, vol. 7947, pp. 35–46. Springer, Heidelberg (2013). https://doi.org/10.1007/978-3-642-39146-0_4

24. Trang, L.H., Huy, T.D., Le, A.N.: Clustering helps to improve price prediction in online booking systems. Int. J. Web Inf. Syst. (2021)

25. Wang, D., Nicolau, J.L.: Price determinants of sharing economy based accommodation rental: a study of listings from 33 cities on airbnb.com. Int. J. Hospital. Manag. **62**, 120–131 (2017)

26. Wang, Y.E., Wei, G.Y., Brooks, D.: Benchmarking TPU, GPU, and CPU platforms for deep learning. arXiv preprint arXiv:1907.10701 (2019)

27. Yang, Y., Mueller, N.J., Croes, R.R.: Market accessibility and hotel prices in the Caribbean: the moderating effect of quality-signaling factors. Tour. Manag. **56**, 40–51 (2016)

28. Yu, H., Wu, J.: Real estate price prediction with regression and classification. CS229 (Machine Learning) Final Project Reports (2016)

29. Zhou, X., Tong, W.: Learning with self-attention for rental market spatial dynamics in the Atlanta metropolitan area. Earth Sci. Inf. **14**(2), 837–845 (2021)

Towards Financial Sentiment Analysis in a South African Landscape

Michelle Terblanche$^{(\boxtimes)}$ ⓘ and Vukosi Marivate$^{(\boxtimes)}$ ⓘ

Department of Computer Science, University of Pretoria, Pretoria, South Africa
vukosi.marivate@cs.up.ac.za

Abstract. Sentiment analysis as a sub-field of natural language processing has received increased attention in the past decade enabling organisations to more effectively manage their reputation through online media monitoring. Many drivers impact reputation, however, this thesis focuses only the aspect of financial performance and explores the gap with regards to financial sentiment analysis in a South African context. Results showed that pre-trained sentiment analysers are least effective for this task and that traditional lexicon-based and machine learning approaches are best suited to predict financial sentiment of news articles. The evaluated methods produced accuracies of 84%–94%. The predicted sentiments correlated quite well with share price and highlighted the potential use of sentiment as an indicator of financial performance. A main contribution of the study was updating an existing sentiment dictionary for financial sentiment analysis. Model generalisation was less acceptable due to the limited amount of training data used. Future work includes expanding the data set to improve general usability and contribute to an open-source financial sentiment analyser for South African data.

Keywords: Financial sentiment analysis · Natural language processing · Corporate reputation · Share price

1 Introduction

Big corporate organisations produce vast amounts of textual information in the form of official financial and non-financial reports, media releases and trading statements. The communication strategy, and hence perception, of an organisation directly impacts it's reputation. One of the industry accepted measures of reputation, *the RepTrak Score*[1], takes into account seven drivers of reputation: products and services, innovation, workplace, citizenship, governance, leadership and performance. The latter is a measure of the financial health of an organisation.

---

[1] https://www.reptrak.com/reputation-intelligence/what-is-it/.

© IFIP International Federation for Information Processing 2021
Published by Springer Nature Switzerland AG 2021
A. Holzinger et al. (Eds.): CD-MAKE 2021, LNCS 12844, pp. 185–202, 2021.
https://doi.org/10.1007/978-3-030-84060-0_12

In the past, historic accounting information formed the basis for financial performance prediction, evolving from statistical models to more sophisticated machine learning models [3]. Subsequent research ventured into the field of qualitative measures such as textual analysis to predict performance [13]. More recent research shows that there is promise in correlating sentiment with financial performance in order to make future predictions [3,7,13,18].

The justification for this study is rooted firstly in the evidence that there is a financial value linked to the reputation of an organisation [12,23]. An improvement in reputation can have in the order of a 6% improvement in the company bottom-line [12]. As a result, reputation risk should form a key component of overall corporate strategy [23].

A further motivation for this study stems from identifying a gap in the South African context with regards to financial sentiment analysis using natural language processing (NLP) techniques. Even though many sentiment analysers are freely available, these models were developed within a given context and relevant to a specific domain and geographical region.

Based on the identified problem and motivation, the following research questions were identified: *What NLP techniques are required to successfully determine the sentiment of financial communication in a South African context?*; *Is there a correlation between the sentiment of financial news and company performance as indicated by share price?*; *How effectively can a narrower sentiment prediction model be applied to a broader scope of finance-related information?* The study only focuses on formal communication channels in the form of online news articles, specifically excluding social media.

2   Sentiment Analysis and Opinion Mining

The terms sentiment analysis and opinion mining are often used interchangeably. The first mention of public opinion analysis dates back to post-World War II and has been one of the fastest developing areas in the last decade. It involves using natural language processing (NLP) techniques to extract and classify subjective information expressed through opinions or through detecting the intended attitude [15,21]. It has been one of the fastest developing areas in the last decade, growing from simple online product reviews to analysing the sentiment from various online platforms such as social media and extending the application to predicting stock markets, tracking polls during elections and disaster management [15]. Research highlighted the following three categories of sentiment analysis:

Open-Source Pre-trained Sentiment Analysers. The *TextBlob* library in **Python** is a simple rule-based sentiment analyser that provides the average sentiment (excluding neutral words) of a text string[2]. *VADER*[3] is another rule-based sentiment analyser specifically trained on social media texts and gener-

---

[2] https://textblob.readthedocs.io/en/dev/.
[3] Valence Aware Dictionary for sEntiment Reasoning.

alises quite well across contexts/domains compared with other sentiment analysers [5]. This analyser outperforms *TextBlob* when predicting sentiment on social media texts [9]. Some of the main reasons are that it takes into account emoticons, capitalization, slang and exclamation marks.

A Simple Dictionary-Based Approach. This method typically uses a dictionary of words/phrases either manually created or automatically generated.

Custom-Built Predictive Models Using Machine Learning. The main techniques generally used involve either 1) traditional models or 2) deep learning models [2]. These are supervised machine learning models and required data sets to be labeled. The traditional models are typically Naive Bayes, logistic regression and support vector machines.

Some of the main challenges in sentiment analysis are language dependency, domain specificity, nature of the topic, negation and the availability of labeled training data [4, 21]. A further challenge in opinion mining from user generated content is to acknowledge the importance of text pre-processing to improve the quality and usability thereof [20].

3   Financial Sentiment Analysis

3.1   Exploiting Typical Financial Headline Structure

A potential way to determine the sentiment of a financial title was explored by introducing the concept that ±30% of such titles follow a hinge structure [24]. The investigation suggested that the hinge, which is typically a word such as *as, amid, after*, splits the sentence into two parts, both parts carrying the same sentiment. If one therefore determines the sentiment of the first part of the sentence, the overall sentiment is inferred. In Fig. 1, the top sentence aims to explain this notion. However, the second sentence shows an example where the part of the sentence following the hinge does not carry the same sentiment as the first part.

Furthermore, it was argued that the verbs hold the key as sentiment carrying words. It was identified, however, that using existing, labeled word lists may still fall short since these lists were created using very domain-specific pieces of text. For e.g. a word such as rise may be listed as positive based on prior usage, however, its use in a new application may indicate it to be negative.

3.2   Existing Approaches for Financial Sentiment Analysis

Existing Popular Financial Sentiment Word Lists. For a lexicon-based approach, a very popular domain-specific (i.e. financial) dictionary is the *Loughran-McDonald sentiment word lists* first created in 2009 [13]. The drive for developing these lists stemmed from the authors showing that a more general dictionary, in this case the *H4N* negative wordlist from the *Harvard Psychological Dictionary*, misclassified the sentiment of financial words quite substantially.

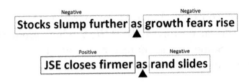

Fig. 1. Hinge concept of financial headlines.

They found that ∼75% of negative words in the aforementioned list are generally not negative in the financial domain. The sentiment categories are negative, positive, uncertainty, weak modal, strong modal, litigious and constraining.

Predictive Models Using Machine Learning. As part of the 11th workshop on Semantic Evaluation (SemEval-2017)[4], one of the tasks was *"Fine-Grained Sentiment Analysis on Financial Microblogs and News"*[5] of which a sub-task was sentiment analysis on news statements and headlines. It was a regression problem and participants had to predict the sentiment in the range −1 to 1 (representing Negative to Positive). The training data provided was annotated in this same range. Table 1 gives a summary of the results and methods for four of the submissions.

Table 1. Summary of the performance of the various annotation methods.

Ranking	Score[1]	Modelling approach
1	0.745	1D convolutional neural network (using word embeddings from GloVe) [14]
4	0.732	Bidirectional Long Short-Term Memory (with early stopping) [17] *Also looked at support vector regression
5	0.711	Ensemble using support vector regression (and gradient boosting regression) [6]
8	0.695	Support vector regression (with word embeddings and lexicon features) [8]

[1] Weighted cosine similarity score

The models used to address the sentiment analysis task range from traditional machine learning to deep learning models with only a ±5% improvement from the latter (Table 1). These results indicate that traditional machine learning models can be used quite successfully for this task to set a baseline for further evaluation. The prediction is still far better than random chance of 50%. An

4 https://alt.qcri.org/semeval2017/.
5 https://alt.qcri.org/semeval2017/task5/.

important observation was the need for domain-specific sentiment lexicons. The challenge provided teams with labeled training data. However, a large number of supervised learning activities start with unlabeled data and a substantial amount of time and effort is required to properly annotate data sets

3.3 Related Work on Financial Sentiment Analysis in the South African Context

Research produced a limited amount of South Africa-related scientific papers on sentiment analysis as a sub-field of natural language processing. Even fewer published results were available on specifically financial sentiment analysis in a South African context.

In 2018, a study on using sentiment analysis to determine alternative indices for tracking consumer confidence (as opposed to making use of surveys) showed high correlation with the traditional consumer confidence indices [19]. These indices are used to better understand current economic conditions as well as to predict future economic activity.

Another study, although not necessarily financial sentiment analysis *per se*, was on measuring the online sentiment of the major banks in South Africa [11]. The data source for this analysis was social media only. Machine learning models were used for both detecting topics and analysing the sentiment of user-generated comments relating to those topics. The main contribution the authors made was to highlight the importance of human validation as part of the process to increase accuracy and precision [11].

Based on the available research, it is deducted that a gap exists for researchers and academics to expand and improve sentiment analysis of online media through natural language processing, especially in the financial domain, in order to increase the knowledge base and pool of technical solutions in the context of South Africa.

4   Sentiment Correlation with Financial Performance

A statistical approach to understanding whether stock market prices follow a trend with the sentiment from news articles relating to the stock/company showed promising results [1]. The method was tested on ∼15 different companies. The study only considered a dictionary-based approach to calculate degrees of positivity, negativity and neutrality. The results showed at 67% correlation between sentiment and share price [1].

A second paper on predicting market trends using sentiment analysis included a broader context through more diverse data [18]. The authors evaluated a predictive model using sentiment attitudes (i.e. Positive and Negative), sentiment emotions (such as joy, anger) as well as common technical drivers of share price. Granger-causality found that only sentiment emotions could potentially be useful indicators [18].

The findings highlight the complexity of share price prediction and the fact that it is determined by a number of factors, of which sentiment could potentially add value. The authors highlighted the need to better understand which stocks are impacted by sentiment to determine to applicability of this proposed method [18].

5   Method

A model development pipeline was designed to answer the research questions and achieve the set objectives. This process flow is given in Fig. 2.

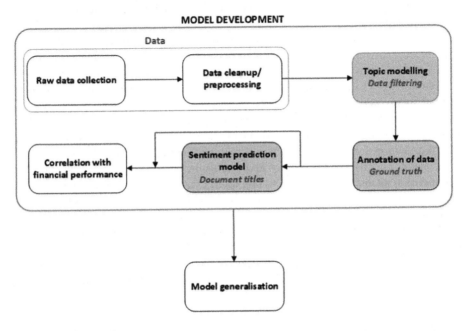

Fig. 2. Process flow for addressing the research questions and objectives.

The following sections give a brief overview of the steps in the model development pipeline.

During the first phase, **Data**, various sources of publicly available textual information was identified and collected. Relevant data from these sources were extracted, cleaned and consolidated. The data sources used for model development are listed in Sect. 6.

5.1   Topic Modelling for Data Filtering

Topic modelling was used to filter the data to specifically extract financial-related documents. Term frequency–inverse document frequency (TF-IDF) was used to determine the word vectors (based on full article content) as input to the topic model. Non-Negative Matrix Factorization (NMF) was used.

5.2   Annotation of Data

Labeling of the data set, comprising only of the financial documents (extracted using topic modelling), was done on document headlines only using four independent annotators. Due to the small size of the data set, the full set of financial documents were labeled. Label options were Positive, Negative and Neutral. The majority label was used as the ground truth sentiment for the relevant documents.

The inter-annotator agreement was calculated using the **_AnnotationTask class_**[6] from NLTK in **Python**. Fleiss' Kappa was used as the statistical measure of inter-rater reliability [10].

5.3   Sentiment Prediction Model

Three sentiment prediction methods were evaluated, compared and the most robust prediction model implemented as the final annotation model.

Existing Rule-Based Approaches: TextBlob and Vader Both _TextBlob_ and **_VADER_** were used to calculate the sentiment of document headlines to understand the usability of pre-trained sentiment analysers on text from South African finance-related articles. Both models predict sentiment as value between −1 and 1 and it was assumed that predicted values between −0.05 and 0.05 are **_Neutral_**.

A Simple Lexicon-Based Approach. For this analysis, the existing _Loughran and McDonald Sentiment Word Lists_ were used as basis [13]. These lists were developed to overcome the fact that more general dictionaries often misclassify financial texts, especially words perceived as negative in a day-to-day context. The following three iterations were performed:

1. Experiment 1
   Base dictionary as updated in 2018[7].
2. Experiment 2
   Base dictionary (Loughran and McDonald Sentiment Word Lists) with added synonyms (using NLTK's Wordnet Interface[8]). These synonyms are given the same sentiment.
3. Experiment 3
   - Base dictionary (as for experiment 2) but without the addition of synonyms for words in the "modal" lists.
   - Manual addition and deletion of words based on the evaluation of a sample of sentiment predictions from this update.

---

[6] https://www.nltk.org/_modules/nltk/metrics/agreement.html.

[7] https://sraf.nd.edu/textual-analysis/resources/.

[8] https://www.nltk.org/howto/wordnet.html.

The concept of a hinge structure (Sect. 3.1) using the words *as, but, amid, after, ahead, while* and *despite* was used. Where hinge words were not present in headings, a *"comma"* was used as a hinge or alternatively the full title was used. Individual words in article headings were lemmatized using multiple lemmas i.e. adjectives, verbs and nouns to ensure maximum chance of matching words in the developed dictionary.

Sentiment was assigned based on the first part of the title (where a hinge was present) alternatively the full sentence was used. This method assumes the sentiment is dictated by the first part, which could be slightly contradictory to the initial hinge structure proposal (Sect. 3.1).

In the case where multiple sentiment-carrying words are present, the first occurring *Positive* or *Negative* word was used as the sentiment of the headline (other sentiments were excluded in this round of the evaluation). Where no sentiment-carrying words were present, the headline was labeled as 'Not detected'. This approach does not take into account context, however, this simple bag-of-words implementation to detect word sentiments was used for the baseline model development.

Feature-Based Approach: XGBoost. A binary classifier was developed, using a traditional machine learning approach, with **Python's** implementation of **XGBoost** (Extreme Gradient Boosting)[9]. It is a boosting algorithm based on an ensemble of decision trees[10].

The following typical cleaning and pre-processing steps were performed: tokenized text into words, converted words to lower case, expanded contractions (e.g. replace can't with can not), removed English stopwords, removed punctuation and lemmatized the words using NLTK's WordNetLemmatizer. Input vectors to the model were then created for the processed document headlines using term frequency inverse document frequency (TF-IDF).

Even though more advanced machine learning models have been used for sentiment classification (Sect. 3.2), it was decided to only evaluate a more traditional machine learning model. The main reason being that the focus of the study was to develop an annotation method in order to set a baseline after which improvements can be investigated.

5.4 Correlation with Financial Performance

The predicted sentiments (from document headlines) and company financial performance (as indicated by share price), over the same time period, were analysed to observe whether patterns can be recognised. Multiple sentiments on a given date were resolved by using the majority sentiment.

Since share price prediction is a complex task and impacted by various factors, it was decided to only illustrate whether a directional correlation can be

---

[9] https://xgboost.readthedocs.io/en/latest/python/index.html.
[10] https://www.datacamp.com/community/tutorials/xgboost-in-python.

observed. For future work, a statistical correlation can be investigated and potentially include additional drivers known to impact a given stock price.

6   Data

The company identified for developing the financial sentiment prediction model is **Sasol**[11].

A variety of data sources were considered for model development and the most relevant were non-official communication in the form of online news articles and Stock Exchange News Service reports, which are company announcements that can have an affect on market movement. These are provided by the Johannesburg Stock Exchange (JSE)[12] and are publicly available[13].

The above-mentioned data was collected for the period April/May 2015 - April/May 2020. The final data set were made up of 7666 online news articles and 168 SENS reports.

7   Model Development Results

7.1   Annotation of Data

Table 2 gives the sentiment distribution for the financial data set based on the majority label from the annotators. The 'None' category was removed.

Table 2. Summary of the sentiment categories of the annotated data.

Sentiment	Count	Percentage
Positive	249	31%
Negative	419	52%
Neutral	141	17%

The financial document data set, after using topic modelling for filtering and removing 'None' labeled documents, consisted of 808 articles (only 33 i.e. 4% were SENS reports).

7.2   Sentiment Prediction Model

Rule-Based Approaches: TextBlob and Vader. The *Loughran and McDonald Sentiment Word Lists* only consider sentiment-carrying words, therefore to compare the various approaches only documents with **Positive** and **Negative** ground truth sentiments were considered (a total of 668 articles). Table 3 shows

---

[11] www.sasol.com.

[12] https://www.jse.co.za/services/market-data/market-announcements.

[13] https://www.sharedata.co.za.

the confusion matrices for using TextBlob and Vader to predict sentiment on document headlines. The overall accuracies were 19% and 51% respectively. The high inaccuracies stem from the majority of headlines being predicted as **Neutral** (i.e. in the range −0.05 to 0.05).

Table 3. Summary of the results using TextBlob and VADER on article headlines.

	TextBlob			VADER		
		Predicted			Predicted	
		Negative	Positive		Negative	Positive
Actual	Negative	68	47	Negative	243	86
	Positive	19	60	Positive	53	99

The superior performance of **VADER** as compared with **TextBlob** is consistent with a previous study on their comparison (Sect. 2) [5]. Furthermore, since **VADER** was trained on social media, the subpar performance on financial headlines is therefore not unexpected.

Lexicon-Based Approach. In Experiment 1, the original word lists (containing 4140 words) were used as is to determine a sentiment based on key words according to hinge structure approach discussed in Sect. 3.1 to observe the baseline accuracy. The method for assigning the sentiment to the headline is as outlined in Sect. 5.3.

The goal of Experiment 2 was to update the word lists with synonyms (of the words in the existing lists) and determine whether it improves prediction accuracy. As part of this experiment, a short list of bi-grams were added based on manual observation where one word was ambiguous (Table 4). The list is not exhaustive and is to indicate the impact of expanding the sentiment dictionary.

Table 4. Bi-grams added to the sentiment dictionary.

Negative	Positive
Record low	New record
Record lows	Record high
Back foot	Record highs
Price halves	Record production
	On track

Thereafter in Experiment 3, random samples were evaluated to update the dictionary from Experiment 2. It was noticed that some of the synonyms added resulted in incorrect predictions and had to be removed again. Also, the synonyms added in this experiment excluded those for the "modal" word lists. Only 4 words were removed from the original dictionary: break, closed, closing and despite. The final dictionary contains 9743 words.

It is recommended, however, that a more robust method be developed to update the dictionary in future since this manual method does not necessarily capture all the required words and may also have redundant words. Furthermore, due to the small size of the data set, manual updates could be performed but will not feasible for large data sets.

Table 5 gives the results for the 3 experiments and highlights the improvement based on the manual dictionary update. After updating the dictionary, the sentiment prediction accuracy improved by 47% compared with the original word lists.

Table 5. Summary of the results of the simple dictionary-based approaches.

Sentiment	Actual count	Experiment 1		Experiment 2		Experiment 3	
		Count	%	Count	%	Count	%
Positive	249	69	28%	123	49%	184	74%
Negative	419	180	43%	323	77%	379	90%
Overall	668	249	37%	446	67%	563	84%

The results from the various experiments highlight the need for not only domain-specific sentiment prediction tools but also region-specific corpora.

The data set is named *LM-SA-2020* representing *Loughran and McDonald Sentiment Word Lists* for *South Africa*.

A future improvement is to assess the sentiment for sentences where multiple sentiment-carrying words are present to evaluate the impact on sentiment prediction accuracy.

From the above results it appears that a simple dictionary based method to annotate the document headlines prove more accurate than pre-trained sentiment analysers.

Feature-Based Approach: XGBoost. For model training, 80% of the data set (of the 668 documents) were used. The accuracy of prediction was 81% $\pm 4.4\%$. The accuracy on the 20% unseen data was also 81%. Table 6 gives the recall and F1-score on the full data set and includes the results for the other approaches for comparison. The overall accuracy for the XGBoost model was 94% using all headlines.

Figure 3 shows the top 20 most important features of the XGBoost classifier. The words flagged are interpretable and useful.

The results given in this section lead to the conclusion that an XGBoost classifier as a simple traditional model performs very well for the given task and is recommended to be used as the sentiment prediction model.

Table 6. Summary of the performance of the various annotation methods.

	Accuracy	Recall		F1-score	
		Pos	Neg	Pos	Neg
Lexicon	84%	74%	90%	80%	89%
TextBlob	19%	24%	16%	34%	27%
Vader	51%	40%	58%	48%	68%
XGBoost	94%	86%	98%	91%	95%

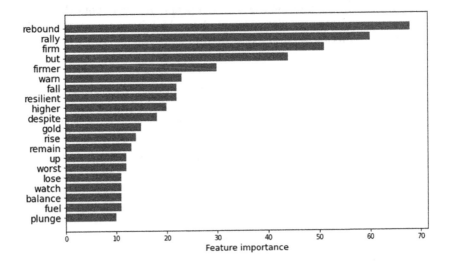

Fig. 3. TF-IDF values as a function of occurrence.

For a potential future improvement, an approach that takes into account the sequence of words in a sentence should be evaluated for e.g. a recurrent neural network (RNN) such as a Long Short Term Memory (LSTM) with attention.

7.3 Sentiment Correlation with Financial Performance

In order to observe whether there is a noticeable trend between sentiment and share price, a time frame of the most recent six months was used. Figure 4 shows this trend. Periods A and B are periods where sentiment improved and was reflected by share price. Similarly Period C stands out through a significant amount of negative sentiments and a severe drop in share price.

The above results show promise that there are indeed periods where sentiment (from financial articles/documents) and share price correlate well.

An additional factor for consideration is the impact of lag when using share price movement and it is recommended to be evaluated in future work.

It is also recommended to expand the sentiment prediction to include additional topics and observe the correlation with share price. An alternative is to

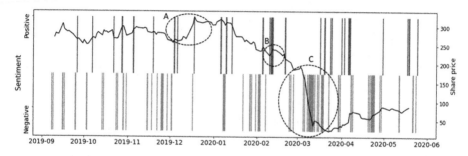

Fig. 4. Sentiment prediction vs. share price for September 2019 – May 2020 using the XGBoost binary classifier.

extract topics according to the seven key drivers that impact reputation (Sect. 1) and apply weightings to an overall reputation score. Lastly, it is recommended to explore using either more fine-grained categories or a continuous scale for sentiment.

8   Model Generalisation

In order to understand how well the models generalise, it was required to use the developed model pipeline on unseen data from a different organisation. For this phase, data for the corporate organisation, ***Anglo American***, was used. The aim was to determine whether language use in financial articles (mostly online news) follow the same pattern for different organisations. This will inform whether such models can be implemented on a larger scale or whether it is company-specific.

As per the pipeline (Sect. 5.3), the following steps to predict sentiment (on headline or the first portion of a document) as well as to understand whether sentiment correlates with financial performance, were performed:

1. Data collection and cleanup/pre-processing (a total of 1758 articles for the period June 2018 – May 2020)
2. Filtering of data for financial documents with topic modelling (a total of 151 articles)
3. Sentiment prediction using document titles:
   - Using the updated dictionary to identify sentiment-carrying keywords (Sect. 7.2)
   - Using the previously developed binary XGBoost classifier based on ***Sasol*** data (without retraining) (Sect. 7.2)
4. Graphically represent daily aggregated sentiments and share price

Table 7 gives the predicted sentiments using the dictionary-based and the XGBoost models.

Table 7. Comparison of sentiment predictions.

	Lexicon-based		XGBoost	
	Count	%	Count	%
Positive	49	32%	43	28%
Negative	74	49%	108	72%
Neutral	20	13%	–	–
Other	8	5%	–	–

Table 8. Comparison of sentiment predictions on *Anglo American* data.

Sentence	XGBoost classifier	Lexicon-based
Aveng execs get R17.7m in bonuses - Moneyweb	Negative	Negative
Sharp (partial) recovery in share prices - Moneyweb	Positive	Positive
JSE tumbles as global growth fears spread | Fin24	Negative	Negative
Anglo American replaces Deloitte with PwC as external auditor after 20 years	Negative	Litigious
Anglo says S. Africa's Eskom a major risk as it mulls Growth - Bloomberg	Negative	Negative
Another major investor leaves the pebble mine | NRDC	Negative	Positive
Mining lobbies and the modern world: new issue of Mine Magazine out now	Positive	Positive
BHP approach to Anglo CEO signals end of Mackenzie era is nearing	Negative	Positive
JSE tracks global markets higher on improved wall street data | Fin24	Positive	Positive
Rand firms as dollar, stocks fall	Negative	Positive
Anglo American delivers 3.5-billion USD profit, declares final dividend	Negative	Neutral
Best mining stocks to buy in 2020 | The Motley fool	Positive	Positive
Rand firmer as dollar falls on rate cut bets	Negative	Positive
The new ministers in charge of the Amazon	Positive	Negative
Anglo American's Cutifani not thinking of retirement as plots coup de grâce - Miningmx	Negative	Negative
Pressure persists for resources stocks | Fin24	Positive	Negative
Markets WRAP: rand closes at R14.73/$ | Fin24	Negative	Neutral
See the top performers on the JSE in 2018 so far	Positive	Litigious
Anglo American seeks to avert revolt over chief's £14.6m pay | Business News | Sky News	Negative	Negative
Rand, stocks slip as investors await big Trump speech	Negative	Negative

From Table 7 it seems that the XGBoost classifier is more biased towards negative sentiments whereas the dictionary-based approach appears more balanced. There is only a 52% agreement between the two models. Since there is no ground truth sentiment labels for the data, it was decided to manually evaluate the predicted sentiments to provide a more informed view. Table 8 is an extract of the headline sentiment predictions using a binary XGBoost classifier as well as a dictionary-based approach.

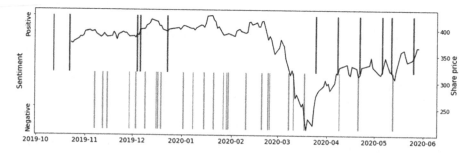

Fig. 5. Sentiment prediction using a dictionary-based approach compared with Anglo American share price movement.

For a better understanding, TF-IDF vectors were determined for the **Anglo American** document headlines and the vocabulary compared with the pre-trained vocabulary. Only 29% of the words in this data set exists in the pre-trained vocabulary.

From the manual inspection it is concluded that the dictionary-based approach predict sentiments more accurately than the XGBoost classifier that was developed using **Sasol** data.

Figure 5 shows the sentiment prediction and share price using a simple dictionary-based approach to identify sentiment-carrying words. It can be seen that there is an upward movement in share price corresponding to more positive sentiments (post April 2020).

Based on the above findings it is surmised that a XGBoost classifier trained on company-specific document titles may be too specific to extend to other industries. However, this can be improved by increasing the size of the data set to improve generalisation.

9   Conclusions and Future Work

Based on the findings it is concluded that natural language processing techniques can be used to predict the sentiment of financial articles in the South African context. Existing off-the-shelf sentiment analysers were evaluated and were found to underperform in predicting sentiment of South African finance-related articles with accuracies just above 50%. Custom models using a simple lexicon-based approach or traditional machine learning such as a binary XGBoost classifier are well suited to the task and produced accuracies of 84% and 94% respectively. These models use document titles only.

Furthermore, an analysis showed there is a good correlation between predicted sentiments and financial performance (as represented by share price). The approach therefore shows promise, and with refinement, can be used to identify at risk periods for an organisation.

Lastly, the sentiment prediction model was evaluated using data from a different company to test how well it generalises. Since there were no ground truth data labels for this, a manual evaluation on a sample of the results was done. The dictionary-based approach the XGBoost classifier were compared and it was concluded that the former was better suited in this case. Sentiment predictions can be improved by increasing the size of the data set used in model development. Despite these shortcomings, a correlation between predicted sentiment and share price was still observed for certain periods. This substantiates the fact that the method has promise.

The main contributions made by this study are as follows: Developed an updated sentiment dictionary suitable for financial articles (the **LM-SA-2020** data set and the accompanying data statement are publicly available) [22].; Setting the foundation for expanding the work to include a broader sentiment prediction model that takes into account various topics and their contribution to overall sentiment as an indication of company reputation.; Progress towards an open-source library for financial sentiment analysis developed on South African data.

The following are some of the main recommendations for future work: A more sophisticated, streamlined process to update/expand the new data set - **LM-SA-2020**.; Improve the model generalisation capability by increasing the size of the data set.; Investigate the impact of share price movement lag on the correlation with sentiment and enhance the understanding on whether there is a causal relationship between sentiment and financial performance.; Expand the sentiment prediction model to include additional topics (over and above financial documents).; Publish an open-source financial sentiment analysis tool that can be used on South African data.; Evaluate the performance of deep learning models - a very recent study indicated that transformers outperformed other sentiment analysis approaches and models in the domain of finance. It therefore warrants further investigation for possible application to this paper [16].

References

1. Chowdhury, S.G., Routh, S., Chakrabarti, S.: News analytics and sentiment analysis to predict stock price trends. Int. J. Comput. Sci. Inf. Technol. **5**(3), 3595–3604 (2014)
2. Dang, N.C., Moreno-García, M.N., De la Prieta, F.: Sentiment analysis based on deep learning: a comparative study. Electronics (Switzerland) **9**(3) (2020). https://doi.org/10.3390/electronics9030483
3. Hajek, P., Olej, V., Myskova, R.: Forecasting corporate financial performance using sentiment in annual reports for stakeholders' decision-making. Technol. Econ. Dev. Econ. **20**(4), 721–738 (2014). https://doi.org/10.3846/20294913.2014.979456
4. Hussein, D.M.E.D.M.: A survey on sentiment analysis challenges. J. King Saud Univ. Eng. Sci. **30**(4), 330–338 (2018). https://doi.org/10.1016/j.jksues.2016.04.002
5. Hutto, C., Gilbert, E.: VADER: a parsimonious rule-based model for sentiment analysis of social media text. In: Proceedings of the 8th International Conference on Weblogs and Social Media, ICWSM 2014, pp. 216–225 (2014)

6. Jiang, M., Lan, M., Wu, Y.: ECNU at SemEval-2017 task 5: an ensemble of regression algorithms with effective features for fine-grained sentiment analysis in financial domain. In: Proceedings of the 11th International Workshop on Semantic Evaluations (SemEval-2017), pp. 888–893 (2017). https://doi.org/10.18653/v1/s17-2152

7. Joshi, K., Bharathi , H.N., Rao, J.: Stock trend prediction using news sentiment analysis. Int. J. Comput. Sci. Inf. Technol. 8(3), 67–76 (2016). https://doi.org/10.5121/ijcsit.2016.8306

8. Kumar, A., Sethi, A., Akhtar, S., Ekbal, A., Biemann, C., Bhattacharyya, P.: IITP-Bat SemEval-2017 task 5: sentiment prediction in financial text. In: Proceedings of the 11th International Workshop on Semantic Evaluations (SemEval-2017), pp. 894–898 (2017). http://nlp.stanford.edu/projects/

9. Kumaresh, N., Bonta, V., Janardhan, N.: A Comprehensive study on lexicon based approaches for sentiment analysis. Asian J. Comput. Sci. Technol. 8(S2), 1–6 (2019). www.rottentomatoes

10. Landis, R.J., Koch, G.G.: The measurement of observer agreement for categorical data. In: Biometrics, pp. 159–174 (1977)

11. Lappeman, J., Clark, R., Evans, J., Sierra-Rubia, L., Gordon, P.: Studying social media sentiment using human validated analysis. MethodsX 7, 100867 (2020). https://doi.org/10.1016/j.mex.2020.100867

12. Lei, Q.: Financial value of reputation: evidence from the ebay auctions of gmail invitations. J. Ind. Econ. 59(3), 422–456 (2011)

13. Loughran, T., McDonald, B.: When is a liability not a liability? Textual analysis, dictionaries, and 10-Ks. J. Financ. 66(1), 35–65 (2011)

14. Mansar, Y., Gatti, L., Ferradans, S., Guerini, M., Staiano, J.: Fortia-FBK at SemEval-2017 task 5: bullish or bearish? Inferring sentiment towards brands from financial news headlines. In: Proceedings of the 11th International Workshop on Semantic Evaluations (SemEval-2017), pp. 817–822 (2017). https://doi.org/10.18653/v1/s17-2138

15. Mäntylä, M.V., Graziotin, D., Kuutila, M.: The evolution of sentiment analysis–a review of research topics, venues, and top cited papers. Comput. Sci. Rev. 27(February), 16–32 (2018). https://doi.org/10.1016/j.cosrev.2017.10.002

16. Mishev, K., Gjorgjevikj, A., Vodenska, I., Chitkushev, L.T., Trajanov, D.: Evaluation of sentiment analysis in finance: from lexicons to transformers. IEEE Access 8, 131662–131682 (2020)

17. Moore, A., Rayson, P.: Lancaster A at SemEval-2017 task 5: evaluation metrics matter: predicting sentiment from financial news headlines. In: Proceedings of the 11th International Workshop on Semantic Evaluations (SemEval-2017), pp. 581–585 (2017). https://doi.org/10.18653/v1/s17-2095

18. Mudinas, A., Zhang, D., Levene, M.: Market trend prediction using sentiment analysis: lessons learned and paths forward (2019). arXiv:1903.05440

19. Odendaal, H., Johannes, N., Reid, M.: Media based sentiment indices as an alternative measure of consumer confidence (2018). A Working paper of the Department of Economics and the Bureau for Economic Research at the University of Stellenbosch. https://towardsdatascience.com/a-new-way-to-sentiment-tag-financial-news-9ac7681836a7. Accessed 13 Mar 2020

20. Petz, G., Karpowicz, M., Fürschuß, H., Auinger, A., Stříteský, V., Holzinger, A.: Reprint of: computational approaches for mining user's opinions on the web 2.0. Inf. Process. Manag. 51(4), 510–519 (2015)

21. Saberi, B., Saad, S.: Sentiment analysis or opinion mining: a review. Int. J. Adv. Sci. Eng. Inf. Technol. **7**(5), 1660–1666 (2017). https://doi.org/10.18517/ijaseit.7. 5.2137
22. Terblanche, M., Marivate, V.: LM-SA-2020 sentiment word list, April 2021. https://doi.org/10.25403/UPresearchdata.14401178
23. Vig, S., Dumicic, K., Klopotan, I.: The impact of reputation on corporate financial performance: median regression approach. Bus. Syst. Res. **8**(2), 40–58 (2017). https://doi.org/10.1515/bsrj-2017-0015
24. Zimmerman, V.: A new way to sentiment-tag financial news (2019). https://towardsdatascience.com/a-new-way-to-sentiment-tag-financial-news-9ac7681836a7. Accessed 13 Feb 2020

Weighted Utility: A Utility Metric Based on the Case-Wise Raters' Perceptions

Andrea Campagner[✉], Enrico Conte, and Federico Cabitza

Universitá degli Studi di Milano-Bicocca, Milan, Italy
`a.campagner@campus.unimib.it`

Abstract. In this article we discuss a novel utility metrics for the evaluation of AI-based decision support systems, which is based on the users' perceptions of the relevance of, and risks associated with, the validation cases. We discuss the relationship between the proposed metric and other previous proposals in the specialist literature; in particular, we show that our metric generalizes the well-known *Net Benefit*. More in general, we make the point for having utility as the prime dimension to optimize machine learning models in critical domains, like the medical one, and to evaluate their potential impact on real-world practices.

Keywords: Utility · Validation · Medical machine learning · Decision support

1 Introduction

Interest in medical AI has grown markedly in the last few years, with a growing number of studies showing how Machine Learning models can achieve performance on par with our clinicians [11,13] in some diagnostic tasks. However, most of these studies were performed in controlled settings, while still few studies have shown significant effects in real practice. In this context, the use of reliable metrics is of paramount importance, as these could be used by both vendors and certification bodies to attest the *validity* of the performance of applications based on ML [3].

Traditional error-based metrics, such as accuracy or AUC, are affected by different types of bias [5,6], mainly due to their susceptibility to label imbalance [16], and are thus not adequate for the above mentioned purpose. Although utility-based [1,10,17] or balanced error-based metrics [2,4,5] address some of these biases, also these metrics ignore other significant and contextual aspects [12]. In this article, we attempt to address these shortcomings by proposing a new utility-based metric, which we call *weighted utility*. This metric generalizes existing efforts by taking into account variations in the impact and relevance of the individual cases on which the ML-based system is trained and evaluated. Finally, we will illustrate the application of our metric in a real-life user study in the field of diagnostic radiology.

© IFIP International Federation for Information Processing 2021
Published by Springer Nature Switzerland AG 2021
A. Holzinger et al. (Eds.): CD-MAKE 2021, LNCS 12844, pp. 203–210, 2021.
https://doi.org/10.1007/978-3-030-84060-0_13

2   Methods

2.1   Weighted Utility Metrics

In this Section, we describe the proposed utility metric and we derive its relationship with other existing utility metrics.

Let $S = \langle (x_1, y_1), ..., (x_m, y_m) \rangle$ be a dataset where $x_i \in X$ is an instance and $y_i \in \{0, 1\}$ is the associated target label (thus, we consider only binary classification problems): generally, we associate *normality* with class 0, and *abnormality* (that is *presence of disease* or *treatment required*) with class 1.

We assume that the evaluated ML model h provides, for each x_i a probabilistic score; in particular, with $h(x_i)$ we denote the probability score that h assigns, for instance x_i, to the positive class (that is $h(x_i) = P(y_i = 1 | x_i, h)$).

Let $\mathbf{r} : X \mapsto [0, 1]$ be a *relevance* function: this function defines, for each instance x_i, "how important it is that the model h correctly classifies x_i". We note that *relevance* could represent multiple properties of instance x_i, for example its complexity or its rarity: we will discuss this aspect further in Sect. 3.

Let $\tau : X \mapsto [0, 1]$ be a probability threshold, which, for instance x, defines the threshold $\tau(x)$ at which one should be maximally undecided between assigning any of the 2 target labels to x. Thus, a probability score $h(x)$ such that $h(x) \geq \tau(x)$ should be interpreted as evidence towards the positive class, while the opposite case (i.e. $h(x) < \tau(x)$) as evidence towards the negative class.

Then the *weighted utility* metrics for dataset S and model h is defined as:

$$wU(\tau, \mathbf{r}, S, h) = \frac{1}{\mathbf{r}(Pos)} \sum_{x_i : y_i = 1} r(x_i) \cdot \mathbf{1}_{h(x_i) \geq \tau(x_i)} \qquad (1)$$

$$- \frac{1}{\mathbf{r}(Pos)} \sum_{x_i : y_i = 0} r(x_i) \cdot \frac{\tau(x_i)}{1 - \tau(x_i)} \mathbf{1}_{h(x_i) \geq \tau(x_i)}. \qquad (2)$$

In what follows, we give an informal explanation of the above expression. We propose to see utility as the difference between the weighted true positive rate and the weighted false positive rate. That is, intuitively a decision support is useful if the number of times it is right in detecting a problem is higher than the number of times it is wrong so. The value of our proposal lies in the concept of *weight*: true positive cases are weighted for their (case-wise) relevance ($r(x_i)$), e.g., complexity and difficulty to detect, as this aspect is perceived by the ground-truth raters. The same logic applies also to the 'false positive' part of the equation; however, to that respect we also consider the risk (i.e., impact, negative importance) associated with giving a wrong advice for positivity (that is in regard to actually negative cases), like e.g., over-diagnosis and over-treatment (τ). The wU metric allows to make all these considerations at the level of single instances: this obviously encompasses the more general case, when the same weights (relevance and positivity risk) are constantly assigned to all of the instances.

Next, we show that our the wU metric represents a natural generalization of the (standardized) Net Benefit. This is defined as [17]:

$$NB(\tau) = TPR_\tau * \pi - (1 - \pi) * \frac{\tau}{1 - \tau} FPR_\tau \qquad sNB(\tau) = \frac{NB(\tau)}{\pi} \qquad (3)$$

where $NB(\tau)$ is the Net Benefit and π is the proportion of positive cases in S. In the following derivations, we assume that in the definition of wU we have $\gamma = 1$.

Theorem 1. *Let, for each x, $\tau(x) = \tilde{\tau}$ (where $\tilde{\tau}$ is a constant) and $\mathbf{r}_1(x) = 1$. Then $wU(\tilde{\tau}, \mathbf{r}_1) = sNB(\tilde{\tau}) = \frac{NB(\tau)}{\pi}$*

Proof. Under the assumptions in the statement it holds that:

$$wU(\tilde{\tau}, \mathbf{r}_1) = \frac{|\{x_i \in S : y_i = 1 \wedge s(x) \geq \tilde{\tau}\}|}{|\{x_i \in S : y_i = 1\}|} - \frac{\frac{\tilde{\tau}}{1 - \tilde{\tau}} |\{x_i \in S : y_i = 0 \wedge s(x) \geq \tilde{\tau}\}|}{|\{x_i \in S : y_i = 1\}|} \qquad (4)$$

$$= TPR_{\tilde{\tau}} - \frac{1}{\pi} \frac{\tilde{\tau}}{1 - \tilde{\tau}} FP_{\tilde{\tau}} = TPR_{\tilde{\tau}} - \frac{1 - \pi}{\pi} \frac{\tilde{\tau}}{1 - \tilde{\tau}} FPR_{\tilde{\tau}} = sNB(\tilde{\tau}) \qquad (5)$$

2.2 Experimental Evaluation

In this Section, we report on a user-based study that we conducted in order to evaluate the viability of the proposed metrics. To this purpose, we involved 13 board-certified radiologists from several Italian hospitals, asking them to annotate a sample of 417 cases randomly extracted from the MRNet dataset[1]. This dataset encompasses 1,370 knee MRI exams performed at the Stanford University Medical Center (with 81% abnormal exams, and in particular 319 Anterior Cruciate Ligament (ACL) tears and 508 meniscal tears).

In the study we used an online questionnaire platform (LimeSurvey, version 3.18[2]) and invited the participants by personal email. As anticipated above, we involved 13 radiologists in a diagnostic task where they were called to discriminate the MRNet cases that were positive, and indicate whether these regarded either ACL or meniscal tears: in particular they had to say whether the presented imaging presented a case of ACL tear (yes/no), and/or a meniscal tear (yes/no). The radiologists were also requested to assess each case in terms of complexity on a 5-level ordinal scale, and the confidence with which they classified the case, on a 6-level ordinal scale. These subjective ratings were then used to define the case-wise *relevance function* $\mathbf{r}$ and the case-wise probability threshold τ.

In order to illustrate the application of the wU metric, we developed a Deep Learning classification model, trained to perform a binary classification task: more precisely, we trained an InceptionV3 Convolutional Neural Network model

---

[1] https://stanfordmlgroup.github.io/competitions/mrnet/.
[2] https://www.limesurvey.org/.

to discriminate between abnormal cases (that is, cases affected by either a meniscal or ACL tear) and normal cases. The training set encompassed a subset of the MRI exams taken from the MRNet dataset that were not given to the radiologists. To this purpose, we randomly selected a subset of 600 individual exams so to obtain a class balanced training set. Each of the images in the training was composed of a variable number of images, depending on the number of slices in the MRI examination. The ML model was then evaluated on the set of 417 images that were given to the radiologists, which was completely disjoint from the training set (to avoid overfitting). Model evaluation was performed using different metrics, namely accuracy, balanced accuracy, AUROC, (standardized) net benefit (at different threshold values), and the wU.

As regards the *relevance* function, we simply used the average reported complexity rating, for each case, maximum normalized so to obtain numbers in $[0, 1]$. As regards the case-wise τ values, we considered three different definitions (we will discuss the semantics behind these three definitions in Sect. 3):

$$- \ \tau_{confidence}(x_i) = \frac{1}{n^\circ \ raters} \left(\sum_{r \ \mathrm{rater}: r(x_i)=1} \frac{c_r(x_i)+1}{2} + \sum_{r \ \mathrm{rater}: r(x_i)=0} \frac{1-c_r(x_i)}{2} \right);$$

$$- \ \tau_{persuasion}(x_i) = \frac{1}{n^\circ \ raters} \left(\sum_{r \ \mathrm{rater}: r(x_i)=0} \frac{c_r(x_i)+1}{2} + \sum_{r \ \mathrm{rater}: r(x_i)=1} \frac{1-c_r(x_i)}{2} \right);$$

$$- \ \tau_{auto-bias}(x_i) = \begin{cases} \frac{d(x_i)}{2} & |\{r \ \mathrm{rater} : r(x_i) = 1\}| \geq |\{r \ \mathrm{rater} : r(x_i) = 0\}| \\ \frac{2-d(x_i)}{2} & otherwise \end{cases}$$

where $r(x_i) \in \{0, 1\}$ is the label annotation reported by rater r for case x_i, $c_r(x_i) \in [0, 1]$ is the (normalized) confidence reported by rater r for their annotation of case x_i, and $d(x_i)$ is the *disagreement rate*. In short, $\tau_{confidence}$, for an instance to be classified as positive, requires the model's probability score to be at least as high as the average of the probabilities expressed by the raters; $\tau_{persuasion}$ requires the model's probability score to be higher than the probability that the raters assigned to the negative class; while for $\tau_{auto-bias}$ the required probability score is defined based on the disagreement among the raters.

3   Results and Discussion

The performance of the raters and of the AI model, in the ROC space, is reported in Fig. 1. The average perceived case complexity was 0.70 (95% C.I [0.69, 0.71], IQR [0.63, 0.77]), the average $\tau_{confidence}$ was 0.72 (95% C.I. [0.71, 0.74], IQR [0.61, 0.87]), the average $\tau_{persuasion}$ was 0.55 (95% C.I. [0.52, 0.58], IQR [0.17, 0.85]), and the average $\tau_{auto-bias}$ was 0.54 (95% C.I [0.51, 0.58], IQR [0.14, 0.86]). The performance of the AI model, in terms of wU (with three different settings of the τ function), and other metrics, is reported in Fig. 2.

Commenting the results, the first observation regards the large differences observed among the different values of the proposed wU metrics, computed according to the three definitions of τ reported in Sect. 2: indeed, we can see that $wU(\tau_{confidence})$ was largely smaller than both $wU(\tau_{auto-bias})$ and $wU(\tau_{persuasion})$, while these latter two were more similar. These numerical differences reflect different semantics underlying the three definitions of τ:

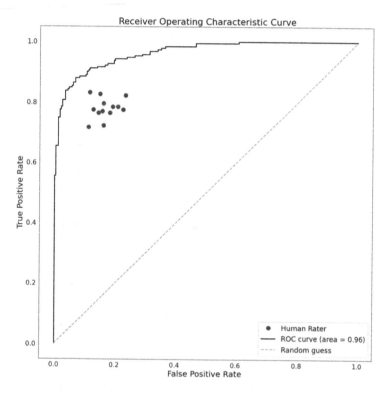

Fig. 1. Performance of the raters and of the AI model, in the ROC space

- $\tau_{confidence}$ acts as a threshold for the probability score of the model: for the classification provided by the ML model to be considered useful, the model's confidence should be at least as high the rater's one.
- On the other hand, the definition of $\tau_{persuasion}$ reflects the fact that to *persuade* a human rater in changing its opinion, the ML model should be very confident in the advice it provides;
- Finally, $\tau_{auto-bias}$ is more directly related to *risk* and to the notion of *automation bias* [8].

Obviously, the proposed approaches to convert the qualitative perceptions of the readers into probability thresholds are only one of the possible approaches to define the τ function in the formulation of the wU. Further research should be devoted at comparing alternative approaches, compared with the ordinal scales adopted in this paper.

As a second observation, we compare the (standardized) Net Benefit and the wU. In Sect. 2, we proved that wU provides a generalization of the Net Benefit by allowing the probability threshold τ to vary with the individual cases, and by attaching a degree of *relevance* to each individual case. The first factor allows to evaluate the costs and benefits of treatment vs non-treatment on an individual, case-wise basis. This provides the wU with an increased level of flexibility, as it

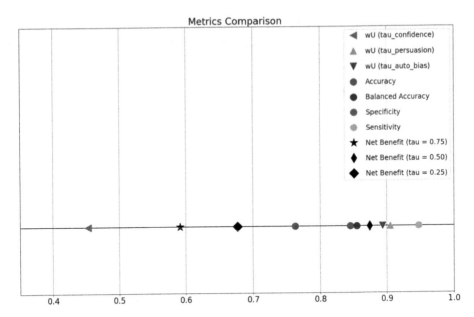

Fig. 2. The performance of the AI model, in terms of the three different versions of wU, and a collection of other pertinent metrics.

allows to differentiate between two cases that, although identical in terms of condition (e.g. same disease and/or stadiation), still differ with respect to the risks of undergoing treatment. The second term, on the other hand, allows to capture case-wise differences in the perceived importance of correctly identifying a case with respect to others. In this paper we focused on *complexity* as a dimension to define relevance; however, other dimensions could be of interest as well: some examples include *rarity*, *severity*, *impact* (if the condition gets undetected), or any combination thereof [15]. In any case, in regard to the mathematical formulation of wU, the relevance factor is an agnostic factor that we introduced with the simple aim of capturing the central notion of *relative importance*.

As a consequence of this increased flexibility, we can easily notice in our exemplificatory study that in no case (i.e. for no risk threshold in the definition of the standardized Net Benefit, and for no *definition* of the τ in the wU) the wU and the Net Benefit were exactly the same. However, the Net Benefit at $\tau = 0.5$, and the wU based on either $\tau_{persuasion}$ or $\tau_{auto-bias}$ were quite similar (i.e. 0.85 vs 0.90 and 0.89, respectively): this can be explained by noting that both $\tau_{persuasion}$ and $\tau_{auto-bias}$ had an average value close to 0.5 (albeit with a relatively large IQR), and the case complexity was stable across the dataset.

Similar comments can be made with respect to the comparison between wU and the considered error rate-based metrics: since the distribution of relevance was significantly skewed towards the positive class (0.75 ± 0.01 vs 0.66 ± 0.01), both $\tau_{persuasion}$ and $\tau_{auto-bias}$ were slightly skewed toward Sensitivity rather than Specificity. In this latter respect, we believe that the relationship between the wU and other error-rate based metrics, including the Matthews Correlation Coefficient [6] and the deep ROC analysis approach [4], should be further investigated, together with the theoretical properties of wU as an evaluation metrics, e.g. with respect to the mathematical framework proposed in [14].

The investigation of such empirical and theoretical properties, in turn, can be important for the development of regulatory standards and tools to evaluate and validate ML model for use in the real world. Indeed, despite the abundance of metrics, reaching consensus on what measure should be used, even in a specific application domain like medicine, has been so far an oft-neglected objective, even within recent recommendations developed for the reporting of prediction models, like TRIPOD [7] or MINIMAR [9]. With this respect, further research should also be devoted at establishing appropriate threshold values for claiming *validity* (so-called *minimum acceptable accuracy* [15]).

Finally, let us consider other possible uses of our proposed metric. Performance metrics can be used for either model training or model selection (equivalently, hyper-parameter optimization). In regard to the latter use case, one could envision the application of utility-based metrics as target metrics for hyper-parameter optimization, so as to represent more clearly the costs and benefits involved in the application of the ML model to be trained: with this respect, the use of wU would allow to more naturally capture the characteristics of the considered cases and the perceptions of the involved actors. By contrast, in the case of model training, it is noteworthy that neither error rate-based nor utility-based metrics are typically used as an optimization target. In their place, so-called *surrogate* metrics are typically used. Therefore, we believe that further research should be aimed at the development of appropriate surrogates for wU.

4   Conclusions

In this paper, we introduced a novel utility metrics, called *weighted Utility* (wU) and discussed its relationships with other existing metrics. The potentiality of the metric was demonstrated proving that it generalizes state-of-the-art metrics like the *Net Benefit* and *Standardized Net Benefit*. The wU metrics allows the description of the same information provided by the above metrics, *but* it is also informed by additional information of the whole clinical process, including information about the individual cases and the perceptions of raters involved in the annotation and decision making process, when compared with other existing metrics. We believe this makes wU measures more indicative of the real usefulness of a classification model when it comes to considering the skills and expectations of the intended users and the kind of decisions these are called to make. Further research is needed to validate this claim.

References

1. Baker, S.G., Cook, N.R., Vickers, A., et al.: Using relative utility curves to evaluate risk prediction. J. R. Statist. Soc. **172**(4), 729–748 (2009)

2. Brodersen, K.H., Ong, C.S., Stephan, K.E., et al.: The balanced accuracy and its posterior distribution. In: Proceedings of ICPR 2010, pp. 3121–3124. IEEE (2010)
3. Cabitza, F., Zeitoun, J.D.: The proof of the pudding: in praise of a culture of real-world validation for medical artificial intelligence. Ann. Transl. Med. **7**(8), 1–9 (2019)
4. Carrington, A.M., Manuel, D.G., Fieguth, P.W., et al.: Deep ROC analysis and AUC as balanced average accuracy to improve model selection, understanding and interpretation (2021). arXiv preprint: arXiv:2103.11357
5. Chicco, D., Jurman, G.: The advantages of the Matthews correlation coefficient (MCC) over f1 score and accuracy in binary classification evaluation. BMC Genomics **21**(1), 1–13 (2020)
6. Chicco, D., Tötsch, N., Jurman, G.: The matthews correlation coefficient (MCC) is more reliable than balanced accuracy, bookmaker informedness, and markedness in two-class confusion matrix evaluation. BioData Mining **14**(1), 1–22 (2021)
7. Collins, G.S., Reitsma, J.B., Altman, D.G., et al.: Transparent reporting of a multi-variable prediction model for individual prognosis or diagnosis (tripod): the tripod statement. BMC Med. **13**(1), 1 (2015)
8. Goddard, K., Roudsari, A., Wyatt, J.C.: Automation bias: a systematic review of frequency, effect mediators, and mitigators. J. Am. Med. Inf. Assoc. **19**(1), 121–127 (2012)
9. Hernandez-Boussard, T., Bozkurt, S., Ioannidis, J.P., et al.: Minimar (minimum information for medical AI reporting): developing reporting standards for artificial intelligence in health care. J. Am. Med. Inf. Assoc. **27**(12), 2011–2015 (2020)
10. Kerr, K.F., Brown, M.D., Zhu, K., et al.: Assessing the clinical impact of risk prediction models with decision curves: guidance for correct interpretation and appropriate use. J. Clin. Oncol. **34**(21), 2534 (2016)
11. Liu, X., Faes, L., Kale, A., et al.: A comparison of deep learning performance against health care professionals in detecting diseases from medical imaging: a systematic review and meta-analysis. The Lancet Digital Health (2019)
12. Oakden-Rayner, L., Dunnmon, J., Carneiro, G., et al.: Hidden stratification causes clinically meaningful failures in machine learning for medical imaging. Proc. ACM CHIL **2020**, 151–159 (2020)
13. Shen, J., Zhang, C.J., Jiang, B., et al.: Artificial intelligence versus clinicians in disease diagnosis: systematic review. JMIR Med. Inf. **7**(3), e10010 (2019)
14. Sokolova, M., Lapalme, G.: A systematic analysis of performance measures for classification tasks. Inf. Process. Manage. **45**(4), 427–437 (2009)
15. Sternini, F., Ravizza, A., Cabitza, F.: How accurate do you want it? Defining minimum required accuracy for medical artificial intelligence. In: Proceedings of E-Health 2020. IADIS (2020)
16. Valverde-Albacete, F.J., Peláez-Moreno, C.: 100% classification accuracy considered harmful: the normalized information transfer factor explains the accuracy paradox. PLOS ONE **9**(1), 1–10 (2014)
17. Vickers, A.J., Van Calster, B., Steyerberg, E.W.: Net benefit approaches to the evaluation of prediction models, molecular markers, and diagnostic tests. BMJ **352**, 1–6 (2016)

Deep Convolutional Neural Network (CNN) Design for Pathology Detection of COVID-19 in Chest X-Ray Images

Narayana Darapaneni[1], Anindya Sil[2], Balaji Kagiti[2], S. Krishna Kumar[2],
N. B. Ramanathan[2], S. B. VasanthaKumara[2], Anwesh Reddy Paduri[2(✉)],
and Abdul Manuf[2]

[1] Northwestern University/Great Learning, Evanston, USA
[2] Great Learning, Bangalore, India
anwesh@greatlearning.in

Abstract. The coronavirus disease 2019 (COVID-19) caused by a novel coronavirus, turned into a pandemic and raised a serious concern to the global healthcare system. The reverse transcription polymerase chain reaction (RT-PCR) is the most widely used diagnostic tool to detect COVID-19. However, this test is time consuming and subject to availability of the test kits during a crisis. An automated method of screening chest x-ray images using convolutional neural network (CNN) Transfer Learning approach has been proposed as a relatively fast and cost-effective, decision support tool to detect pulmonary pathology due to COVID-19. In this study we have used Kaggle dataset with chest x-ray images of normal and pneumonia cases. We have added COVID-19 x-ray images from 5 different open-source datasets. The images were pre-processed based on the position of radiography images and greyscale was applied and subsequently the images were used for training. After consolidation, COVID-19 images comprised only 5% of the dataset. To address the class imbalance, we have used dynamic image augmentation technique to reduce the bias. We have then explored custom CNN and VGG-16, InceptionNet-V3, MobileNet-V2, ResNet-50, and DarkNet-53 transfer learning approaches to classify COVID-19, other pneumonia and normal x-ray images and compared their performances. So far, we have achieved the best score of F1 score 0.95, sensitivity 95% and specificity 95% for COVID-19 class with Darknet-53 feature extractor. Darknet-53 classifier is part of the state-of-the-art object detection algorithm named Yolo-v3. We have also done a McNemar-Bowker post-hoc test to compare Darknet-53 performance with the next best ResNet-50. This test suggests that Darknet-53 is significantly better skilled than ResNet-50 in differentiating COVID-19 from other pneumonia in chest x-ray images.

Keywords: COVID-19 · CNN · Transfer learning · VGG-16 · ResNet-50 · InceptionNet-V3 · MobileNet-V2 · DarkNet-53 · McNemar-Bowker test

© IFIP International Federation for Information Processing 2021
Published by Springer Nature Switzerland AG 2021
A. Holzinger et al. (Eds.): CD-MAKE 2021, LNCS 12844, pp. 211–223, 2021.
https://doi.org/10.1007/978-3-030-84060-0_14

1 Introduction

More than 1 million adults are hospitalized with pneumonia and around 50,000 die from the disease every year in the US alone. Chest X-rays are currently the best available method for diagnosing pneumonia, playing a crucial role in clinical care [8] and epidemiological studies [7]. However, detecting pneumonia in chest X-rays is a challenging task that relies on the availability of expert radiologists. X-rays produced worldwide are analyzed visually on scan-by-scan basis. It requires a relatively high degree of accuracy. It is time-consuming, expensive and is prone to manual bias or wrong interpretation. Errors and delays in these diagnostic methods still contribute to a large number of patient deaths in hospitals, making these errors one of the largest causes of death along with heart disease and cancer. Detecting pneumonia in chest radiography [21] can be difficult for radiologists. The appearance of pneumonia in X-ray images is often vague, can overlap with other diagnoses [28], and can mimic many other benign abnormalities. These discrepancies cause considerable variability among radiologists in the diagnosis of pneumonia. Deep learning is a machine learning technique that teaches computers to do what comes naturally to humans: learn by example. In deep learning, a computer model learns to perform classification tasks directly from images, text, or sound. Deep learning models can achieve state-of-the-art accuracy, sometimes exceeding human-level performance [26]. Models are trained by using a large set of labeled data and neural network architectures that contain many layers. Convolutional Neural Network (ConvNet/CNN) [3, 27] is a Deep Learning Algorithm which could absorb an entire image, and assign importance (learnable weights and biases) to numerous features inside the image and have the ability to distinguish one from the other. CNN is capable of correctly catching the Spatial and Temporal dependencies in an image through the use of learnable filters. Due to the reduced number of parameters involved and the reusability of weights, this architecture provides superior fitting to the image dataset. In other words, the network may be trained to better recognise the image's sophistication. When compared to other classification algorithms, the amount of pre-processing required by a CNN is significantly less. Prior to the CNN the traditional methods for image classification had to do a lot of hand engineered feature engineering and also based on the problem, the feature engineering should be varied and it proved to be a time consuming and expensive approach and also heavily dependent on the expert's domain knowledge. While filters are hand-engineered in basic approaches, CNN can learn these filters/characteristics with adequate training. The convolutional layers in this case act as feature extractors and then the pooling layer reduces the dimensions [9]. The main advantage of the CNN especially while processing the images are the reduced need for the feature Engineering.

COVID-19 is caused by a new type of coronavirus. The symptoms of the infection include fever, cough, shortness of breath, and diarrhea. In more severe cases, COVID-19 can cause pneumonia and even death. The COVID-19 pandemic continues to have a devastating effect on the health and well-being of the global population. A critical step in the fight against COVID-19 is effective screening of infected patients. The main screening method used for detecting COVID-19 cases is polymerase chain [1, 2] reaction (PCR) testing, which can detect SARSCoV-2 RNA from respiratory specimens. While PCR testing is the gold standard as it is highly sensitive, it is a very time-consuming, laborious, and complicated manual process that is in short supply. This is also a very risky procedure since the health care fraternity could come in direct contact with infected people and get infected themselves. It was found in early studies that patients present abnormalities in chest radiography images that are characteristic of those infected with COVID-19. Motivated by this, a number of artificial intelligence (AI) systems based on deep learning have been proposed and results have been shown to be quite promising in terms of accuracy in detecting patients infected with COVID-19 using chest radiography images [3]. These developed AI systems have been closed source and unavailable to the research community for deeper understanding and extension, and unavailable for public access and use. AI based diagnostic systems that can aid radiologists to more rapidly and accurately interpret radiography images to detect COVID-19 cases is highly desired [12]. This method is also cost effective and contactless. Hence reduces the risk of infection of health care fraternity.

1.1 Initial Goals

- To develop a solution based on CNN, which will classify the X-Ray images into: Normal, Pneumonia or COVID-19
- Develop a model with: COVID-19 sensitivity $\geq 80\%$ and specificity $\geq 80\%$.

As a first step in this study after the pre-processing of images, the team designed a custom CNN algorithm to do multi-class classification of pathology using Tensorflow 2.0. The result obtained was sub-optimal with around **35%** accuracy, which did not improve even after tuning of network and hyperparameters. Hence team shifted towards CNN based Transfer Learning methods and explored the following algorithms in sequence for rest of the project namely: VGG-16 [20], InceptionNet-V3 [22], MobileNet-V2 [31], ResNet-50 [23], and DarkNet-53 [10] based training on the consolidated data set. Also added trainable fully connected layer (between 1–3 layers) to the respective Transfer learning network and performed hyperparameter tuning for each algorithm.

2 Data Sources

Based on the above goal, open source datasets comprising of the Chest X-Ray images and the respective metadata for Normal, Pneumonia and COVID-19 were acquired from Kaggle, Cohen [5, 14] and other sources listed below:

The Normal and Pneumonia X-Rays have relatively equal distribution. However the number of COVID-19 images are less leading a Data Imbalance problem. The X-Ray data is restricted to very few geographies Ex – North America & Europe. All the input images were of good quality 1024 pixels. The radiography images were taken in AP (Anterior Posterior), PA(Posterior Anterior) and AP Supine positions which is ideal for analyzing and training the chest pathology.

Refer Kaggle dataset [14] for Normal and Pneumonia [14]. The origin of the Kaggle dataset is [15] COVID-19 dataset reference: [12, 16–18].

3   Data Integration and Preprocessing

The below data pre-processing steps were followed on the X-Ray images before it was consumed for modelling [6].

A) As a first step, using the metadata file from the respective source, only the following conditions were selected for analysis. COVID-19, Normal, SARS, MERS, Streptococcus, Klebsiella, Chlamydophila, Legionella, E.Coli, Lung Opacity. Other cases are rejected.

B) Further to simplify the analysis, [SARS through Lung Opacity] conditions were grouped as Pneumonia based on the advice of medical domain experts.

C) While studying X-Ray images with the help of Radiologists or machines, the position of the X-Ray images becomes an important factor for learning. The quality of information in the X-Ray varies based on the positions PA vs AP. PA is most preferred in terms of Quality over AP, Lateral & Decubitus. The present algorithms accept both PA and AP. However Lateral images and Decubitus are avoided for analysis.

- **AP** – Anterior – Posterior (Back facing X-Ray film)
- **PA** – Posterior – Anterior (Chest facing X-Ray film)
- **AP Supine** – AP Sleeping with face upwards
- **Decubitus** - Decubitus means lying down; thus, this projection is made with the patient lying on their side and the x-ray beam horizontal (parallel) to the floor.

D) Accordingly, only "PA", "AP", "AP Supine", "AP semi erect", "AP erect" type of images are selected for analysis. Decubitus & Lateral images are not useful for training and hence filtered out.

E) The images were converted to Gray scale in case they are in RGB or BGR format. The image resolution and the channels are preserved as is.

F) The data was split into Train and Test data sets using the metadata file.

In this use case, the data is highly imbalanced for COVID-19 class due to lack of certified X-Ray data. The other two classes namely : Pneumonia and Normal X-Rays are relatively balanced. Hence, Data Augmentation methods need to be applied to ensure that we obtain a generalized model.

3.1   Applying Transfer Learning for Classifying X-Ray Images

The following Transfer learning methods were applied on the processed data to classify X-Ray images into Normal, Pneumonia and COVID-19.

1. VGGNet-16 using pre-trained ImageNet weights
2. ResNet-50 using pre-trained ImageNet weights
3. InceptionNet-V3 using pre-trained ImageNet weights
4. MobileNet-V2 using pre-trained ImageNet weights

For each of these Transfer Learning networks, a trainable fully connected layer (between 1–3 layers) to the respective Transfer learning network and performed hyper-parameter tuning for each algorithm. The maximum overall accuracy reached was 91%. The COVID-19 F1 score obtained was 0.82 (Table 1) (Fig. 1).

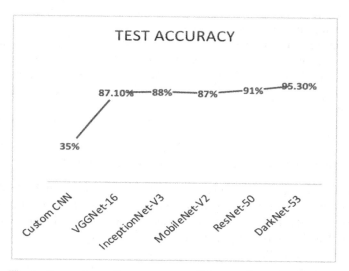

Fig. 1. Comparison of test accuracy and F1-score for the x-ray dataset

Table 1. Performance comparison of transfer learning algorithms.

Input Image size = 448 pixels

Model	Precision Weighted	Recall Weighted	F1-score Weighted	COVID-19 Precision	COVID-19 Recall	COVID-19 F1-Score	Train Acc.%	Overall Test Acc %	Train Loss	Test Loss
VGG-16	0.88	0.87	0.87	1	0.45	0.62	85.1	87.1	0.39	0.33
InceptionNet-V3	0.88	0.88	0.88	1	0.35	0.52	85.7	88	0.37	0.36
MobileNet-V2	0.88	0.87	0.87	1	0.39	0.56	86.4	87	0.34	0.38
ResNet-50	0.91	0.91	0.91	1	0.7	0.82	88.5	91	0.3	0.25

In the above experiments it is evident that ResNet-50 delivered the best score for COVID-19 class. The ResNet-50 Residual block with ReLU activation provided the best results.

The authors in the work [4] used DarkNet-19 Transfer Learning using Binary and Multi-class classification for COVID-19 classification. The dataset used was limited – 500 Normal, 500 Pneumonia and 125 COVID-19 images for training and does not use any type of image augmentation.

In our study we are using DarkNet-53 Transfer Learning network comprising of 53 Convolutional layers. The improvements upon its predecessor Darknet-19 include the use of residual connections, as well as more layers. Our approach was to propose a model which is reliable and robust compared to the studies done so far on COVID-19 detection. Hence, In the proposed method, our training data set was relatively large, comprising of 7966 normal, 5475 Pneumonia and 517 COVID-19 images. We took advantage of the high resolution input images (with 448 pixels) for training. We consolidated COVID-19 datasets from 3 continents (North America, Europe & Asia) based on the available sources. Even with this data consolidation, since the COVID-19 images were relatively less compared to other 2 data sets, we used dynamic data augmentation techniques before training. Some examples of the augmentation techniques included rotation, applying zoom, width shift and height shift of existing images. The image augmentation was applied dynamically at the time of training using the above mentioned aspects. Deliberately, image brightness was avoided as this could manipulate the pathology of the x-ray images. Also, image flip was avoided as the lung physiology would be manipulated.

The test data comprised of 885 Normal, 594 Pneumonia and 100 COVID-19 images.

3.2 Improving Model Performance Using Darknet-53(Yolo-V3)

In order to improve the Classification metrics further, the team researched and explored other classification and detection algorithms like YOLO-v3 [10]. Yolo-v3 uses DarkNet-53[10] as the classification algorithm before detection of subjects. From the literature it was evident that DarkNet-53 uses LeakyRelu for activations and provided much better metrics on ImageNet [10].

Based on the intuition mentioned above, the team researched the open source for Yolo-v3/DarkNet-53 TensorFlow implementations. Separating out the DarkNet-53 from Yolo-v3 was an initial challenge. This was resolved by experimenting with a few open source implementations of DarkNet-53 from GitHub [11]. The architecture was re-used from the open source implementation. In the Yolo-v3 author's website [32] DarkNet-53 pre-trained ImageNet weights were available in CUDA format [19]. In addition to extracting the architecture the second challenge was to convert the DarkNet-53 CUDA weights into TensorFlow compatible format. After converting the weights from CUDA to TensorFlow compatible format, the weights were loaded into the DarkNet-53 architecture.

Team used the above two resources and performed tuning & tweaking of the source code, to arrive at a reliable baseline of DarkNet-53 Transfer Learning architecture implementation loaded with ImageNet weights. Refer the architecture image derived from the open source implementation using TensorFlow 2.0 (Fig. 2).

The team used the baseline DarkNet-53 architecture, added trainable fully connected layers (between 1–3), applied Dynamic Image Augmentation & performed hyper parameter tuning to do a series of ~25 tests using the raw image data. Eventually selected the fully connected network and hyper parameters based on the best F1-score of COVID-19 class and best overall accuracy on test data.

The best model had approximately 42 million non-trainable parameters and 38K trainable parameters with 2 fully connected dense layers and SGD optimizer with learning rate of 1e-3 (Fig. 3).

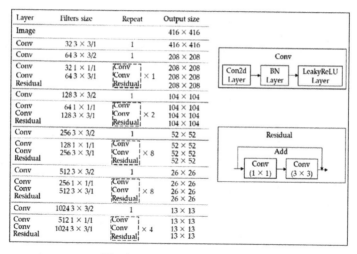

Fig. 2. DarkNet-53 architecture

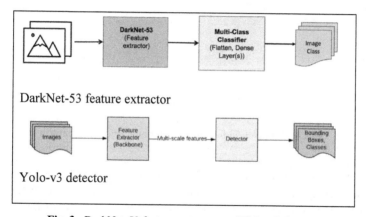

Fig. 3. DarkNet-53 feature extractor and Yolo-v3 detector

4 Results

Among all the Transfer Learning methods, DarkNet-53 which was so far only used for Image Object Detection along with Yolo-v3 by the AI community, is now proven to be very useful and accurate in the Medical X-Ray pattern classification. This model is robust and reliable since it is trained on large data set using image augmentation. Using DarkNet-53, the accuracy can vary for other tests (unseen data) between 94.5% to 96.2% at 95% confidence interval.

This model is quite robust to different position of X-Rays as it is trained with AP, PA, AP Supine type of radiography images. Lateral, Decubitus or CT images are not supported by this model.

These results were obtained by changing the images to grey scale. Due to limitations of the computing resources, the image resolution was changed from 1024 × 1024 pixels

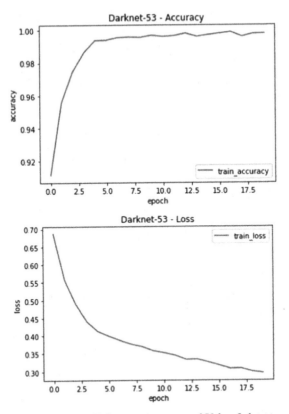

Fig. 4. DarkNet-53 feature extractor and Yolo-v3 detector

to 448 × 448 pixels. If the limitations of computing resource is addressed, then we could train/test with higher resolution. This can lead to better accuracy and COVID F1-score. Among various experiments we conducted, we observed that use of high resolution images yields better results.

The model is trained with well distributed dataset of Pneumonia and Normal radiography images comprise 95% of training images. COVID-19 images comprises of 5% of training images. Overall this dataset is highly imbalanced to classify COVID-19 class reliably. This class imbalance is compensated using Dynamic Image Augmentation in TensorFlow 2.0. With Image Augmentation the COVID-19 Precision & Recall achieved is 95%. As and when more certified COVID-19 data set with the prescribed standards is available, the accuracy, Precision and Recall can be further improved by training the model with more COVID-19 Radiography images.

After achieving a jump in the accuracy from ResNet-50 to DarkNet-53, a number of Hyperparameter Tuning experiments were conducted to find the best local optima for the training loss. These experiments ranged from using fully connected layers (between 1–3 layers), applying BatchNormalization or Dropout in the Fully connected layers, using different optimizers with different learning rate (Adam and SGD), applying different dynamic image augmentation techniques and training with higher number of epochs.

Table 2. Multi-class classification using DarkNet-53 model & Hyperparameter tuning.

Test Run ID	Model	Hyper-params	Img Aug? (Yes/No)	Precision WTD	Recall WTD	F1-score WTD	COVID-19 Precision	COVID-19 Recall	COVID-19 F1-Score	Train Acc %	Test Acc %	Train Loss	Test Loss
4	Darknet-53	2 Dense layers Adam LR=0.00001	No	0.94	0.94	0.94	0.95	0.91	0.93	99.73	94.1	0.299	0.69
5	Darknet-53	2 Dense layers Adam LR=0.00001	Yes	0.96	0.9	0.93	1	0.79	0.88	99.17	94.68	0.344	0.515
7	Darknet-53	2 Dense layers Adam LR=0.000001	No	0.94	0.94	0.94	0.95	0.87	0.91	99.91	94	0.418	0.715
14	Darknet-53	3 dense layers (1024, 512, 3) 2 Batch Norm 2 Dropout 0.6 Adam LR=0.00001	Yes	0.94	0.94	0.94	1	0.67	0.8	95.84	94	0.557	0.586
19	Darknet-53	1 Dense Layer Adam LR=0.00001	Yes	0.96	0.96	0.96	1	0.85	0.92	98.97	96	0.333	0.503
21	Darknet-53	2 Dense layers SGD LR=0.001, decay=0.0005, momentum=0.9	Yes	0.95	0.95	0.95	0.94	0.96	0.95	99.19	95.3	0.442	0.56

As described in Table 2, Test ID #21 gave best results with fully connected network comprising of 2 dense layers with SGD optimizer and learning rate of 1e-3. Along with this Dynamic Image Augmentation was applied using TensorFlow 2.0 and the network was trained with 20 epochs. The training was done on 13,958 radiology images (7966 normal, 5475 Pneumonia and 517 COVID-19) and testing was done on an out-of-sample data set of 1579 radiology images (885 Normal, 594 Pneumonia and 100 COVID-19).

4.1 Model Performance Evaluation Using Statistical Methods

Classification of images using ImageNet transfer learning algorithms namely ResNet-50 and DarkNet-53 have yielded the best results in this scenario. Overall accuracy is much better in Darknet-53 over Resnet-50 by ~4%. The COVID-19 F1-score is marginally improved by 0.1 in Darknet-53 over Resnet-50. However It is difficult to exactly pinpoint which classes are contributing significantly to these improvements. Hence we conducted McNemar's [29] paired test to verify if any class (Es) are contributing significantly to the results. Since there is less control on the distribution of the test data, a non-parametric test should be used in this scenario.

We are using statistical methods like McNemar-Bowker test statistic to compare the model results specifically to determine the model performance. McNemar's test is paired & non-parametric. The McNemar's test is checking if the disagreements between two cases match. McNemar's test is a type of homogeneity test for contingency tables. In terms of comparing two binary classification algorithms, the test is commenting whether the two models disagree in the same way (or not). It does not commenting on whether one model is more or less accurate or error prone than another. Further, Bowker's test is used to paired test with 3 or more categories.

H0: Null Hypothesis – There is no difference in the error count between the 2 models. Both models are performing in the same way.

H1: Alternate Hypothesis – There is significant difference in the errors between the 2 models (Table 3).

Table 3. Confusion matrix based on results from 2 models

	Model 2 correct	Model 2 incorrect
Model 1 correct	Yes/Yes	Yes/No
Model 1 incorrect	No/Yes	No/No

The McNemar's test statistic is calculated as:

Statistic = (Yes/No – No/Yes)^2/ (Yes/No + No/Yes)

The Table 4 confusion matrix/cross tab is arrived prior to the McNemar-Bowker test based on the results of ResNet-50 and Darknet-53.

Table 4. Confusion matrix based on ResNet-50 and Darknet-53 results

	Darknet-53				
		COVID-19	Normal	Pneumonia	All
ResNet-50	COVID-19	69	1	1	71
	Normal	5	876	47	928
	Pneumonia	17	46	517	580
	All	**91**	**923**	**565**	1579

Table 5. McNemar-Bowker statistic based on Table 4

Category 1	Category 2	chi2 value	p-unadj.	p-adj.
Normal	Pneumonia	0.0108	0.9174	1
Normal	COVID-19	2.6667	0.1025	0.3075
Pneumonia	COVID-19	14.2222	0.0002	0.0006

Inference: p-value for 'COVID-19' versus 'Pneumonia' comparison is less than the cut-off value of 0.05. Hence we will reject the null hypothesis. This signifies that errors are dissimilar between Resnet-50 and Darknet-53.

Based on the results from Tables 4 and 5, it can be concluded that Darknet-53 is significantly contributing in differentiating between 'Pneumonia' and 'COVID-19' images. Hence, Darknet-53 model is better skilled and efficient than ResNet-50 model in detecting COVID-19 cases which is our area of interest in this study.

5 Discussion and Conclusion

AI based diagnostic systems that can aid radiologists to more rapidly and accurately interpret radiography images to detect COVID-19 cases is highly desired. This method is also cost effective and contactless. Hence reduces the risk of infection among health care fraternity. Since COVID-19 virus infects the lungs as the primary organ, this method can help in detection of certain asymptomatic cases and early treatment can be started to arrest the infection.

The Transfer Learning methods provide very accurate results in identifying the pathology in X-Rays. This can be very good aid to Radiologists in confirming the COVID lung infection. This model can be used a Remote diagnostic tool (Ex: in places where there are lack of expert Radiologists).

When we analysed the results of the experiments closely with different Transfer Learning methods, we can infer that inclusion of residual blocks in ResNet-50 significantly improves both accuracy (~3%) and F1-score weighted (~0.03) compared to our experiments with InceptionNet-v3. With DarkNet-53, the overall accuracy (~4%) and F1-Score weighted (~0.04) is further improved compared to ResNet-50. The important aspect of this improvement can be attributed to Residual network block usage and usage of Leaky Relu activation in the entire DarkNet-53 design. LeakyReLU is a variant of the ReLU operation that is used to avoid neurons from dying. Unlike ReLU and sigmoid activation functions, which have zero value in the negative part of their derivatives, LeakyReLU features a tiny epsilon value to avoid the problem of dying neurons. This is demonstrated in the loss graph in Fig. 4. We can infer that the vanishing gradients is addressed effectively by use of LeakyReLU activations.

Hence, we can conclude that even with very less samples of COVID-19 X-Ray set DarkNet-53 is able to detect the pathology accurately up to 95%. DarkNet-53 is very robust and can be used for trials in the real-world pathology detection in X-Rays.

5.1 Future Directions

In order to reduce the bias, improve accuracy and also improve the robustness of the model for use in real world scenarios and also in production, we need to include more number of COVID Radiography images spread across multiple days of the infection. Also, we need to include additional COVID data from more variety of data sources. At present, only a small set of sample is used from SIRM [18] for Train and Test.

Including COVID images from variety of Geographies Ex: Africa, South America, Australia etc. will help in reducing model bias and improve model generalization across world's prominent populations which are presently affected.

The present data set used in this paper do not have annotated bounding boxes or segmentation to detect the affected areas in COVID-19 X-Rays. The data source [13] contains the COVID-19 lung x-ray data with bounding boxes and segmentation information. After detection of COVID-19 pathology, U-Net [25] can be used to do plot the affected areas using bounding boxes or with segmentation areas. This will be of great use to the medical fraternity to analyze the affected areas in lungs and provide targeted treatment to affected patients.

5.2 Limitations

This model is trained with only AP, PA, AP Supine type of radiography images. Lateral, Decubitus X-Ray images are not supported. These images have to be filtered out as these kind of images provide only Right or Left lung image which is not suitable for Machine Learning. Further CT images are not supported by this model. A separate DarkNet-53 model needs to be built to detect COVID-19 using CT training samples.

Due to the nature of the COVID-19 virus, the infection spreads in lungs quite rapidly across multiple days [24]. Unlike Pneumonia, the COVID infection in lungs across multiple days, specifically from day-1 through day-10 is very significant among patients. Hence the samples taken across multiple days will eventually prove to be very robust in detecting the lung infection. In this study, the COVID-19 radiography images fed during training is a mixture of lung images taken during early onset of COVID or late onset of COVID. The distribution of these COVID X-Ray images is not annotated based on patient's day of infection (Ex: COVID day-1, day-2....day-10) when the X-Ray was taken. Hence the model may fail to accurately detect COVID in early stages as there may be very weak clues or low lung infection for the disease. In order to make the model even more robust, COVID-19 lung x-ray data of patients spread across multiple days will need to be fed during the training. This will provide the much needed data variety to COVID images during training. This will also ensure during out-of-sample testing, that the detection will be accurate.

References

1. Corman, V.M., et al.: others, Detection of 2019 novel coronavirus (2019-nCoV) by real-time RT-PCR. Eurosurveillance **25**, 2000045 (2020)
2. Maguolo, G., Nanni, L.: A critic evaluation of methods for COVID-19 automatic detection from X-Ray images (2020). arXiv [eess.IV]
3. Narin, A., Kaya, C., Pamuk, Z.: Automatic detection of Coronavirus disease (COVID-19) using X-ray images and deep convolutional neural networks (2020). arXiv [eess.IV]
4. Ozturk, T., Talo, M., Yildirim, E.A., Baloglu, U.B., Yildirim, O., Rajendra Acharya, U.: Automated detection of COVID-19 cases using deep neural networks with X-ray images. Comput. Biol. Med. **121**, 103792 (2020)
5. Cohen, J.P., Morrison, P., Dao, L., Roth, K., Duong, T.Q., Ghassemi, M.: COVID-19 Image Data Collection: Prospective Predictions Are the Future (2020). arXiv [q-bio.QM]
6. Wang, L., Lin, Z.Q., Wong, A.: COVID-Net: a tailored deep convolutional neural network design for detection of COVID-19 cases from chest X-ray images. Sci. Rep. **10**(1), 19549 (2020)
7. Ng, M.-Y., et al.: Imaging profile of the COVID-19 infection: Radiologic findings and literature review. Radiol. Cardiothorac. Imaging. **2**(1), 200034 (2020)
8. Abiyev, R.H., Maaitah, M.K.S.: Deep convolutional neural networks for chest diseases detection. J. Healthc. Eng. **2018**, 1–11 (2018)
9. Shin, H.-C., et al.: Deep convolutional neural networks for computer-aided detection: CNN architectures, dataset characteristics and transfer learning. IEEE Trans. Med. Imaging **35**(5), 1285–1298 (2016)
10. Redmon, J., Farhadi, A.: YOLOv3: An Incremental Improvement (2018). arXiv [cs.CV]
11. Convert Darknet-53 ImageNet weights and architecture from NVIDIA CUDA framework to Tensor Flow 2.0: Github repository - https://github.com/qqwweee/keras-yolo3

12. COVID-19 Image Data collection from Cohen et al.: [10] - https://github.com/ieee8023/covid-chestxray-dataset
13. COVID-19 datasets with Bounding box and Segmentation information: https://github.com/GeneralBlockchain
14. Kaggle RSNA Pneumonia detection challenge dataset: https://www.kaggle.com/c/rsna-pneumonia-detection-challenge
15. Data Source reference for Kaggle RSNA Pneumonia detection: https://nihcc.app.box.com/v/ChestXray-NIHCC
16. Figure 1 COVID-19 Chest X-ray Dataset Initiative - https://github.com/agchung/Figure1-COVID-chestxray-dataset
17. Actualmed COVID-19 Chest X-ray Dataset Initiative - https://github.com/agchung/Actualmed-COVID-chestxray-dataset
18. Italian Society of Medical and Interventional Radiology (SIRM). https://www.kaggle.com/tawsifurrahman/covid19-radiography-database
19. DarkNet-53 ImageNet weights in CUDA format. https://pjreddie.com/darknet/imagenet/#darknet53
20. Simonyan, K., Zisserman, A.: Very deep convolutional networks for large-scale image recognition (2014). arXiv [cs.CV]
21. Darapaneni, N., et al.: COVID 19 severity of pneumonia analysis using chest X rays. In: 2020 IEEE 15th International Conference on Industrial and Information Systems (ICIIS), pp. 381–386 (2020)
22. Szegedy, C., Vanhoucke, V., Ioffe, S., Shlens, J., Wojna, Z.: Rethinking the inception architecture for computer vision. In: 2016 IEEE Conference on Computer Vision and Pattern Recognition (CVPR) (2016)
23. He, K., Zhang, X., Ren, S., Sun, J.: Deep residual learning for image recognition (2015). arXiv [cs.CV]
24. Dubey, R., Sen, K.K., Panda, S., Goyal, M., Menon, S.M., Arora, R.: Potential of Conventional Chest Radiography as a diagnostic and follow up imaging tool in COVID pandemic era Ejmcm.com. https://www.ejmcm.com/article_7273_4828eb9ff564ab3bc54bcfee5973b8b6.pdf. Accessed 10 Apr 2021
25. Ronneberger, O., Fischer, P., Brox, T.: U-Net: Convolutional Networks for Biomedical Image Segmentation (2015). arXiv [cs.CV]
26. Darapaneni, N., et al.: Inception C-net(IC-net): altered inception module for detection of covid-19 and pneumonia using chest X-rays. In: 2020 IEEE 15th International Conference on Industrial and Information Systems (ICIIS), pp. 393–398 (2020)
27. Albawi, S., Mohammed, T.A., Al-Zawi, S.: Understanding of a convolutional neural network. Int. Conf. Eng. Technol. (ICET) **2017**, 1–6 (2017). https://doi.org/10.1109/ICEngTechnol.2017.8308186
28. Neuman, M.I., et al.: Variability in the interpretation of chest radiographs for the diagnosis of pneumonia in children. J. Hosp. Med. **7**(4), 294–298 (2012)
29. Wikipedia reference for Evaluating McNemar test statistic. https://en.wikipedia.org/wiki/McNemar%27s_test
30. Reference book for Bowkers test – Basic Biostatistics for medical and Bio-Medical practioners (second edition) (2019). https://www.sciencedirect.com/book/9780128170847/basic-biostatistics-for-medical-and-biomedical-practitioners
31. Sandler, M., Howard, A., Zhu, M., Zhmoginov, A., Chen, L.-C.: MobileNetV2: Inverted residuals and linear bottlenecks (2018). arXiv [cs.CV]
32. Yolo-v3 author PJ Reddie's website used for downloading DarkNet-53 weights. https://pjreddie.com/darknet/imagenet/#darknet53

Anomaly Detection for Skin Lesion Images Using Replicator Neural Networks

Fabrizio Nunnari[1]([✉])(ⅈD), Hasan Md Tusfiqur Alam[1](ⅈD), and Daniel Sonntag[1,2](ⅈD)

[1] German Research Center for Artificial Intelligence (DFKI) Saarland Informatics Campus D3 2, Saarbrücken, Germany
{fabrizio.nunnari,hasan_md_tusfiqur.alam,daniel.sonntag}@dfki.de
[2] Oldenburg University, Oldenburg, Germany

Abstract. This paper presents an investigation on the task of anomaly detection for images of skin lesions. The goal is to provide a decision support system with an extra filtering layer to inform users if a classifier should not be used for a given sample. We tested anomaly detectors based on autoencoders and three discrimination methods: feature vector distance, replicator neural networks, and support vector data description fine-tuning. Results show that neural-based detectors can perfectly discriminate between skin lesions and open world images, but class discrimination cannot easily be accomplished and requires further investigation.

Keywords: Skin cancer · Anomaly detection · Autoencoders · Replicator neural networks · SVDD

1 Introduction

Clinical decision support systems (CDSS) for skin cancer detection, based on deep neural networks, have proven to be effective and in some cases surpass human performances [1,2,9,14].

To foster research in this direction, from 2016 on, the International Society for Digital Imaging of the Skin[1] organizes the ISIC[2] challenge for the development of computer vision systems supporting clinical decision in the field of skin lesions. The tasks considered in the past editions include classification [6,7], lesion segmentation, and feature extraction [5].

The 2019 edition[3] contained, as an implicit task, anomaly detection. The training dataset provided for the ISIC 2019 challenge included images pertaining to 8 classes of skin lesions. However, the test dataset contained also images pertaining to none of those categories, named the *unknown* (UNK) class. In other words, as the training set was providing material for 8 known classes, the test phase asked for a classification into 9 classes (see Fig. 1).

---

[1] https://isdis.org/.

[2] https://www.isic-archive.com/.

[3] https://challenge.isic-archive.com/landing/2019.

© IFIP International Federation for Information Processing 2021
Published by Springer Nature Switzerland AG 2021
A. Holzinger et al. (Eds.): CD-MAKE 2021, LNCS 12844, pp. 225–240, 2021.
https://doi.org/10.1007/978-3-030-84060-0_15

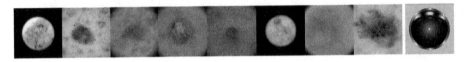

Fig. 1. A sample for each of the nine classes in the ISIC 2019 dataset. From left to right: Melanoma, Melanocytic nevus, Basal cell carcinoma, Actinic keratosis, Benign keratosis, Dermatofibroma, Vascular lesion, and Squamous cell carcinoma, followed by a sample of the test set clearly belonging to the UNK class.

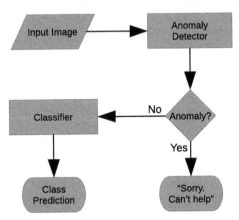

Fig. 2. The classification chain based on the concatenation of an anomaly detector and a standard classifier.

One approach to solve this problem would be to inject random pictures of other known skin pathologies into the training, or random pictures from the real world, and mark them as UNK. However, the choice of such extra training images would be arbitrary and possibly not reflect the selection criteria used for the preparation of the test set.

An alternative approach would be to chain two models: the first dedicated to performing *anomaly detection*, followed by a classification model (see Fig. 2). Hence, a new sample would be first filtered by the anomaly detector. If detected as not-pertaining to any of the 8 classes, it would be marked as UNK, or continue through the classification model otherwise.

In general, anomaly detection, in our approach also known as 1-class classification, is the task of discriminating if a given sample pertains to the same distribution of a reference set. Such a pre-filtering strategy would help circumventing the critical limitation of classifiers, which are unable to output choices beyond the closed-list of classes provided at training time.

Here, the purpose would be to enhance clinical decision support systems to provide answers like "I cannot take a decision: this system was not prepared for this kind of input image". Another possible application would be of an automatic filtering during the automated collection of images, for example, from the web.

Despite the potential advantages of anomaly detection in the field of skin lesions, from the results of the ISIC 2019 challenge, it emerges that none of the participants was able to reach satisfactory specificity for the UNK class, with some of the participants ignoring the problem as a whole.

Hence, in this paper, we report on a post-challenge investigation that we conducted to measure the effectiveness of deep-learning-based anomaly detection on skin lesion images.

From a survey on the ISIC 2019 reports, it looks like all of the participants addressed the problem of anomaly detection through a statistical analysis of the softmax output of their classifiers. The work we present here seems to be the first one to tackle the problem of anomaly detection using deep neural networks configured as autoencoders. Our results do not show major gains in classification performance, i.e., discrimination methods based on feature vector distance, Replicator Neural Networks, and Support Vector Data Description do not perform as good as they do on other domains. Nevertheless, we contribute with several hints when dealing with anomaly detection for (skin lesion) images and an investigation methodology that could be used as starting point for future work in this field or related imaging task.

2 Related Work

Anomaly detection (aka 1-class classification, outlier detection, novelty detection) refers to the task of discriminating between samples pertaining to a reference *target* distribution and samples coming from whatever kind of other distribution, and identify them as *anomalies*, or *outliers*. See Chandola et al. [3] for a comprehensive review.

Anomaly detection presents distinct problem complexities compared to the majority of analytical and learning problems. Pang et al. [20] discuss some unique problem complexities like unknowness, heterogeneous anomaly classes, rarity and class imbalance and the diverseness in the types of anomaly that results in largely unsolved challenges.

The One-Class SVM [22] is a popular solution for anomaly detection based on the SVM method. The drawback is that it doesn't scale with the number of features, and is thus not applicable to CNN-driven image classification, where the number of features describing a sample before the softmax stage is above 1000.

When using CNN-based classificators, an approach that reaches state-of-the-art performance comes from Lee et al. [17], who proposed a method for detecting out of distribution (OOD) samples where class conditional Gaussian distributions with respect to the features of the deep models are obtained under Gaussian discriminant analysis. Then, the confidence score are obtained by using the Mahalanobis distance metric. Their method considers both the final softmax scores and the intermediate features of internal hidden layers.

In the context of dermatoscopy, Li et al. [18] proposed a non-parametric deep isolation forest (DeepIF) as a modification of the method from Lee et al. [17] in order to take into account the huge intra-class diversity of skin disease images. With this approach they reach an average 0.71 ROC on intra-class discrimination on the HAM10000 dataset [24].

As a new approach, the tests reported in this paper use Replicator Neural Networks [12], which are based on the training of an autoencoder on the target set and a measurement of the reconstruction error between an input image and the encoded-decoded output image. The hypothesis is that an autoencoder "specialized" in compressing and decompressing a certain type of images will show a higher reconstruction error if applied to images never used during the training phase.

Additionally, we test the effectiveness of the deep support vector data description (SVDD) technique proposed by Ruff et al. [21], who used neural-based anomaly detectors on images of digits as well as on open space images. The SVDD optimization technique is a post-training, fine-tuning technique increasing the accuracy of the detection through an analysis of the internal feature vector of the autoencoder.

A closer look at the results of the ISIC2019 challenge[4] (see Table 1) denotes that the classification for the UNK class was poor, and in some cases the problem was ignored as a whole. For the UNK class, only four teams reached a sensitivity above 0.1.

Table 1. Results of the top 10 performers of the ISIC2019 challenge. The *Acc.* column refers to the Balanced Multiclass Accuracy (i.e. average sensitivity among all classes) which is the main ranking metric of the challenge.

Team	Acc.	Ext. data	UNK acc.	UNK sens.	UNK spec.	UNK AUC
DAISY lab	0.636	Yes	0.808	0.002	0.999	0.808
DysionAI	0.606	No	0.798	0.179	0.946	0.562
AImage lab	0.592	No	0.808	0.004	0.999	0.502
DermaCode	0.578	No	0.807	0.012	0.997	0.642
Nurithm labs	0.569	Yes	0.806	0.002	0.997	0.551
Torus actions	0.563	No	0.808	0.000	1.000	0.500
BIT deeper	0.558	No	0.729	0.390	0.810	0.705
SYSU-MIA-Group	0.557	No	0.801	0.272	0.920	0.600
MelanoNorm_IITRopar	0.546	No	0.802	0.004	0.992	0.496
MH_team	0.544	No	0.799	0.118	0.961	0.556

For example, the first in the rank (DAISYLab) [10], who reached a balanced multiclass accuracy of 0.636, achieved only 0.002 sensitivity for the UNK class. Their strategy was to train directly a classifier on 9 classes, injecting in the training set a collection of 2334 images from other datasets, including healthy skin.

Among the best performers MH.team (ranked 10th with accuracy 0.544) performed a post-prediction analysis using the minimum, maximum and standard deviation of the softmax output of each sample. By cross-validating on 7

---

[4] https://challenge.isic-archive.com/leaderboards/2019.

classes against the others (eight times), they manually selected the discrimation thresholds. With this approach they reached 0.118 sensitivity for UNK.

DysionAI (ranked 2nd with 0.607 accuracy) achieved an UNK sensitivity of 0.179 by training as 9-class classification with 0 images for UNK class. During prediction, they assign the input sample to UNK if its softmax probability is greater than a threshold set to 0.35.

The SYSU-MIA-Group (8th with 0.557 accuracy) computed the entropy of a softmax prediction on 8 classes. They interpret entropy as the inverse of confidence when the classification network makes a prediction. If the confidence is below a certain threshold, the sample is marked as UNK. The threshold was manually set during internal tests by using two under-represented classes (AK and VASC) as UNK class. With this approach they reached 0.272 sensitivity for UNK.

Finally, the highest sensitivity for the UNK class (0.390) was achieved by the BITDeeper team (7th with 0.557 accuracy). They trained a multi-class classifier in parallel with a multi-label classifier (actually implemented via 8 binary classifiers) on the 8 known classes. The output for the UNK class is computed as a class-wise combination of the 8 softmax (multi-class) and the 8 sigmoid (multi-label) outputs. However, the choice of the combination formula and its parameter values is not explicitly motivated.

3   Method

The goal is to build an anomaly detection system that, given the image of a skin lesion as input, outputs a binary decision stating whether the input pertains to the *target* distribution, i.e., the same class of images on which the model was trained (negative case), or it is an outlier (positive case).

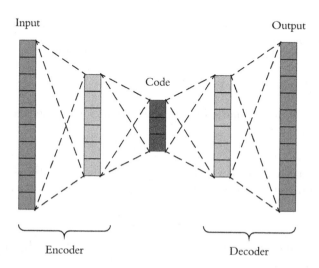

Fig. 3. The autoencoder architecture used to train the anomaly detection model.

As already introduced, we build an anomaly detection system based on a deep convolutional neural network autoencoder. The configuration, training, and testing procedures work as follows:

1. **Configure.** Figure 3 shows the general structure of an autoencoder. Our goal is to configure an autoencoder based on a convolutional architecture composed by a sequence Conv : $[D_c :] F [: D_d]$: Deconv, where F is a central dense layer with the *code* or *features* of input images, while the (optional) D_c and D_d are dense layers connecting the last convolution stage to F and the same to the first deconvolution stage;

2. **Train** the autoencoder f using a *target* set S_{train} of m images, where $f(x; w)$ takes as input an image x and the encoder weights w and outputs another image after encoding and decoding steps. The objective function for training the autoencoder is:

$$\min_{W} \frac{1}{m} \sum_{i=1}^{m} \|f(x_i; W) - x_i\|^2 \tag{1}$$

where $x_i \in S_{train}$ is an input image and W are the initial pre-trained parameters (weights) of the deep autoencoder. In other words, the goal is to minimize the l^2-norm computed on the pixel-wise difference between the original and the reconstructed image. After training, W^* are parameters of the trained model;

3. Test method l^2-**norm.** Given $\phi(x; w)$ the function that computes the feature vector of an image x for the weights w, find the center c of the hypersphere for the training set in the feature space:

$$c = \frac{\sum \phi(x, W^*)}{m}, x \in S_{train} \tag{2}$$

and d_{std} as the standard deviation of the l^2-norm between the feature vector of every sample and the center:

$$d_{std} = \sqrt{\frac{\sum \|\phi(x; W^*) - c\|^2}{m}}, x \in S_{train} \tag{3}$$

Test using the discrimination formula that marks a sample x as *anomaly* if

$$\|\phi(x; W^*) - c\|^2 > d_{std} * T \tag{4}$$

where $T > 0$ is a multiplier which sets the "threshold" for the discrimination.

4. Test method **Err.** Define the reconstruction error E of an image x as:

$$E(x) = \|x - f(x; W^*)\|^2 \tag{5}$$

Mean and standard deviation of the reconstruction error E of train set images are used to determine the binary classification:

$$E_m = \frac{\sum E(x)}{m}, x \in S_{train} \tag{6}$$

$$E_{std} = \sqrt{\frac{\sum (x - E_m)^2}{m}}, x \in S_{train} \tag{7}$$

Test using the discrimination formula:

$$\|E(x) - E_m\|^2 > E_{std} * T \tag{8}$$

5. Test method **SVDD**. Fine-tune the `Conv` stage using the Deep Support Vector Data Description (SVDD) method [21], which consists of training further the `Conv : [D_c :] F` part of the model with the following objective:

$$\min_W \frac{1}{m} \sum_{i=1}^{m} \|\phi(x_i; W) - c\|^2 + \frac{\lambda}{2} \sum_{l=1}^{L} \|\mathbf{W}^l\|_F^2 \tag{9}$$

where c is the center of the learned hypersphere that represents the training set in the feature space, $L \in \mathbb{N}$ is total number of hidden layers and $\lambda > 0$ is the weight decay regularization parameter.

Then, test using the same formulas of method l^2-norm (Eqs. 2, 3, and 4).

Architecture Configuration. We used two backbone CNN architectures for our tests, where the plain convolution stage was used as encoder and its transpose for the decoding part. The first backbone CNN architecture is VGG16 [23], which has proven to be sufficiently accurate in the classification skin lesions during previous ISIC challenges as well as still relatively fast to train. The second architecture is LeNet [15], which was successfully used by Ruff et al. [21] in the anomaly detection applied to the MINST [16] and CIFAR-10[5] datasets.

We tried both networks together with several configurations for the internal dense layers (hence, the number of features describing an image) and optionally the optimization method SVDD. As an additional hyper-parameter, we optionally frozen the parameters of both the `Conv` and `Deconv` stages instead of training the whole autoencoder. We also tried a combination of freezing the encoder and training the decoder together with the dense layers, but we did not observe any significant improvement, hence, results on this combination will not be reported

Dataset. Training stages were performed on the ISIC2019 dataset (S), which consists of 25331 images pertaining to 8 classes. Table 2 shows the class frequencies. To conduct our studies, we selected the *nevus* (S^{NV}) as target class, as it contains the highest number of samples. The dataset S^{NV} was further split into S_{train}^{NV}, S_{val}^{NV}, and S_{test}^{NV}, where the two last subsets included 2500 images each.

---

[5] https://www.cs.toronto.edu/~kriz/cifar.html.

Table 2. Class frequency for the ISIC2019 dataset.

Lesion	MEL	NV	BCC	AK	BKL	DF	VASC	SCC	Tot
Pct.	17.8%	50.8%	13.1%	3.4%	10.4%	1.0%	1.0%	2.5%	100%
Count	4522	12875	3323	867	2624	239	253	628	25331

Training. It has to be noted that while training for the ISIC2019, using randomly initialized weights couldn't converge. We had to use a *double transfer* approach. First, a classifier based on the VGG16 architecture was initialized with the weights computed for the ImageNet dataset [8]. Second, the dense layers were substituted with a 2X 2048 nodes dense layers, followed by a final 8-level softmax output and the model trained on an S_{train} set. This model scored 0.91 accuracy and 0.53 sensitivity in the ISIC 2019 challenge. The resulting weights were then used to initialize both the Conv and Deconv stages of the VGG16-based autoencoder.

After initialization, we also distinguished between training the full autoencoder or only the internal dense layers (All vs. Dense-only).

The structure of SVDD is identical to the encoder part of the autoencoder along with the final representation layer and the initial weights of SVDD architecure are transferred from the trained autoencoder part and further optimization is done using the objective function 9.

Testing. We tested our architectures using three test sets. The first T_{7cls} is composed by the union of S_{test}^{NV} with the remaining seven classes of the ISIC 2019 set ($S - S^{NV}$), for a total of 4154 samples. As the nevus class is already contained in the S_{train}^{NV} set, the goal was to discriminate from nevus as target and melanoma as anomaly. The second test set $T_{MedNode}$ is the MedNode dataset [11], which contains 100 images for nevi and 70 melanomas. Finally, the third test set T_{coco} is composed by the union of S_{test}^{NV} with a selection of 4989 random images from the COCO dataset [19]. The goal here is to set a baseline for the discrimination between skin lesion images and random "outside-world" ones.

4   Results

Table 3 show the test results for several combination of hyperparameters and test sets. The positive case (i.e., high sensitivity) is associated with the capability of detecting an anomaly.

In addition to the reference CNN architecture (base arch.) and the configuration of the dense layers (dense layers), we test the difference between training the whole autoencoder vs. training only the internal dense layers (trained layers) and use different norm and three discrimination methods: feature vector distance, (Err, l^2, and SVDD). The AUC is computed considering all of the samples of the test set, and gives an indication on the capability of the method into discriminating between the target and the anomaly classes. However, the

Table 3. A selection of the tests of different architectures against other 7 classes and COCO datasets.

Test #	Base arch.	Dense nodes	Train layers	Test method	Test set	AUC	T = 1			T = 3		
							acc.	spec.	sens.	acc.	spec.	sens.
1	LeNet	128	All	Err	7cls	0.49	0.39	0.15	0.86	0.35	0.03	0.98
2	LeNet	128	All	Err	Coco	1	0.96	1	0.86	0.99	1	0.98
3	LeNet	128	All	SVDD	7cls	0.49	0.38	0.12	0.88	0.35	0.3	0.97
4	LeNet	128	All	SVDD	coco	1	0.7	1	0	0.99	1	0.97
5	VGG16	1960:1960:1960	Dense	Err	7cls	0.51	0.45	0.34	0.66	0.34	0	1
6	VGG16	1960:1960:1960	Dense	Err	Coco	1	0.9	1	0.66	1	1	1
7	VGG16	1960:1960:1960	Dense	SVDD	7cls	0.49	0.39	0.15	0.85	0.37	0.07	0.94
8	VGG16	1960:1960:1960	Dense	SVDD	Coco	0.99	0.96	1	0.85	0.98	1	0.94
9	VGG16	1960:1960:1960	All	Err	7cls	0.49	0.4	0.17	0.82	0.36	0.04	0.96
10	VGG16	1960:1960:1960	All	Err	Coco	1	0.95	1	0.82	0.99	1	0.96
11	VGG16	1960:1960:1960	All	SVDD	7cls	0.5	0.66	1	0	0.34	0	1
12	VGG16	1960:1960:1960	All	SVDD	Coco	1	0.7	1	0	1	1	1
13	VGG16	1960X2:980:1960X2	All	Err	Coco	0.95	0.91	0.97	0.77	0.39	0.1	0.99
14	VGG16	3920	All	Err	Coco	0.93	0.88	0.91	0.83	0.63	0.47	0.96
15	VGG16	490	All	SVDD	Coco	0.92	0.87	0.89	0.84	0.59	0.41	0.96
16	VGG16	980	All	SVDD	Coco	0.92	0.87	0.89	0.84	0.58	0.4	0.96
17	VGG16	147	All	Err	Coco	0.92	0.87	0.88	0.83	0.58	0.4	0.96
18	VGG16	1960:980:1960	All	Err	Coco	0.9	0.86	0.87	0.82	0.5	0.29	0.95
19	VGG16	1960:do(0.5):980:1960	All	Err	Coco	0.9	0.86	0.87	0.82	0.5	0.29	0.95
20	VGG16	1960:1960:1960	All	l^2-norm	7cls	0.5	0.41	0.22	0.77	0.34	0	0.99
21	VGG16	1960:1960:1960	All	l^2-norm	Coco	0.73	0.34	0.16	0.77	0.30	0	0.99

AUC does *not* suggest what would be a proper distance (or error) threshold value T for deploying the system in real settings.

Hence, with reference to Eq. 8, we tested the performances of the anomaly detector using two threshold T values: 1 and 3. With $T = 1$, our hypothesis is that the hypersphere including the target samples would be very narrow, thus including only some of the target samples, but no anomaly samples. Differently, with $T = 3$, which for normal distributions would include 99.7% of the samples, our hypothesis is to have a discriminator which retain most of the target samples at a risk of missing many anomalies.

The top section of Table 3 reports results for the LeNet architecture. From the AUC measurement, we can see that the network is perfectly able to detect COCO classes, but the discrimination with 7cls fails ($AUC \simeq 0.5$). This is reflected in the high sensitivity couple with a very low specificity.

Therefore, we configured a more powerful autoencoder, based on the VGG16 architecture, experimenting with several configurations for the internal dense layers. In Table 3, lines from 13 to 19 show the test results for several combinations of dense layers. Such configurations where not able to reach AUC 1.0 even on the COCO dataset. Lines 20–21 show the results for the l^2-norm method, which was, too, unable to reach AUC 1.0 on the S_{coco} test set.

The perfect detection of COCO images is achieved by the dense nodes configuration $L_{d_c} = 1960 : L_f = 1960 : L_{d_d} = 1960$ (Table 3, lines 5–12). However, in spite of the similarity with the original classifier (Conv : 2048 : 2048 :

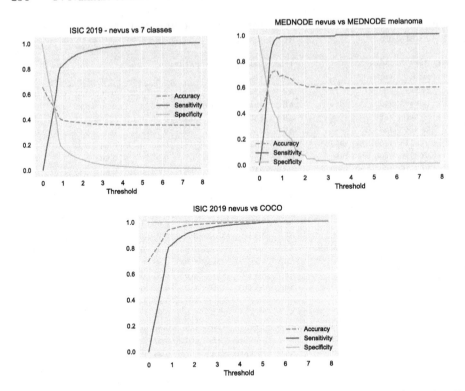

Fig. 4. Performance metrics as function of the threshold for tests 5 and 6 (training all layers), and the same architecture tested on the $T_{MedNode}$ test set.

softmax), the test on the 7cls dataset lead to an $AUC \simeq 0.5$. It can be seen that tests on thresholds 1 and 3 lead (in some cases) to opposite results in terms of sensitivity and specificity.

To better understand the behaviour of the discriminator as function of the threshold T, we computed the quality metrics for different values of T, ranging from 0 to 8 with increments of 0.1. This last procedure is also essential for fixing the T parameter to a value that should include most (or better all) of the samples of the target distribution, and be ready to intercept anomalies in a real application scenario, such as an online web service, where input samples can come from unpredictable distributions.

Figures 4 and 5 show the results for tests number 5–6 and 9–10, respectively (The other configurations show a similar behaviour). The top-left plots show the variation when testing NV against the other 7 classes. As the threshold increases, the sensitivity (i.e., the capability to detect an anomaly reaches 1.0). However, the specificity drops to 0.0, meaning that the system is not able to discriminate at all. The accuracy reflects this behaviour and converges to the class proportions ratio. The top-right plots show the same behaviour when trying to discriminate against the melanoma class in the MedNode dataset. Finally, the bottom plots

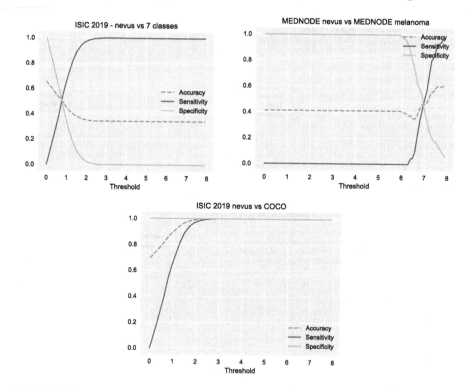

Fig. 5. Performance metrics as function of the threshold for tests 9 and 10 (training only dense layers), and the same architecture tested on the $T_{MedNode}$ test set.

show positive results when testing against the COCO dataset. By setting $T = 6$ for full training, and $T = 3$ for only-dense layers training, we reach accuracy 1.0. When comparing the two configurations, it means that by training only the dense layers, the target samples are closer to the center of the hypershpere, potentially meaning that the discrimination among classes can be more difficult.

To better inspect the behaviour when applying the SVDD technique, we plotted the metrics variation for tests 11–12 (which correspond to the non-SVDD test 5–6 of Fig. 4). Figure 6 shows that when testing against 7-classes and against melanoma, around $T = 1.5$ there is a sudden inversion between specificity and sensitivity. It suggests that, as is the purpose of SVDD, the sample features space is contracted towards the center of the target hypersphere, reducing the range of the distribution. However, this leads to poor results also when testing against the COCO images, meaning that also the feature vectors of fairly different images are collapsing together with the target lesion images. The coherence of this last results with other configurations, led us to mark the SVDD method as ineffective for the skin lesion domain.

The discrimination between targets and anomalies is based on the measurement of the error E between the original and the reconstructed image. Here, the

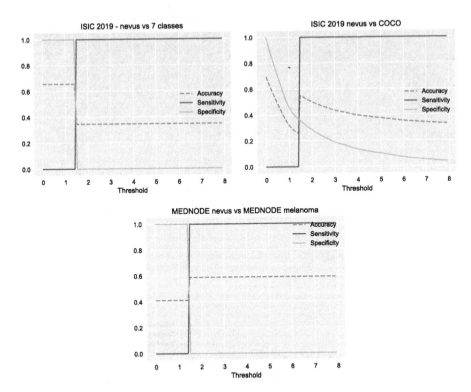

Fig. 6. Performance metrics as function of the threshold for tests 11 and 12 (training all layers, plus SVDD), and the same architecture tested on the $T_{MedNode}$ test set.

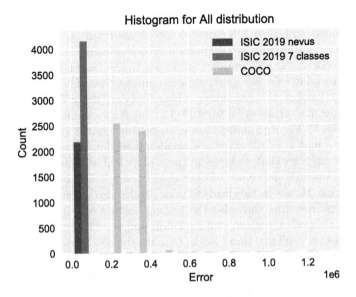

Fig. 7. Distribution of the MSEs for the CNN configuration used in tests 5 and 6.

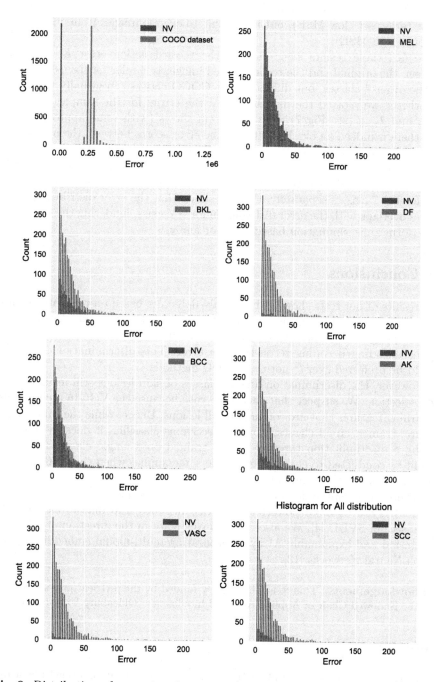

Fig. 8. Distribution of reconstruction errors for nevus and all of the other classes separately.

idea is that during the training the autoencoder specializes in encoding images of the target set (low MSE), but is not able to encode images from other distributions (high MSE).

So far, these results suggests that the reconstruction error E, measured between the original and the reconstructed image, is similar for nevus as well as for the other 7 classes, but differs for the COCO classes. To visually verify this hypothesis, we plotted the distribution of the errors for the samples for S_{test}^{NV}, T_{7cls}, and T_{coco} (see Fig. 7). The histogram shows (with some approximation) that there is indeed an overlap between the error scores between the nevus class and the other 7 classes, while samples of the COCO dataset are well distanced. Finally, to understand if there would be the possibility to discriminate between the nevus class and any other the 7 other classes, we plotted the error distribution for the 7 classes separately. Figure 8 shows that the error distribution of all classes overlaps with the error distribution for NV class, hindering the capability to perform a discrimination based on error analysis.

5   Conclusions

The results of our tests show that anomaly detectors based on replicator neural networks, initially trained as autoencoders, can distinguish skin lesions from random images of the outside world very well when the discrimination is based on the encoding/decoding reconstruction error. This discrimination technique should be preferred over l^2-norm or SVDD methods.

However, the discrimination among classes of skin lesions still leads to random selection. We suspect that this is the case because the VGG16 architecture is learning features that are common to all lesions. Hence, while the same architecture, trained on all classes, can be effective as classifier, it doesn't allow for setting a discrimination threshold when trained on a single class. More tests should be conducted to check whether the same applies when changing the target class, from nevus to any of the other seven.

Future work can be done in several directions: i) explore more hyperparameters, ii) try with more powerful networks, iii) solve the limitations recently addressed on SVDD [4]. We also aim to investigate in the direction of information fusion and explainable AI by incorporating multi-modal embeddings with Graph Neural Networks [13].

Acknowledgements. This research is partly funded by the pAItient project (BMG) and the Endowed Chair of Applied Artificial Intelligence (Oldenburg University).

References

1. Brinker, T.J., Hekler, A., Enk, A.H., et al.: Deep learning outperformed 136 of 157 dermatologists in a head-to-head dermoscopic melanoma image classification task. Eur. J. Cancer **113**, 47–54 (2019)

2. Celebi, M.E., Codella, N., Halpern, A.: Dermoscopy image analysis: overview and future directions. IEEE J. Biomed. Health Inf. **23**(2), 474–478 (2019)
3. Chandola, V., Banerjee, A., Kumar, V.: Anomaly detection: a survey. ACM Comput. Surv. **41**(3), 15:1-15:58 (2009). https://doi.org/10.1145/1541880.1541882
4. Chong, P., Ruff, L., Kloft, M., Binder, A.: Simple and effective prevention of mode collapse in deep one-class classification. In: 2020 International Joint Conference on Neural Networks (IJCNN), July 2020. http://dx.doi.org/10.1109/IJCNN48605.2020.9207209
5. Codella, N., et al.: Skin Lesion Analysis Toward Melanoma Detection 2018: A Challenge Hosted by the International Skin Imaging Collaboration (ISIC), February 2019. arXiv:1902.03368
6. Codella, N.C.F., et al.: Skin Lesion Analysis Toward Melanoma Detection: A Challenge at the 2017 International Symposium on Biomedical Imaging (ISBI), Hosted by the International Skin Imaging Collaboration (ISIC), October 2017. arXiv:1710.05006
7. Codella, N.C.F., et al.: Skin lesion analysis toward melanoma detection: a challenge at the 2017 International symposium on biomedical imaging (ISBI), hosted by the international skin imaging collaboration (ISIC). In: 2018 IEEE 15th International Symposium on Biomedical Imaging (ISBI 2018), pp. 168–172. IEEE, Washington, DC, April 2018. https://ieeexplore.ieee.org/document/8363547/
8. Deng, J., Dong, W., Socher, R., Li, L.J., Li, K., Li, F.-F.: ImageNet: a large-scale hierarchical image database. In: 2009 IEEE Conference on Computer Vision and Pattern Recognition, pp. 248–255. IEEE, Miami, FL, June 2009. http://ieeexplore.ieee.org/document/5206848/
9. Esteva, A., et al.: Dermatologist-level classification of skin cancer with deep neural networks. Nature **542**, 115 (2017). https://doi.org/10.1038/nature21056
10. Gessert, N., Nielsen, M., Shaikh, M., Werner, R., Schlaefer, A.: Skin lesion classification using ensembles of multi-resolution EfficientNets with meta data. MethodsX **7**, 100864 (2020)
11. Giotis, I., Molders, N., Land, S., Biehl, M., Jonkman, M.F., Petkov, N.: MED-NODE: a computer-assisted melanoma diagnosis system using non-dermoscopic images. Expert Syst. App. **42**(19), 6578–6585 (2015)
12. Hawkins, S., He, H., Williams, G., Baxter, R.: Outlier detection using replicator neural networks. In: Kambayashi, Y., Winiwarter, W., Arikawa, M. (eds.) DaWaK 2002. LNCS, vol. 2454, pp. 170–180. Springer, Berlin, Heidelberg (2002). https://doi.org/10.1007/3-540-46145-0_17
13. Holzinger, A., Malle, B., Saranti, A., Pfeifer, B.: Towards multi-modal causability with graph neural networks enabling information fusion for explainable AI. Inf. Fusion **71**, 28–37 (2021)
14. Kawahara, J., Hamarneh, G.: Visual Diagnosis of Dermatological Disorders: Human and Machine Performance, June 2019. arXiv:1906.01256
15. LeCun, Y., et al.: Backpropagation applied to handwritten zip code recognition. Neural Comput. **1**(4), 541–551 (1989)
16. Lecun, Y., Bottou, L., Bengio, Y., Haffner, P.: Gradient-based learning applied to document recognition. Proc. IEEE **86**(11), 2278–2324 (1998)
17. Lee, K., Lee, K., Lee, H., Shin, J.: A simple unified framework for detecting out-of-distribution samples and adversarial attacks (2018)
18. Li, X., Lu, Y., Desrosiers, C., Liu, X.: Out-of-distribution detection for skin lesion images with deep isolation forest. In: Liu, M., Yan, P., Lian, C., Cao, X. (eds.) MLMI 2020. LNCS, vol. 12436, pp. 91–100. Springer, Cham (2020). https://doi.org/10.1007/978-3-030-59861-7_10

19. Lin, T.Y., et al.: Microsoft COCO: common objects in context. In: Fleet, D., Pajdla, T., Schiele, B., Tuytelaars, T. (eds.) ECCV 2014. LNCS, vol. 8693, pp. 740–755. Springer International Publishing, Cham (2014). https://doi.org/10.1007/978-3-319-10602-1_48

20. Pang, G., Shen, C., Cao, L., Hengel, A.V.D.: Deep learning for anomaly detection. ACM Comput. Surv. **54**(2), 1–38 (2021)

21. Ruff, L., et al.: Deep one-class classification. In: Proceedings of the 35th International Conference on Machine Learning, vol. 80, pp. 4393–4402. PMLR, Stockholm, Sweden, July 2018. http://proceedings.mlr.press/v80/ruff18a.html

22. Schölkopf, B., Platt, J.C., Shawe-Taylor, J., Smola, A.J., Williamson, R.C.: Estimating the support of a high-dimensional distribution. Neural Comput. **13**(7), 1443–1471 (2001)

23. Simonyan, K., Zisserman, A.: Very Deep Convolutional Networks for Large-Scale Image Recognition, September 2014. arXiv:1409.1556

24. Tschandl, P., Rosendahl, C., Kittler, H.: The HAM10000 dataset, a large collection of multi-source dermatoscopic images of common pigmented skin lesions. Sci. Data **5**(1), 180161 (2018). https://doi.org/10.1038/sdata.2018.161

On the Overlap Between Grad-CAM Saliency Maps and Explainable Visual Features in Skin Cancer Images

Fabrizio Nunnari[1]([envelope]) [iD], Md Abdul Kadir[1,2] [iD], and Daniel Sonntag[1,2] [iD]

[1] German Research Center for Artificial Intelligence (DFKI), Saarland Informatics Campus D3 2, Saarbrücken, Germany
{fabrizio.nunnari,md_abdul.kadir,daniel.sonntag}@dfki.de
[2] Oldenburg University, Oldenburg, Germany

Abstract. Dermatologists recognize melanomas by inspecting images in which they identify human-comprehensible visual features. In this paper, we investigate to what extent such features correspond to the saliency areas identified on CNNs trained for classification. Our experiments, conducted on two neural architectures characterized by different depth and different resolution of the last convolutional layer, quantify to what extent thresholded Grad-CAM saliency maps can be used to identify visual features of skin cancer. We found that the best threshold value, i.e., the threshold at which we can measure the highest Jaccard index, varies significantly among features; ranging from 0.3 to 0.7. In addition, we measured Jaccard indices as high as 0.143, which is almost 50% of the performance of state-of-the-art architectures specialized in feature mask prediction at pixel-level, such as U-Net. Finally, a breakdown test between malignancy and classification correctness shows that higher resolution saliency maps could help doctors in spotting wrong classifications.

Keywords: Skin cancer · Visual features · Explainable AI · Saliency maps

1 Introduction

The recognition of skin cancer from digital pictures is a task that has received much attention in the last years [3,22,23]. Many evaluations show that convolutional neural networks (CNNs) are capable of distinguishing between malignant skin cancer and benign lesions with higher accuracy than experienced practitioners [2,11].

When neural networks are employed as classification models, decisions come by default without a human-comprehensible explanation—an issue affecting the adoption of neural networks in medical applications both for legal reasons as well as for the lack of trust in such systems. On the contrary, dermatologists

© IFIP International Federation for Information Processing 2021
Published by Springer Nature Switzerland AG 2021
A. Holzinger et al. (Eds.): CD-MAKE 2021, LNCS 12844, pp. 241–253, 2021.
https://doi.org/10.1007/978-3-030-84060-0_16

Fig. 1. The picture of a melanoma (ISIC_0000013, from the picture archive of the International Skin Imaging Collaboration) and its segmentation, followed by annotations for globules, pigment network, streaks, and the union of the three.

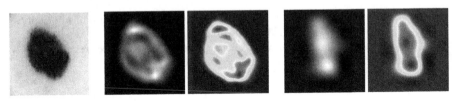

Fig. 2. The picture of a correctly classified melanoma (ISIC_0000013) (left), the saliency map and its colored heatmap computed on a VGG16 (middle) and on a RESNET50 model (right). (Color figure online)

diagnose skin cancer in the basis of widely recognized *visual features*, i.e., areas of the skin, or regions of interest (ROIs), characterized by well-defined visual patterns associated with medical concepts [21]. Figure 1 shows some examples. Such visual features can be present in both benign as well as malignant lesions. When their visual presence is significant, clinical guidelines suggest to assume malignancy.

New algorithms in the field of eXplainable Artificial Intelligence (XAI) allow for the extraction of saliency maps from classification models; Grad-CAM [27] is one of the most popular algorithms. Saliency maps are images that indicate the pixels areas contributing to a certain classification decision. Saliency maps are normally encoded as greyscale images or converted to heatmaps for visual inspection. Figure 2 provides an example.

Intuitively, it can be expected and observed an overlap between regions with high saliency and regions of interest (that practitioners would identify as signs of malignancy) occurs. However, this relationship has never been investigated in detail.

In this paper, we present a study measuring to what extent saliency maps can be used to identify visual features of skin lesions. In particular, we investigate the behavior of the involved deep learning architectures on actual data, in order to extract reference measurement values, reference thresholds, and to identify the limits of this approach.

The remainder of this paper first describes in Sect. 2 related work in the field of feature extraction and XAI. In Sect. 3, we describe the two classification models used for our experiments, able to discriminate (among others) between nevus and melanoma. Section 4 describes the skin lesion images and masking data used for the experiments. Then, Sect. 5 describes the first experiment, aiming to find the best threshold value maximizing the overlap between saliency maps and ground truth regions of interest, which is not granted to be the usual 0.5. The second experiment (Sect. 6), investigates the difference in overlpping when distinguishing between correctly vs. wrongly classified lesions, showing significant differences. Finally, Sect. 7 discusses the results of the experiments and Sect. 8 describes future work.

2   Related Work

The experiments presented in this paper are conducted on the dataset presented for the Task 2 (feature extraction) of the ISIC 2018 challenge (https://challenge2018.isic-archive.com) [5]. The dataset contains 2386 dermoscopy images, all of them annotated with binary masks highlighting the presence of five "features" at pixel-level, i.e., patterns on skin lesions unanimously recognized as indicators of potential malignancy [21]. Namely, the are: globules, streaks, pigment network, negative network, and milia-like cysts.

The performances on the feature extraction task are measured using the Jaccard index. Given two 2-color (black-white) image masks of the same resolution, the index J returns the ratio between the count of the common white pixels (correctly classified) and the union of all the white pixels. This is also known as the "intersection over union" ratio. To give a reference on the performances of the best feature extraction models, the first three ranks of the ISIC 2018 challenge were taken by the NMN-Team, with three approaches reaching $J = 0.307$, 0.305, and 0.304 respectively [18]. One of the goals of the analysis presented in this paper is to assess to what extent thresholded saliency maps can identify skin lesion features, to compare performances with the best ISIC 2018 systems, and to provide a reference performance baseline of a XAI-based system.

The recent experiments [8] on skin cancer detection focus on image classification only: they cannot produce explanations. Esteva et al. [11] also show that Inception v3 works very well in skin lesion detection and outperforms doctors. Even though the algorithm outperforms doctors, it cannot explain its decision accurately. Han et al. [13] fine-tuned the ResNet-152 model for cutaneous tumor detection. The classification performance of the network was comparable with that of 16 dermatologists, and they also exploit Grad-CAM to explain the classifications. There is a lot of success in explaining algorithmic output, but to increase the performance of an explainable model, we need a way to evaluate the quality of an explanation [16].

Concerning visual explanation techniques, GradCAM [27] is an analytical technique that, applied to convolutional neural classifiers, is able to highlight the areas of a picture contributing to the classification choice. The method is fast, as it needs just one forward and one backward propagation step, and then builds the saliency map from an analysis of the activation values of an intermediate chosen convolutional layer. Best XAI results are obtained by analyzing the last convolutional layer of a network [10,28]; however, the results are often low-resolution images.

A method that provides higher-resolution images is RISE [25], which is based on a stochastic approach. Input images are iteratively altered via random noise, and the final saliency map is composed by accumulating the partial estimations. However, its application requires much more computational power, as it needs to run hundreds of thousands of prediction cycles. Additionally, from a set of initial tests, it seems that RISE is not able to highlight regions of interest of skin lesion images with the same reliability as on pictures of real-world objects. The experimentation using RISE (together with other visual XAI variants like Grad-CAM++ [4] and SmoothGrad [30]) is deferred to future work.

Arun et al. [1] measured the overlapping between saliency maps and human-traced ground truth, but in the domain of chest X-rays, and used only very deep networks (InceptionV3 and DenseNet121), which provide very low resolution maps. Interestingly, they found the best XAI method being XRAI [19], which we plan to include in future work using also the evaluation methodology suggested by Sun et al. [31].

Several works focus on the use of saliency map to perform lesion segmentation (i.e., distinguish the lessioned from the healthy skin area) before passing it to a classifier. Among them, Gonzalez-Diaz proposes DermaKNet [12], which follows several pre-processing steps before the classification of skin lesion. In the first step, it creates a segmentation mask and applies it to the dermoscopic image. Secondly, it creates a structure segmentation mask to identify the structure of the dermoscopic image. After masking, the original segmented image and some nonvisual metadata are fed into a convolutional neural network for classification. Khan et al. [20] propose a channel enhancing technique to increase the contrast of lesion area. As a result, there is an improvement in the quality of the segmentation mask. Jahanifar et al. [17] also propose a modified DRFI (Discriminative Regional Feature Integration) technique for a similar task for multi-level segmentation task. By combining multiple segmentation masks, they produce a more accurate mask. During the generation of the mask, they use a threshold value of 0.5, but they did not provide a reason for which they choose this value.

In our work, we rather focus on the specific degree of overlap between the saliency maps and the visual features that dermatologists search for a diagnosis.

3   Classification Architectures and Models

In this section, we describe the two classification models used for our experiments. The two models are based on two different architectures: VGG16 [29] and RESNET50 [14]. We selected these two architectures to monitor the saliency map generation process according to two important model differences: the classification performances and the resolution of the last convolutional layer. Interestingly, while the RESNET50 architecture definitely results in the better classifier, the resolution of its last convolution layer (`res5c_branch2c`) is limited to 8×8 pixels, which is much less than the 28×28 pixels resolution of the last convolution layer (`block5_conv3`) in the VGG16 architecture. Our first hypothesis is that although more prone to classification errors, the VGG16 could still deliver better "visual explanations" because of its higher resolution.

Following a transfer learning approach, both the VGG16 and the RESNET50 models are pre-trained on the Imagenet dataset [9], and their final layers are substituted with randomly initialized fully connected layers of 2048 nodes and a final 8-level softmax. Then, both models are trained using 20k images of the ISIC2019 challenge (https://challenge2019.isic-archive.com) [6,7,33], which contains 8 classes (MEL, NV, BCC, AK, BLK, DF, VASC, SCC). VGG16 was configured with an input resolution of 450 × 450 pixels, while RESNET50 at 227 × 227 pixels.

The models are tested on 2529 held-out images, and we report the following class specific metrics.

For VGG16, accuracies are MEL: 0.845, NV: 0.827, BCC: 0.934, AK: 0.964, BLK: 0.912, DF: 0.991, VASC: 0.995, SCC: 0.977, and the class sensitivities are MEL: 0.659, NV: 0.755, BCC: 0.783, AK: 0.593, BLK: 0.626, DF: 0.826, VASC: 0.880, and SCC: 0.661. Overall accuracy is 0.722 and average balanced accuracy (mean sensitivity, the metric for the ISIC challenge) is 0.723.

For RESNET 50, accuracies are MEL: 0.873, NV: 0.857, BCC: 0.947, AK: 0.972, BLK: 0.920, DF: 0.992, VASC: 0.995, SCC: 0.977, and the class sensitivities are MEL: 0.675, NV: 0.812, BCC: 0.834, AK: 0.628, BLK: 0.672, DF: 0.739, VASC: 0.800, and SCC: 0.726. Overall accuracy is 0.767 and the average balanced accuracy is 0.736

As a reference, the 2nd placed at the ISIC challenge, which is the best approach not using external data (therefore comparable to our approach), measured an average balanced accuracy of 0.753 [34].

All of the model training and testing was performed using our Toolkit for Interactive Machine Learning (TIML) [24], which operates on top of the Keras and Tensorflow frameworks.

4   Data Preparation

To generate the saliency maps for our experiments, we run the two classification models (VGG16 and RESNET50) on the images of the ISIC Challenge 2018 Task 2 (see related work). The dataset contains 2386 RGB skin lesion images (519 melanomas, 1867 nevi), each associated to five ground truth black-white feature maps: globules, streaks, pigment network, negative network, and milia-like cysts. As an additional feature, we compute the pixels-wise *union* of all the features (see Fig. 1). The resolution of the ground truth feature maps is consistent with their corresponding colored picture.

Table 1. Counts of non-black masks for each feature class on the 2386 total samples.

Globules	Mil.	Neg. net.	Pig. net.	Streaks	Union
601	574	188	1502	98	1963

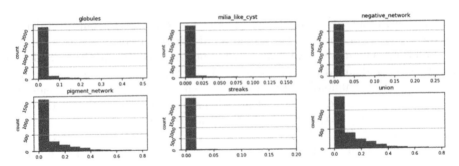

Fig. 3. Distribution of the measured Jaccard indices (horizontal axis), computed on $S_V^{0.5}$.

Fig. 4. VGG16: the saliency map for sample ISIC_0000013 thresholded for 0.0 = all-white, 0.1, 0.2, ..., 1.0 = all-black. Corresponding Jaccard indices between ground truth and union feature are 0.064, 0.150, 0.171, 0.189, 0.221, 0.242, 0.243, 0.216, 0.137, 0.059, and 0.000.

Some of the ground truth feature maps are completely black, as the dermatologists did not find any region of the corresponding class during the annotation. As can be seen in Table 1, only 1963 pictures have at least one non-black feature map. In our experiments, we ignore the skin lesion samples with no features.

The *generation of the saliency maps* consists of running the Grad-CAM algorithm [27] on each skin lesion picture with non-black union mask. The saliency is generated for the predicted class. We repeat the procedure for both the VGG16 and the RESNET50 models, generating the S_V and S_R greyscale picture sets, where $\|S_V\| = \|S_R\| = 1963$. Saliency maps have a resolution of 24×24 pixels for S_V and 8×8 for S_R, and their pixels are normalized in the range $[0, 1]$.

To *compare the saliency maps with ground truth maps*, we scaled up S_V and S_R to the resolution of the original images using a nearest neighbour filter. Figure 3 shows the histogram distribution of the Jaccard indices J computed between the features class (plus union) and S_V at threshold 0.5 ($S_V^{0.5}$). We can observe that all distributions are strongly right skewed, and all Js are mostly below 0.2, with the exception of a peak in performance for the pigment network class. A similar profile could be observed for $S_R^{0.5}$. The next step is to investigate whether 0.5 is the best value to use for thresholding saliency maps.

5 First Experiment

With the first experiment we aim at identifying the threshold value that leads to a maximization of the overlap between saliency maps and ground truth. To do so, we converted each saliency map into 11 binary maps using thresholds from 0.0 to 1.0 with steps of 0.1. For example, for VGG16, we define 11 sets $\mathcal{S}_V^t, t \in 0.0, 0.1, ..., 0.9, 1.0$. Figures 4 and 5 show examples of this threshold process. Then, we proceed by computing the Jaccard indices J between the ground truth and all of the processed saliencies $\mathcal{S}_V^x$ and $\mathcal{S}_R^x$.

Tables 2 and 3 report the summary of the *threshold analysis* for VGG16 and RESNET50, respectively, on which we report the threshold leading to the highest average Jaccard index.

Table 2. For VGG16, the best performing masking threshold together with corresponding Jaccard index data.

Feature	Best Thr.	J-min	J-mean (SD)	J-max
Globules	0.600	0.000	0.067 (0.078)	0.428
Mil.	0.600	0.000	0.019 (0.032)	0.236
Neg. net.	0.400	0.000	0.044 (0.048)	0.201
Pig. net.	0.500	0.000	0.141 (0.146)	0.797
Streaks	0.700	0.000	0.062 (0.062)	0.271
Union	0.500	0.000	0.132 (0.137)	0.784

Fig. 5. RESNET50: the saliency map for sample ISIC_0000013 thresholded for 0.0 = all-white, 0.1, 0.2, ..., 1.0 = all-black. Corresponding Jaccard indices between ground truth and union feature are 0.064, 0.104, 0.156, 0.174, 0.174, 0.158, 0.127, 0.101, 0.072, 0.009, and 0.000.

Table 3. For RESNET50, the best performing masking threshold together with corresponding Jaccard index data.

Feature	Best Thr.	J-min	J-mean (SD)	J-max
Globules	0.500	0.000	0.079 (0.090)	0.591
Mil.	0.700	0.000	0.032 (0.043)	0.288
Neg. net.	0.600	0.000	0.100 (0.102)	0.526
Pig. net.	0.300	0.000	0.133 (0.134)	0.720
Streaks	0.600	0.000	0.041 (0.050)	0.265
Union	0.300	0.000	0.136 (0.130)	0.720

For VGG16, among the features classes, the best threshold ranges between 0.4 and 0.7. The minimum J index is 0.0 on all categories, meaning that among all samples there is always at least one map with zero-overlap with the ground truth. The highest average $(J = 0.141)$ and maximum $(J = 0.797)$ belong to the pigmented network class. The union of all features lowers the scores to average $J = 0.132$ and max $J = 0.784$ at threshold 0.5.

When switching to RESNET50, the best thresholds range between 0.3 and 0.7. With respect to VGG16, pigmented network and streaks present the worse performance, while the average J increases for the other three classes. Overall, the union class has slightly higher average performance (average $J = 0.136$) at threshold 0.3.

Surprisingly, the Jaccard indices measured with the RESNET50 maps, which have a resolution limited to 8×8 pixels, are comparable to the ones extracted from the VGG16 models (24×24 pixels). The second hypothesis is that the lower resolution of the RESNET50 maps is compensated by the higher accuracy of the classification model, i.e., a better overall overlap.

6   Second Experiment

We proceed with a deeper analysis by further diving the samples into Melanoma and Nevus, and into correctly vs. wrongly classified samples. The goal is to observe the correlation between the measured J and the correctness of the classification. Here, the Jaccard indices are calculated using the *union* feature and using the best threshold identified in the first experiment, hence on $\mathcal{S}_V^{0.5}$ and $\mathcal{S}_R^{0.3}$. Tables 4 and 5 report the results for VGG16 and RESNET50, respectively.

For VGG16, we can observe that the mean J for correctly classified melanomas (0.135) is similar to the union class average (0.132). However, when melanomas are wrongly classified, the Jaccard index drops to 0.086, meaning that the saliency maps diverges from the ground truth. This could effectively help doctors is spotting a wrong classification. The idea is that: if the classifier tells the doctor that the sample is a melanoma, but then the reported saliency areas diverge a lot from what would be manually marked, then doctors can be more easily induced to think that the system is mis-classifying the image. For correctly classified nevi, the average J (0.134) is also similar to the full class

Table 4. For VGG16, statistics for the union feature as measured by splitting the $\mathcal{S}_V^{0.5}$ dataset in MELanoma and NeVus, either correctly or wrongly classified.

Feature	Count	Best Thr.	J-mean (SD)	J-max
MEL-correct	279	0.500	0.135 (0.108)	0.553
MEL-wrong	158	0.500	0.086 (0.089)	0.495
NV-correct	1165	0.500	0.134 (0.147)	0.784
NV-wrong	361	0.500	0.143 (0.145)	0.666

Table 5. For RESNET50, statistics for the union feature as measured by splitting the $S_R^{0.3}$ dataset MELanoma and NeVus, either correctly or wrongly classified.

Feature	Count	Best Thr.	J-mean (SD)	J-max
MEL-correct	314	0.200	0.114 (0.109)	0.564
MEL-wrong	123	0.400	0.132 (0.120)	0.554
NV-correct	1259	0.400	0.144 (0.135)	0.706
NV-wrong	267	0.300	0.127 (0.120)	0.517

Table 6. Results of a Mann-Whitney U-test on the Jaccard indices between correctly and wrongly classified classes.

Model	Vs.	U	p-value
VGG16	MEL-cor vs MEL-wr	28747.5	**1.2E − 7**
VGG16	NV-cor vs NV-wr	172027.0	**0.038**
RESNET50	MEL-cor vs MEL-wr	49693.5	0.8620
RESNET50	NV-cor vs NV-wr	151981.5	0.393

average (0.132), and for wrongly classified nevi the average J increases to 0.143. This suggests that, for nevi, doctors can better rely of the suggested saliency areas, which helps them in identifying the true area of interest.

To verify if these differences between correct vs. wrong classification are statistically significant, we ran a set of tests on the J indices measured on all items. As the distributions are not normal, we used the Mann-Whitney U-test. Table 6, top, shows that for the VGG16 maps the difference between the two conditions is statistically significant for $\alpha = 0.05$. The same tests are inconclusive for the RESNET50 model (Table 6, bottom), for which we couldn't identify a statistical significance.

7   Discussion

Our experiments show that the generation of features masks from threshold saliency maps performs, on the union of the features, at maximum $J = 0.136$. Among the five features, only Pigment Network reaches the same level of accuracy of the union class. This value is less that the half with respect to state-of-the-art networks specialized for pixel-level classification such as U-Net [26] or pyramid pooling [18]. Nevertheless, when considering the union of all the features, threshold saliency maps could still be a valid alternative to ad-hoc pixel-level feature extraction when dedicated features data sets are not available. In fact, the creation of ground truth datasets for feature extraction requires a considerable amount of work, involving experts in tracing the contour of regions of interests, or labeling super-pixels [5]. This is a huge annotation overhead when compared to labeling images with their diagnose class.

The value of the threshold to reach the best J index varies among datasets and features. Since it is not possible to analytically foresee the best threshold of a given dataset, we suggest the development of interactive exploratory visual interfaces, where dermatologists can autonomously control the saliency threshold value in an interactive fashion for exploration.

Our second hypothesis, that higher resolution saliency maps would lead to a higher Jaccard index than lower resolution ones, cannot be confirmed. However, from a decomposition between classes and correctness of classification, it appears that, for higher resolution maps (24×24 pixels on VGG16), saliency maps overlap much better with ground truth features when the classifier is correctly classifying a melanoma ($J = 0.135$) and performance drops when the prediction is incorrect ($J = 0.086$).

In summary, it seems that for the VGG16 model, in case of misclassification of melanoma, the saliency maps have the tendency to draw the attention of the observer to areas that they would rather ignore, thus inducing doctors to question the choice of the machine. This holds only for the VGG16 architecture, whereas this is not true in case of a low the resolution maps produced by RESNET50 (8×8 pixels), thus supporting our first hypothesis (i.e., higher resolution layers deliver better visual explanations).

8   Conclusions and Future Work

In this paper we presented an investigation on how saliency maps (an explainable AI technique) could be used to identify regions of interest in the diagnosis of skin cancer.

Our experiments show that thresholded saliency maps extracted from classifiers perform, in terms of Jaccard index, almost the half w.r.t. deep neural networks specialized for mask prediction. This applies only when using architectures with high resolution saliency. On the contrary, very deep architectures, usually characterized by very low resolution at the last convolution layer, would lead to the generation of maps with less explanatory power.

The long term goal of this research is the development of an interactive reinforcement learning approach involving human practitioners and their feedback to improve attribute detection. Due to the existence of uncertainty and incompleteness in data, the traditional approach of data-driven algorithms fails. In such a scenario, the "human-in-the-loop" approach can retrain a classification model to increase performance based on the knowledge of domain experts [15]. Starting from a base classifier, trained from a wide set of labeled images, whenever a dermatologist recognizes a wrong classification, he or she provides the correct class and marks the image with the regions of interests (features) that he or she recognizes. The human feedback could then be used to improve the automatic classification performance by comparing the human feedback with the saliency map of the CNN. The measured discrepancy between the two maps could be used to fine-tune the architecture towards higher accuracy [32].

Further, we would like to investigate on better options for thresholding. In this paper, a global threshold, in the range of 0.0 to 1.0, was simultaneously searched and applied to all the saliency map. This allows for an "emersion" of the most relevant region of interests of a *global* scale. However, there might be regions of saliency below the global threshold which are relevant with respect to the *local* surrounding area. To spot local maxima, we could split the maps into tiles, or super-pixels, and iteratively identify multiple local threshold values based on the range of saliency values of each region.

Finally, the current implementation of Grad-CAM returns saliency maps whose range $[0, 1]$ is filled by stretching the range of activation values of the target convolution layer. Each saliency map is forced to use the full activation range, independent of other samples. In so doing, regions of interests are "forced" to emerge, even when the activation values of the inner layer are lower when compared to other images. As future work, we could consider performing saliency normalization according to global statistics (mean and variance) on the tested set.

Acknowledgements. This research is partly funded by the pAItient project (BMG) and the Endowed Chair of Applied Artificial Intelligence (Oldenburg University).

References

1. Arun, N., et al.: Assessing the (un)trustworthiness of saliency maps for localizing abnormalities in medical imaging (2020)
2. Brinker, T.J., Hekler, A., Enk, A.H., et al.: Deep learning outperformed 136 of 157 dermatologists in a head-to-head dermoscopic melanoma image classification task. Eur. J. Cancer **113**, 47–54 (2019). https://doi.org/10.1016/j.ejca.2019.04.001
3. Brinker, T.J., Hekler, A., Utikal, J.S., et al.: Skin cancer classification using convolutional neural networks: systematic review. J. Med. Internet Res. **20**(10) (2018). https://doi.org/10.2196/11936
4. Chattopadhay, A., Sarkar, A., Howlader, P., Balasubramanian, V.N.: Grad-CAM++: generalized gradient-based visual explanations for deep convolutional networks. In: 2018 IEEE Winter Conference on Applications of Computer Vision (WACV), March 2018. https://doi.org/10.1109/wacv.2018.00097
5. Codella, N., Rotemberg, V., Tschandl, P., et al.: Skin lesion analysis toward Melanoma detection 2018: a challenge hosted by the international skin imaging collaboration (ISIC), February 2019. arXiv: 1902.03368
6. Codella, N.C.F., Gutman, D., Celebi, M.E., et al.: Skin lesion analysis toward melanoma detection: a challenge at the 2017 international symposium on biomedical imaging (ISBI), hosted by the international skin imaging collaboration (ISIC), October 2017. arXiv: 1710.05006
7. Combalia, M., Codella, N.C.F., Rotemberg, V., et al.: BCN20000: dermoscopic lesions in the wild. arXiv:1908.02288 [cs, eess], August 2019. arXiv: 1908.02288
8. Curiel-Lewandrowski, C., et al.: Artificial intelligence approach in Melanoma. In: Fisher, D.E., Bastian, B.C. (eds.) Melanoma, pp. 1–31. Springer, New York (2019). https://doi.org/10.1007/978-1-4614-7322-0_43-1
9. Deng, J., Dong, W., Socher, R., et al.: ImageNet: a large-scale hierarchical image database. In: 2009 IEEE Conference on Computer Vision and Pattern Recognition, pp. 248–255. IEEE, Miami, June 2009. https://doi.org/10.1109/CVPR.2009.5206848

10. Donahue, J., Jia, Y., Vinyals, O., et al.: DeCAF: a deep convolutional activation feature for generic visual recognition. In: Proceedings of the 31st International Conference on Machine Learning. Proceedings of Machine Learning Research, vol. 32, pp. 647–655. PMLR, Bejing, June 2014
11. Esteva, A., Kuprel, B., Novoa, R.A., et al.: Dermatologist-level classification of skin cancer with deep neural networks. Nature **542**, 115 (2017). https://doi.org/10.1038/nature21056
12. Gonzalez-Diaz, I.: DermaKNet: incorporating the knowledge of dermatologists to convolutional neural networks for skin lesion diagnosis. IEEE J. Biomed. Health Inform. **23**(2), 547–559 (2019). https://doi.org/10.1109/JBHI.2018.2806962
13. Han, S.S., Kim, M.S., Lim, W., Park, G.H., Park, I., Chang, S.E.: Classification of the clinical images for benign and malignant cutaneous tumors using a deep learning algorithm. J. Invest. Dermatol. **138**(7), 1529–1538 (2018)
14. He, K., Zhang, X., Ren, S., Sun, J.: Deep residual learning for image recognition. In: The IEEE Conference on Computer Vision and Pattern Recognition (CVPR), June 2016
15. Holzinger, A.: Interactive machine learning for health informatics: when do we need the human-in-the-loop? Brain Inform. **3**(2), 119–131 (2016)
16. Holzinger, A., Carrington, A., Müller, H.: Measuring the quality of explanations: the system causability scale (SCS). KI - Künstliche Intelligenz **34**(2), 193–198 (2020)
17. Jahanifar, M., Tajeddin, N.Z., Asl, B.M., Gooya, A.: Supervised saliency map driven segmentation of lesions in dermoscopic images. IEEE J. Biomed. Health Inform. **23**(2), 509–518 (2019). https://doi.org/10.1109/JBHI.2018.2839647
18. Jahanifar, M., Tajeddin, N.Z., Koohbanani, N.A., et al.: Segmentation of skin lesions and their attributes using multi-scale convolutional neural networks and domain specific augmentations (2018)
19. Kapishnikov, A., Bolukbasi, T., Viegas, F., Terry, M.: XRAI: better attributions through regions. In: Proceedings of the IEEE/CVF International Conference on Computer Vision (ICCV), October 2019
20. Khan, M.A., et al.: Construction of saliency map and hybrid set of features for efficient segmentation and classification of skin lesion. Microscopy Res. Tech. **82**(6), 741–763 (2019). https://doi.org/10.1002/jemt.23220
21. Mishra, N.K., Celebi, M.E.: An overview of Melanoma detection in dermoscopy images using image processing and machine learning, Janurary 2016. arXiv: 1601.07843
22. Nunnari, F., Bhuvaneshwara, C., Ezema, A.O., Sonntag, D.: A study on the fusion of pixels and patient metadata in CNN-based classification of skin lesion images. In: Holzinger, A., Kieseberg, P., Tjoa, A., Weippl, E. (eds.) CD-MAKE 2020. LNCS, vol. 12279, pp. 191–208. Springer, Cham (2020). https://doi.org/10.1007/978-3-030-57321-8_11
23. Nunnari, F., Sonntag, D.: A CNN toolbox for skin cancer classification. CoRR abs/1908.08187 (2019)
24. Nunnari, F., Sonntag, D.: A software toolbox for deploying deep learning decision support systems with XAI capabilities. In: Companion of the 2021 ACM SIGCHI Symposium on Engineering Interactive Computing Systems. EICS 2021, pp. 44–49, Association for Computing Machinery, New York (2021). https://doi.org/10.1145/3459926.3464753
25. Petsiuk, V., Das, A., Saenko, K.: RISE: randomized input sampling for explanation of black-box models. In: Proceedings of the British Machine Vision Conference (BMVC) (2018)

26. Ronneberger, O., Fischer, P., Brox, T.: U-net: convolutional networks for biomedical image segmentation. In: Navab, N., Hornegger, J., Wells, W.M., Frangi, A.F. (eds.) MICCAI 2015. LNCS, vol. 9351, pp. 234–241. Springer, Cham (2015). https://doi.org/10.1007/978-3-319-24574-4_28

27. Selvaraju, R.R., Cogswell, M., Das, A., et al.: Grad-CAM: visual explanations from deep networks via gradient-based localization. In: The IEEE International Conference on Computer Vision (ICCV), October 2017

28. Sharif Razavian, A., Azizpour, H., Sullivan, J., Carlsson, S.: CNN features off-the-shelf: an astounding baseline for recognition. In: The IEEE Conference on Computer Vision and Pattern Recognition (CVPR) Workshops, June 2014

29. Simonyan, K., Zisserman, A.: Very deep convolutional networks for large-scale image recognition, September 2014. arXiv:1409.1556

30. Smilkov, D., Thorat, N., Kim, B., et al.: SmoothGrad: removing noise by adding noise (2017)

31. Sun, J., Chakraborti, T., Noble, J.A.: A comparative study of explainer modules applied to automated skin lesion classification. In: Atzmüller, M., Kliegr, T., Schmid, U. (eds.) Proceedings of the First International Workshop on Explainable and Interpretable Machine Learning (XI-ML 2020) Co-located with the 43rd German Conference on Artificial Intelligence (KI 2020), Bamberg, Germany, 21 September 2020 (Virtual Workshop). CEUR Workshop Proceedings, vol. 2796. CEUR-WS.org (2020). http://ceur-ws.org/Vol-2796/xi-ml-2020_sun.pdf

32. Teso, S.: Toward faithful explanatory active learning with self-explainable neural nets. Interact. Adapt. Learn. **2444**, 13 (2019)

33. Tschandl, P., Rosendahl, C., Kittler, H.: The HAM10000 dataset, a large collection of multi-source dermatoscopic images of common pigmented skin lesions. Sci. Data **5**(1) (2018). https://doi.org/10.1038/sdata.2018.161

34. Zhou, S., Zhuang, Y., Meng, R.: Multi-category skin lesion diagnosis using dermoscopy images and deep CNN ensembles (2019)

From Explainable to Reliable Artificial Intelligence

Sara Narteni$^{(\boxtimes)}$ ⓘ, Melissa Ferretti, Vanessa Orani ⓘ, Ivan Vaccari ⓘ,
Enrico Cambiaso ⓘ, and Maurizio Mongelli ⓘ

Consiglio Nazionale delle Ricerche - Institute of Electronics,
Information Engineering and Telecommunications (CNR-IEIIT), Genoa, Italy
{sara.narteni,melissa.ferretti,vanessa.orani,
ivan.vaccari,enrico.cambiaso,maurizio.mongelli}@ieiit.cnr.it

Abstract. Artificial Intelligence systems are characterized by always less interactions with humans today, leading to autonomous decision-making processes. In this context, erroneous predictions can have severe consequences. As a solution, we design and develop a set of methods derived from eXplainable AI models. The aim is to define "safety regions" in the feature space where false negatives (e.g., in a mobility scenario, prediction of no collision, but collision in reality) tend to zero. We test and compare the proposed algorithms on two different datasets (physical fatigue and vehicle platooning) and achieve quite different conclusions in terms of results that strongly depend on the level of noise in the dataset rather than on the algorithms at hand.

Keywords: Reliable AI · Logic Learning Machine · Skope rules

1 Introduction

Artificial Intelligence is a very wide discipline which is undergoing an unprecedented development in recent years. Algorithmic decision-making is now ubiquitous, with always less human intervention, even in critical contexts such as automotive, finance or healthcare. For this reason, there is a need for an "Algorithmic Audit" [21] facing the legal, ethical and safety issues derived from such a growth: technology experts and policy makers should cooperate in order to make AI trustworthy and responsible for users [23]. To this effort, regulation is being developed, stating the requirements that AI systems should follow to achieve such goals. Between that legislation, we must remark the European GDPR[1], introduced in 2018, which states the need of a *"right to explanation"* when dealing with automated systems. This has paved the way to the development of a subfield of AI, referred to as eXplainable AI (XAI), aiming to provide humans with understanding and trust in models outcomes. Hence, XAI models often come in the form of intelligible rules, being simpler and generally less accurate

---

[1] https://gdpr.eu/tag/gdpr/.

© IFIP International Federation for Information Processing 2021
Published by Springer Nature Switzerland AG 2021
A. Holzinger et al. (Eds.): CD-MAKE 2021, LNCS 12844, pp. 255–273, 2021.
https://doi.org/10.1007/978-3-030-84060-0_17

than more sophisticated models (such as those of deep learning) [35], but with the enormous advantage of being interpretable.

Another point of view to trustworthy AI is identifying and handling assurance under uncertainties in AI systems [11]. This means improving reliability of prediction confidence. The topic remains a significant challenge in machine learning, as learning algorithms proliferate into difficult real-world pattern recognition applications. The intrinsic statistical error introduced by any machine learning algorithm may lead to criticism by safety engineers. This is corroborated even more by the intrinsic instability of deep learning in the presence of malicious noise [8,39]. The topic has received a great interest from industry [20], in particular in the automotive [38] and avionics [9] sectors. In this context, the conformal predictions framework [3] studies methodologies to associate reliable measures of confidence with pattern recognition settings including classification, regression, and clustering.

Keeping in mind these emerging research directions, our work shows how global rule-based XAI can be used as a warranty of reliability. In particular, we give the following contributions:

- We define reliability from outside (Sect. 5.1) and reliability from inside (Sect. 5.2) methodologies, through which Logic Learning Machine characteristic value ranking becomes an instrument to achieve "safety regions" in the feature space with zero statistical error.
- We show how intelligible rules (Logic Learning Machine and Skope-Rules), when trained with zero error, can be joined and then perturbed on their most important features to obtain more complex "safety regions" (Sect. 5.3).
- We apply the proposed approaches on two different datasets, concerning different kinds of problems, and demonstrate how our methods may perform differently according to the data (Sect. 6).

2   Related Work

In the era of massive automation, a big effort must be put on developing ML/AI algorithms that should never fail when producing their outcomes: erroneous predictions may lead to severe consequences in many safety-critical fields [2]. Many different approaches have been carried out to this purpose, which will be summarized in the following subsections.

2.1   Safety Engineering-Based Methods

In the context of autonomous driving, safety assessment has been studied in recent years by considering typical safety engineering approaches (safety-by-design, safe fail, safety margins) and extending them to ML paradigm [25,40], with major focus on neural networks and the most advanced Deep Learning solutions. These certification approaches include formal verification [37], transparent implementation [1], uncertainty estimation [22], error detection [16], domain generalization [43] and adversarial approaches based on data perturbation and corruption [13,17]. Furthermore, AI certification may rely on training data quality as in [7], where authors introduced metrics such as scenario coverage for ensuring that the data used in training has possibly covered all important scenarios. Also, [15] proposed a Feature Space Partitioning Tree (FSPT) method which splits the feature space into multiple parts with different training data densities, in order to identify those where there is lack of training samples. Another work [33] adopted the same safety engineering approach to identify safety hazards related to each different phase of a typical ML pipeline and propose product-oriented (i.e. technical requirements) and process-oriented (i.e. processes to be followed) methodologies for the mitigation of such risks. In [36], authors focus on autonomous driving and review the existing machine learning safety assurance methods, categorizing them by following the system's life-cycle. Here, DNNs are massively recurrent in all the collected works, with no mention to XAI. Nowadays, most autonomous systems are based on Deep Neural Networks (DNNs), since they guarantee very accurate performance on high-dimensional data. A lot of literature exists on safety of deep models: in [13], a DNN analyzer based on abstract interpretation is introduced to enhance reliability. Safety engineering approaches are also adopted in healthcare [4] to assess Convolutional Neural Networks safety for pattern recognition using a medical device, combining the known approach of error correcting memory with the introduction of default values to use in case of uncorrectable errors. Safety of DL models is also considered in [12] by using Bayesian neural networks to quantify uncertainty of CNN models in image segmentation tasks.

Moreover, some methods integrate safety assurance into reinforcement learning (RL) framework, by making predictions to guide the agent towards safe decisions [19].

2.2   Classification with Abstension

A different branch of methodologies to achieve reliability of AI consists in allowing classifiers to abstain from making predictions when they are considered uncertain according to a given loss function. Classification with abstension is achieved in [41], where a pointwise-competitive selective classification method was introduced to look for classifiers that minimize the true risk by using a selection function with the property of abstaining from predictions if the empirical risk minimizer does not agree with the true risk minimizer. Moreover, in [10] authors developed an innovative approach for classification with abstension, based on learning a predictor and the abstaining function simultaneously. Another solution is to perform a three-way decision, where an "uncertain" category is added to the task, being chosen if its cost is lower than providing a clear decision: such an approach is showing promising results either when used *a posteriori* either when embedded in the training of traditional ML [5]. However, the evaluation of such abstension-based methodologies needs to be based on a trade-off between accuracy of prediction and the rate of abstention, which cannot be too high to have useful models. In contrast, our XAI-based methods to handle uncertainty do not need such consideration.

2.3   Explainable AI-Based Methods

While AI systems certification is widely investigated for black-box deep learning models, it's not the same for explainable AI (XAI) models. Many XAI techniques are now available [2] with application in critical systems, e.g. in medicine [18]. In [34], the role of XAI is recognized as a way to achieve the verification of the system and the legislation compliance, but the proposed framework is based on explanations of black-boxes. Only a few works exist on the usage of XAI methods to address reliability in autonomous driving [26–29] or medicine [14]. Based on this, we investigate the role of global rule-based models and apply them to vehicle platooning and physical fatigue detection cases.

3   Logic Learning Machine

Logic Learning Machine (LLM) is an innovative global explainable supervised method; it is an efficient implementation of Switching Neural Networks [30]. LLM has the aim of building a classifier $g(x)$ described by a set of rules structured as follows: if *<premise>* then *<consequence>*. The *<premise>* is a logical product ($\wedge$) of conditions on the input features, whereas *<consequence>* corresponds to the output class. The model is built by following a three-step process:

1. *Discretization and Latticization*: each variable is transformed into a string of binary data in a proper Boolean lattice, using the inverse only-one code binarization. All the strings are then concatenated in one unique large string per each sample.
2. *Shadow Clustering*: a set of binary values, called implicants, are generated, allowing the identification of groups of points associated with a specific class.
3. *Rule Generation*: all the implicants are transformed into a set of simple conditions and eventually combined into a collection of intelligible rules.

An implicant is defined as a binary string in a Boolean lattice that uniquely determines a group of points associated with a given class. It is straightforward to derive from an implicant an intelligible rule having in its premise a logical product of threshold conditions based on the cutoffs obtained during the discretization step. In LLM all the implicants are generated via Shadow Clustering by looking at the whole training set: in this way, resulting rules can overlap and represent different relevant aspects of the underlying problem [31,32].

3.1   Feature and Value Ranking

Being a rule-based method, it is possible to inspect LLM results through feature and value ranking.

Consider a set of m rules $\mathbf{r}_k, k = 1, \ldots, m$, each including d_k conditions $c_{l_k}, l_k = 1_k, \ldots, d_k$. Let $X_1, \ldots, X_n$ be the input variables, s.t. $X_j = x_j \in \mathcal{X} \subseteq \mathbb{R}$ $\forall j = 1, \ldots, n$. Let also $\hat{y}$ be the class assigned by the rule and y_j the real output of the $j - th$ instance.

A condition c_{l_k} involving the variable X_j, can assume one of the following forms [29]:

$$X_j > s, \quad X_j \leq t, \quad s < X_j \leq t, \tag{1}$$

being $s, t \in \mathcal{X}$.

For each rule generated by the algorithm, it is possible to define a confusion matrix associated to the rule. It is made up of four indices: $TP(\mathbf{r}_k)$ and $FP(\mathbf{r}_k)$, defined as the number of instances (x_j, y_j) that satisfy all the conditions in rule $\mathbf{r}_k$ with $\hat{y} = y_j$ and $\hat{y} \neq y_j$ respectively; $TN(\mathbf{r}_k)$ and $FN(\mathbf{r}_k)$, defined as the number of examples (x_j, y_j) which do not satisfy at least one condition in rule $\mathbf{r}_k$, with $\hat{y} \neq y_j$ and $\hat{y} = y_j$, respectively.

Consequently, the following useful metrics can be derived [6]:

$$C(\mathbf{r}_k) = \frac{TP(\mathbf{r}_k)}{TP(\mathbf{r}_k) + FN(\mathbf{r}_k)} \tag{2}$$

$$E(\mathbf{r}_k) = \frac{FP(\mathbf{r}_k)}{TN(\mathbf{r}_k) + FP(\mathbf{r}_k)} \tag{3}$$

The covering $C(\mathbf{r}_k)$ is adopted as a measure of relevance for a rule $\mathbf{r}_k$; as a matter of fact, the greater is the covering, the higher is the generality of the corresponding rule. The error $E(\mathbf{r}_k)$ is a measure of how many data are wrongly covered by the rule. Both covering and error are used to define feature ranking and the subsequent value ranking.

Feature ranking (FR) provides a way to rank the features included into the rules according to a measure of relevance. In order to obtain such measure of relevance $R(c_{l_k})$ for a condition, we consider the rule $\mathbf{r}_k$ in which condition c_{l_k} occurs, and the same rule without condition c_{l_k}, denoted as $\mathbf{r}'_k$. Since the premise part of $\mathbf{r}'_k$ is less stringent, we obtain that $E(\mathbf{r}'_k) \geq E(\mathbf{r}_k)$, thus the quantity $R(c_{l_k}) = (E(\mathbf{r}'_k) - E(\mathbf{r}_k))C(\mathbf{r}_k)$ can be used as a measure of relevance for the condition of interest c_{l_k}. Each condition c_{l_k} refers to a specific variable X_j and is verified by some values $\nu_j \in \mathcal{X}$. In this way, a measure of relevance $R_{\hat{y}}(\nu_j)$ for every value assumed by X_j is derived by the following Eq. 4 [29]:

$$R_{\hat{y}}(\nu_j) = 1 - \prod_k (1 - R(c_{l_k})) \tag{4}$$

where the product is computed on the rules $\mathbf{r}_k$ that include a condition c_{l_k} verified when $X_j = \nu_j$. Since the measure of relevance $R_{\hat{y}}(\nu_j)$ takes values in $[0, 1]$, it can be interpreted as the probability that value ν_j occurs to predict $\hat{y}$. The same argument can be extended to intervals $I \subseteq \mathcal{X}$, thus giving rise to *Value Ranking (VR)*. Relevance scores are then ordered, thus giving evidence of the most sensitive interval of the feature with respect to each class.

4   Skope-Rules

Another global explainable supervised method is Skope-Rules[2], a Python machine learning module built on top of scikit-learn. Like LLM, Skope-Rules is an interpretable rule-based model consisting of a series of **if** <*premise*> **then** <*consequence*> rules; the difference between the two models lies in the way these rules are generated, selected and finally filtered. The three-step process for rules generation in Skope-Rules is as follows:

1. *Bagging estimator training*: rules generation is done from a set of decision trees and/or regressors. Each path or sub-path of a branch of a tree is transformed into a decision rule. Trees are trained to predict the output class of interest. This ensures that the splits are made in such a way as to guarantee that they are meant for the prediction task.
2. *Performance filtering*: from this set of rules an initial screening is carried out based on precision and recall thresholds.
3. *Semantic deduplication*: the last filter applied for the choice of rules is based on a criterion of similarity between terms, whereby term is meant the feature associated with the comparison operator with which it appears in the rule. The measure of similarity of two rules is determined by how many terms they have in common.

5   Reliability Assessment Methods

Considering a binary classification problem, we refer to the positive class ($y = 1$) as the unsafe one. In contrast, class $y = 0$ is referred to as the safe class. Based on this, we call "safety regions" those regions in the feature space where false negatives tend to zero. In this work, we developed three different methods to look for such regions.

5.1   Reliability from Outside

Let X be a $D \times N$ matrix of all the input vectors $x_i \in \mathbb{R}^N$, with the total number of features N and $i \in [1, D]$. Let $g(x_i) = y$ be the function describing the LLM classification. For binary classifications, we consider $g(x_i) = 1$ for the positive class, while $g(x_i) = 0$ for the other. Let D_1 be the number of instances belonging to class $y = 1$ and D_0 the number of instances in class $y = 0$, so that $D_1 + D_0 = D$.

Let N^{FR} be the number of the most significant features obtained through the feature ranking for class $y = 1$. For each feature $j \in [1, N^{FR}]$, we can use the LLM value ranking to define the most significant interval for $y = 1$ as $[s_j, t_j]$. Our method consists in expanding such intervals as follows: $[s_j - \delta_{s_j} \cdot s_j, t_j + \delta_{t_j} \cdot t_j]$.

Being $\Delta = (\delta_1, , \delta_{N^{FR}})$ a matrix, with $\delta_j = (\delta_{s_j}, \delta_{t_j})$, the optimal Δ is computed through the following optimization problem. Let $\mathcal{P}(\Delta)$ be the hyper-rectangle under the expanded intervals and let $\mathcal{V}(\mathcal{P}(\Delta))$ be the inherent volume.

---

[2] https://github.com/scikit-learn-contrib/skope-rules.

Then, the optimization problem identifies the best fit from the outside of class $y = 1$, namely, it finds the most suitable shape, in terms of rule-based intervals, of safe points around the unsafe ones. It is as follows:

$$\Delta^* = arg \min_{\Delta:N_1=D_1} \mathcal{V}(\mathcal{P}(\Delta)) \tag{5}$$

being N_1 the number of elements in X classified as $y = 1$ and included into $\mathcal{V}(\mathcal{P}(\Delta))$.

For instance, if we fix $N^{FR}=2$, the hyper-rectangle $\mathcal{P}$ becomes a rectangle $\mathcal{S}$. The optimization process let us find out the matrix $\Delta^* = (\delta_1^*, \delta_2^*)$. The related optimal intervals are $I_1 = (s_1 - \delta_{s_1}^* \cdot s_1, t_1 + \delta_{t_1}^* \cdot t_1), I_2 = (s_2 - \delta_{s_2}^* \cdot s_2, t_2 + \delta_{t_2}^* \cdot t_2)$, corresponding to the features $j = 1$ and $j = 2$ respectively: their logical union ($\vee$) defines a surface $\mathcal{S}$.

Then, the "safety region" is defined as the complementary bi-dimensional surface of $\mathcal{S}$, which can be written as follows:

$$\mathcal{S}_1 = ((-\infty, s_1 - \delta_{s_1}^* \cdot s_1) \vee (t_1 + \delta_{t_1}^* \cdot t_1, \infty)) \wedge$$
$$((-\infty, s_2 - \delta_{s_2}^* \cdot s_2) \vee (t_2 + \delta_{t_2}^* \cdot t_2, \infty)) \tag{6}$$

5.2 Reliability from Inside

An alternative way to perform the same search for "safety regions" consists in considering the N^{FR} most important features for safe ($y = 0$) class instead and reducing their most relevant intervals (again, provided by LLM value ranking) until the obtained region only contains true negative instances.

In this case, with the same notation as for the previous definition (Sect. 5.1), the reduced intervals are: $[s_j + \delta_{s_j} \cdot s_j, t_j - \delta_{t_j} \cdot t_j]$. Being Δ defined in the same way as for Eq. 5 and $\mathcal{P}_0$ the hyper-rectangle under the reduced intervals, the optimal Δ is found by enlarging as much as possible the hyper-rectangle from inside the non-fatigue class, until a fatigued point is reached. It is as follows:

$$\Delta^* = arg \max_{\Delta:N_1=0} \mathcal{V}(\mathcal{P}_0(\Delta)) \tag{7}$$

For $N^{FR} = 2$, the "safety region" is the following rectangle $\mathcal{S}_0$:

$$\mathcal{S}_0 = (s_1 + \delta_{s_1}^* \cdot s_1, t_1 - \delta_{t_1}^* \cdot t_1) \vee (s_2 + \delta_{s_2}^* \cdot s_2, t_2 - \delta_{t_2}^* \cdot t_2) \tag{8}$$

5.3 Rules with Zero Error

As the sharp angularity of hyper-rectangles may be not fine enough to follow the potential complex shapes of the boundaries between the classes, a more refined approach would ask for more complex separators, still preserving the zero statistical error constraint and by starting from the available rule baseline.

Given a rule-based model, it can be trained so to define a set of m rules $\mathbf{r}_k, k = 1, \ldots, m$ denoted by $E(\mathbf{r}_k) = 0 \ \forall k \in [1, m]$. Suppose that this procedure provides a set of m^0 rules $\mathbf{r}_k^0$, $k = 1, \ldots, m^0$ for the safe class ($y = 0$). Also, let $c_{l_k}^0$, $l_k^0 = (1, \ldots, d_k^0)$ be the set of d_k^0 conditions inside of each rule $\mathbf{r}_k^0$. We can join all the obtained rules $\mathbf{r}_k^0$ in logical OR operation ($\vee$), thus building a new predictor $\hat{r}$. Our goal is to assess its ability of classifying new test set data with statistical zero error (FNR $= 0$). This implies to further tune $\hat{r}$, by tuning a subset of its conditions $c_{l_k}^0$, chosen as those containing the first N^{FR} features obtained from the rules feature ranking for class $y = 0$. In mathematical terms, for each feature $j \in [1, N^{FR}]$, we add the thresholds of the chosen conditions by applying $\boldsymbol{\delta} = (\delta_s, \delta_t)$, being δ_s and δ_t the perturbation applied to s and t thresholds, respectively, as defined in Eq. 1. Let $\hat{r}(\boldsymbol{\delta})$ be the resulting perturbed predictor, our goal is then to find the optimal $\boldsymbol{\delta}$ as follows:

$$\boldsymbol{\delta}^* = arg \max_{\boldsymbol{\delta}: E(\hat{r}(\boldsymbol{\delta}))=0} C(\hat{r}(\boldsymbol{\delta})) \tag{9}$$

This procedure can be applied to any rule-based model, provided that it is possible to train it with zero error.

As regards the LLM model, zero error classification (for the safe class) is readily available by the shadow clustering adopted by LLM. The clustering process is applied with the further constraint of building clusters without superposition of points of more than one class [27] (LLM 0%, in the following).

In the case of Skope-Rules (Sect. 4), the same zero error for safe class rules can be obtained by training the model with *precision_min* parameter fixed to 1.

6   Applications and Results

The methods described in the previous Sect. 5 have been applied and tested on two different classification problems: physical fatigue detection in working task simulation (Sect. 6.1) and collision detection in vehicle platooning (Sect. 6.2).

6.1   Physical Fatigue

The data used in this test phase belong to an open-source dataset[3]. Data were collected through wearable sensors, i.e. Inertial Movement Units (IMUs), from 15 participants who were asked to perform a simulation of an industrial task for 180 min and provide a fatigue level every 10 min using RPE [42]. According to such scale, RPE$\geq$13 corresponds to a fatigued state (class $y = 1$), otherwise to non-fatigued (class $y = 0$). From sensors raw data, a list of features is derived (see Table 2 in [24]). We removed heart-rate related features as well as gender, since it is not numerical, and standardized data by applying z-score transformation.

We then trained LLM model with standard 5% maximum error allowed for rules on a 67% training set. We evaluated it on a 33% test set using common metrics, namely an accuracy of 82%, sensitivity of 71%, specificity of 95% and F1-score of 0.81.

---

[3] https://github.com/zahrame/FatigueManagement.github.io/tree/master/Data.

Reliability from Outside. In order to test this method, we considered the first two most important intervals for fatigued class that we got from LLM value ranking: *back rotation position in sagittal plane* > 0.03 and *wrist jerk coefficient of variation* > 0.03. We applied the optimization algorithm (Eq. 5) on such intervals and obtained $\delta^*_{s_1} = -13, \delta^*_{s_2} = 28$. For such values, we got FNR $= 0$ and TNR $= 0.20$. Therefore, the "safety region", which we call "non-fatigue region" in this context, can be expressed as follows (for brevity, let f_1 and f_2 be the two above mentioned features):

$$\mathcal{S}_1 = ((f_1 \in (-\infty, 0.42)) \wedge (f_2 \in (-\infty, -0.81))$$

The resulting region was then validated in order to take into account that the involved feature values should vary in a limited range, so to reflect real human movement capabilities and correspond to proper execution of the task. In general, we cannot assume that a subject who stays still will not ever get fatigued, but the nature of the task in which the subject is involved should provide indications on the ranges of parameters assessing the required movements. Since the dataset documentation does not drive in this direction and the inherent literature lacks of standard ranges, we chose to consider maximum and minimum values for the features based on two age groups (age $\leq$ 40 and age > 40). This helps to highlight the further stratification readily available from the sensitivity analysis.

Doing so, we were able to redefine two "non-fatigue regions" by limiting the previous one according to the ranges we found; such new regions are expressed as follows:

$$\mathcal{S}_1 = ((f_1 \in (-2.52, 0.42)) \wedge (f_2 \in (-1.78, -0.81)) \text{ for } age \leq 40 \text{ y.o}$$
$$\mathcal{S}_1 = ((f_1 \in (-1.86, 0.42)) \wedge (f_2 \in (-2.0, -0.81)) \text{ for } age > 40 \text{ y.o}$$

In Fig. 1 a visual representation of the obtained regions is provided.

Reliability from Inside. We considered the problem of identifying non-fatigue regions starting from the non-fatigued class too, thus adopting the reliability from inside approach. The value ranking shown *back rotation position in sagittal plane* $\leq$ 0.03 and *chest acceleration mean* > −0.47 as the two most relevant intervals for predicting non-fatigued class. On such conditions, we applied the optimization problem (Eq. 7), which led us to individuate $\delta^*_{t_1} = 57, \delta^*_{s_2} = 8.78$. For these values, we got FNR $= 0$ and TNR $= 0.06$. The "non-fatigued region" $\mathcal{S}_0$ is then found (with f_1 and f_2 being *back rotation position in sagittal plane* and *chest acceleration mean* respectively):

$$\mathcal{S}_0 = (f_1 \in (-\infty, -1.68) \vee f2 \in (3.65, \infty))$$

Just as for the outside approach, we limited such region in function of the two group ages (up to and over 40 years old). This procedure redefines $\mathcal{S}_0$ for the two age groups as follows (see Fig. 2 for the graphical representation):

$$\mathcal{S}_0 = (f_1 \in (-2.52, -1.68) \vee f2 \in (3.65, 3.99)) \text{ for } age \leq 40 \text{ y.o.}$$
$$\mathcal{S}_0 = (f_1 \in (-1.86, -1.68) \vee f2 \in (3.65, 3.99)) \text{ for } age > 40 \text{ y.o.}$$

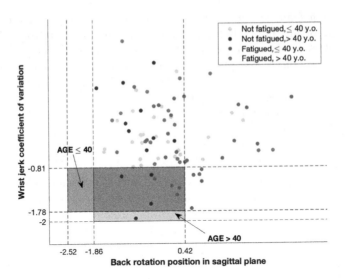

Fig. 1. Scatter plot of the first two features (back rotation position in sagittal plane and wrist jerk coefficient of variation) with representations of the "non-fatigue region" (FNR = 0) individuated for age ≤ 40 group (pink) and age > 40 (violet). (Color figure online)

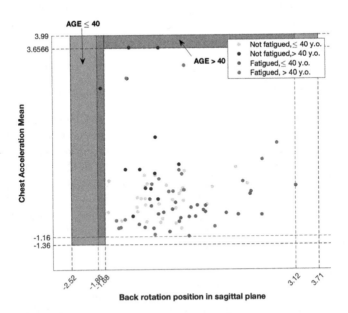

Fig. 2. Scatter plot of the first two features (back rotation position in sagittal plane, Chest Acceleration Mean) from value ranking of non-fatigued class, with representations of the "non-fatigue regions" (FNR = 0) based on the age group (violet for age ≤40, pink otherwise) (Color figure online)

Zero Error LLM. Both the previous approaches have the limitation of individuating optimal solutions to the identification of "non-fatigue regions" characterized by relatively low values of TNR, i.e. number of instances included in such surfaces.

In order to assess if such values could be increased, we trained the LLM 0% and built a new predictor by joining the first four highest coverage rules in logical OR (see below).

if $(0.51 <$ HipACCMean ≤ 1.98 **and** ChestACCcoefficientofvariation ≤ 1.11 **and** -1.73 $<$ averagestepdistance ≤ 0.81 **and** backrotationpositioninsagplane $\leq$ 0.52) $\vee$
(WristjerkMean > 0.55 **and** -1.35 $<$ Back rotation position in sag plane $\leq$ 0.04) $\vee$
(-1.73 $<$ averagestepdistance $\leq$ -0.22 **and** backrotationpositioninsagplane $\leq$ -0.25 **and** -0.44 $<$ numberofsteps ≤ 3.75 **and** -1.73 $<$ Wristjerkcoefficientofvariation ≤ 0.55) $\vee$
(ChestxpostureMean > -0.033 **and** HipzpostureMean > 0.43 **and** WristACCMean > -0.83 **and** -0.88 $<$ backrotationpositioninsagplane ≤ 0.29) **then** non-fatigued

By evaluating the joining before any perturbation, we got FNR $= 0.06$ and TNR $= 0.75$. To further decrease the FNR, we conducted the optimization process described in Eq. 9 by tuning the thresholds for the first $N^{FR} = 2$ features from non-fatigued feature ranking, namely *HipACCMean* and *WristjerkMean*. We obtained $\delta_{s_1}^* = 1.848$ and $\delta_{t_2}^* = 0.027$ for such features respectively: these thresholds perturbations brought FNR $= 0.02$, with TNR $= 0.42$.

Skope-Rules. To ensure that we obtained rules with zero errors on the non-fatigue classification task, we trained several models with a *precision_min* $= 1$, where *precision_min* is the parameter that defines the minimum precision of a rule to be selected in the *performance filtering*. Trained models differ in *n_estimators* and *max_depth_duplication*, where *n_estimators* is the number of base estimators to use for prediction and *max_depth_duplication* is the maximum depth of the decision tree for *semantic deduplication* (Sect. 4). For each model thus obtained, we calculated precision and recall by varying the number of rules applied (from 2 up to the maximum number of rules generated by the model) and then chose the one that maximised precision and recall. This led us to use a model trained with the following parameters:

1. *n_estimators* $= 200$
2. *precision_min* $= 1$
3. *max_depth_duplication* $= 5$

We then chose the first 3 rules generated by this model which correspond to the following logical OR ($\vee$):

if (backrotationpositioninsagplane ≤ 0.08 **and** HipjerkMean > -1.03 **and** HipACCcoefficientofvariation ≤ 0.75 **and** HipypostureMean ≤ 1.12 **and** HipzpostureMean > -1.78) $\vee$

(backrotationpositioninsagplane $\leq$ 0.17 **and** Wristjerkcoefficientofvariation $\leq$
0.05 **and** HipACCMean > -0.47) $\lor$
(backrotationpositioninsagplane $\leq$ 0.22 **and** Wristjerkcoefficientofvariation $\leq$
0.06 **and** HipACCMean > -0.10 **and** ChestjerkMean > -1.36) **then**
non-fatigued

This new predictor, before applying any perturbation, leads to FNR $= 0.11$ and TNR $= 0.69$. As in the previous case, let's see what happens in terms of FN by perturbing two features. The features we are going to perturb are *backrotationpositioninsagplane* and *Wristjerkcoefficientofvariation* and they are respectively the first and second most present features in the rules derived from the performance filtering (Sect. 4). To carry out the perturbation we used the procedure as described in Sect. 5.3, applying the method of Eq. 9 and perturbing only the most restrictive thresholds when the same features appeared in more than one rule. This leads us to the following suboptimal solution, with an FNR $= 0.07$ and TNR $= 0.67$, corresponding to $\delta_{t1} = 1.717$ for *backrotationpositioninsagplane* and $\delta_{t2} = 15.845$ for *Wristjerkcoefficientofvariation*.

6.2   Vehicle Platooning

Vehicle platooning is one of the most important challenges in autonomous driving, dealing with a trade-off between performance and safety. In our analysis we considered a scenario of cooperative adaptive cruise control (CACC) as described in [27], where the platoon is in a steady state of speed and reciprocal intervehicular distance when a braking force is applied by the leader of the platoon. For the application of our safety assessment methods we used simulation data generated by Plexe simulator[4]. For each of the 4744 generated samples, 5 features were computed within the following ranges: the number of vehicles, $N \in [3, 8]$ the braking force $F_0 \in [-8, -1] \times 10^3$ N the Packet Error Rate $PER \in [0.2, 0.5]$ the initial distance between vehicles $d(0) \in [4, 9]$ m (supposed equal for all of them); the initial speed $v(0) \in [10; 90]$km/h. The system registers a collision when distance between two vehicles is lower than 2 m.

Applying the default LLM with maximum error of 5% on a 30% test set, we obtained 85,9% of accuracy, 75.4% sensitivity, 86.8% specificity and 0.46 F1-score. We then performed the safety analysis to find out regions were collisions are avoided with no error.

Reliability from Outside. From the value ranking for the collision class ($y = 1$), we obtained $PER > 0.43$ and $F_0 \leq -7.50 \times 10^3$N as the first two most important intervals. We then applied the optimization approach as in Eq. 5 and found $\delta^*_{s_1} = -0.034$, $\delta^*_{t_2} = -0.416$, which correspond to reach FNR $= 0$ with TNR $= 0.34$. Thus, according to the definition in Eq. 6, the safety region we obtain is the following:

$$\mathcal{S}_1 = ((PER \in (0.2, 0.4154)) \land (F_0 \in (-4.37, -1) \times 10^3)$$

---

[4] https://github.com/mopamopa/Platooning.

A visual representation of such region is in Fig. 3. Also, we performed a search for safety regions by considering three features, including the third most important interval from value ranking too, i.e. $N > 6$. We got $\delta^*_{s_1} = -0.184, \delta^*_{t_2} = -0.166$ and $\delta^*_{s_3} = -0.1$ with FNR = 0 and TNR = 0.19. In this case, the safety region is tridimensional, corresponding to the following volume (Fig. 4):

$$\mathcal{V}_1 = ((PER \in (0.2, 0.3509)) \wedge (F_0 \in (-6.255, -1) \times 10^3) \wedge (N \in (3, 5.4)$$

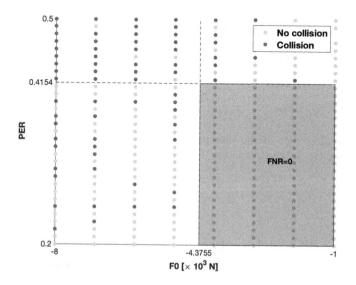

Fig. 3. Scatter plot of the first two features (PER and F0) with representations of the safety region

Reliability from Inside. Following the optimization approach in Eq. 7, we first chose the first two intervals from the value ranking of the safe class ($y = 0$): $PER \leq 0.33$ and $F_0 > -3.50 \times 10^3$N. Then, we computed the optimal threshold perturbations $\delta^*_{t_1} = 0.356, \delta^*_{s_2} = 0.686$, for which we got FNR = 0 with TNR = 0.13. The safety region is then individuated by the following surface (Fig. 5):

$$\mathcal{S}_0 = (PER \in (0.2, 0.2125) \vee F_0 \in (-1.1001, -1) \times 10^3)$$

Zero Error LLM. By lowering the LLM maximum error allowed to 0% we were able to look for more complex safety regions. After training the LLM model with 0% error, we joined the first 4 rules for safe class with the highest coverage. This corresponded to the following logical OR ($\vee$):

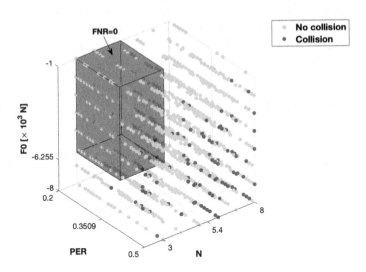

Fig. 4. 3D scatter plot of the first three features (PER,F0,N): the safety region is represented by the volume (in violet) (Color figure online)

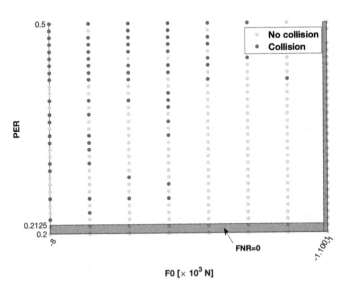

Fig. 5. Scatter plot of the first two features (PER and F0) for safe class with representations of the safety region

$$\textbf{if } (N \leq 5 \textbf{ and } v(0) \leq 54.50) \vee$$
$$(PER \leq 0.295 \textbf{ and } N \leq 7 \textbf{ and } v(0) \leq 86.50) \vee$$
$$(v(0) \leq 28.50 \textbf{ and } PER \leq 0.445) \vee$$
$$(v(0) \leq 28.50 \textbf{ and } N \leq 6 \textbf{ and } d(0) \leq 7.86) \textbf{ then } \text{safe}$$

This new predictor, before applying any perturbation, leads to FNR $= 0.05$ and TNR $= 0.55$. We then exploited the feature ranking to individuate which features we should tune in order to lower FNR as much as possible. The two most influent features resulted to be $v(0)$ and PER in this case. Then, by applying the method in Eq. 9 we perturbed such features: in this case, we were able to achieve only a suboptimal solution, with FNR $= 0.02$ and TNR $= 0.45$, corresponding to $\delta_{t1} = 0.000877$ for $v(0)$ and $\delta_{t2} = 0.277$ for PER. Where the same feature was present in more than one joined rule, we perturbed only the most stringent threshold.

Skope-Rules. As explained above for the Physical Fatigue case, also for Platooning, we trained different models by varying the parameters *n_estimators* and *max_depth_duplication*. Again, we chose to set *precision_min* $= 1$ to obtain rules with zero errors on the non-collision classification task. Again, for each model thus obtained, we calculated precision and recall by varying the number of rules applied (from 2 up to the maximum number of rules generated by the model) and then chose the one that maximised precision and recall. This led us to use a model trained with the following parameters:

1. *n_estimators* $= 75$
2. *precision_min* $= 1$
3. *max_depth_duplication* $= 2$

We then chose the first 4 rules generated by this model which correspond to the following logical OR ($\vee$):

$$\textbf{if } (PER \leq 0.41 \textbf{ and } v(0) \leq 45.5) \vee$$
$$(N \leq 7.5 \textbf{ and} F0 > -7.5 \textbf{ and} PER \leq 0.32) \vee$$
$$(N \leq 5.5 \textbf{ and } v(0) \leq 54.5) \vee$$
$$(F(0) > -4.5 \textbf{ and } PER \leq 0.41 \textbf{ and } v(0) > 64.5) \textbf{ then } \text{safe}$$

This new predictor, before applying any perturbation, leads to FNR $= 0.04$ and TNR $= 0.57$. To compare the results obtained previously for the Zero Error LLM, we decided again to perturb two features in the same way as described above (applying the method of Eq. 9 and perturbing only the most restrictive thresholds). In this case, the first and second most present features in the rules derived from the performance filtering (Sect. 4) are *v0* and *PER*, the same obtained from the LLM ranking. This leads us to the following suboptimal solution, with an FNR $= 0.02$ and TNR $= 0.52$, corresponding to $\delta_{t1} = -0.649$ for *v(0)* and $\delta_{t2} = -0.172$ for *PER*.

6.3   Discussion

From a comparison between the obtained results on the two datasets, we can notice that inferring reliability from the available rules is highly dependent on the structure of the data under analysis. The inside-outside (Sects. 5.2, 5.1) methods show flexibility in looking at the feature space, alternating good results (outside in platooning in two dimensions), surprising results (outside in platooning in three dimensions is outperformed by the same in two dimensions) and bad results (inside in platooning in two dimensions). The outside approach finds larger (higher TNR) safety regions than the inside one both in fatigue and platooning. Inside-outside may be even joined together when the feature ranking agrees on the most important features for the available classes. As this happens in the platooning case, we may consider the safety regions involving PER and F_0 (Figs. 3 and 5), and, by visual analysis of the overlap of such regions (see Fig. 6), we could join them to find a larger and more complex (in terms of rules) safety region.

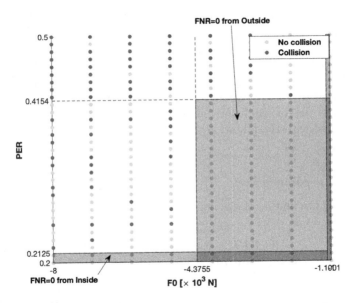

Fig. 6. Scatter plot of the two most important features in vehicle platooning LLM classification (PER and F_0), with representation of the safety regions found with Inside (pink area) and Outside (blue area) methods: the overlap of such regions defines a new safety region, where TNR reaches higher values (Color figure online)

On the other hand, due to the similarity of the adopted rules optimization approach (Sect. 5.3), we can compare the results of LLM 0% and Skope-Rules. Since we were dealing with more complex profiles than rectangles, results have shown an increase of TNR for both the models on the two datasets. However, in the physical fatigue test case, the LLM 0% starts by a much lower FNR (0.06) than Skope (0.11) before perturbation, reaching a sub-optimal solution after tuning; in contrast, Skope achieves a suboptimal solution too, with a FNR (0.07) that is surprisingly higher than the corresponding value of LLM 0% before optimization (0.06). As regards the vehicle platooning problem, results are more consistent in the two algorithms, showing the same FNR and a higher TNR with Skope.

7 Conclusions and Future Works

In this work, we have studied how XAI models can represent a solution towards safety assurance in predictive analytics. We first focused on a global rule-based model, the LLM, and demonstrated how its characteric value ranking property can be exploited for the design of "safety regions" in the features space with zero statistical error. This was achieved by developing our innovative "reliability from outside" and "reliability from inside" methodologies. Then, we used a third method to optimize more complex rule profiles and applied it to LLM 0% and Skope-Rules.

Data and code are available at the following Github repository: https://github.com/saranrt95/safety-from-valueranking.

By testing and comparing our proposed methodologies on problem instances of different nature (physical fatigue and vehicle platooning), we have also shown how their performance varies between the datasets.

Future works may extend the testing through cross-validation in the presence of a large amount of data, including the adoption of data augmentation techniques and the experimentation on benchmark datasets. The characterization of the placement of the points deserves further study to understand the optimal covering of the safety regions. The translation of deep learning logic into rules with further design of safety envelope is another topic we are going to pursue in the near future.

References

1. Adebayo, J., et al.: Sanity checks for saliency maps. arXiv preprint arXiv:1810.03292 (2018)
2. Arrieta, A.B., et al.: Explainable artificial intelligence (XAI): concepts, taxonomies, opportunities and challenges toward responsible AI. Inf. Fusion **58**, 82–115 (2020)
3. Balasubramanian, V.N., Ho, S., Vovk, V.: Conformal Prediction for Reliable Machine Learning, 1st edn. Morgan Kaufmann Elsevier (2014)
4. Becker, U.: Increasing safety of neural networks in medical devices. In: Romanovsky, A., Troubitsyna, E., Gashi, I., Schoitsch, E., Bitsch, F. (eds.) SAFE-COMP 2019. LNCS, vol. 11699, pp. 127–136. Springer, Cham (2019). https://doi.org/10.1007/978-3-030-26250-1_10
5. Campagner, A., Cabitza, F., Ciucci, D.: Three-way decision for handling uncertainty in machine learning: a narrative review. In: Bello, R., Miao, D., Falcon, R., Nakata, M., Rosete, A., Ciucci, D. (eds.) IJCRS 2020. LNCS (LNAI), vol. 12179, pp. 137–152. Springer, Cham (2020). https://doi.org/10.1007/978-3-030-52705-1_10
6. Cangelosi, D., et al.: Logic learning machine creates explicit and stable rules stratifying neuroblastoma patients. BMC Bioinform. **14**(7), 1–20 (2013)

7. Cheng, C.H., et al.: Towards dependability metrics for neural networks (2018)
8. Clavière, A., Asselin, E., Garion, C., Pagetti, C.: Safety verification of neural network controlled systems. arXiv preprint arXiv:2011.05174 (2020)
9. Cluzeau, J., et al.: Concepts of design assurance for neural networks CoDANN. Standard, European Union Aviation Safety Agency, Daedalean, AG, March 2020. https://www.easa.europa.eu/sites/default/files/dfu/EASA-DDLN-Concepts-of-Design-Assurance-for-Neural-Networks-CoDANN.pdf
10. Cortes, C., et al.: Boosting with abstention. In: Lee, D., Sugiyama, M., Luxburg, U., Guyon, I., Garnett, R. (eds.) Advances in Neural Information Processing Systems, vol. 29. Curran Associates, Inc. (2016). https://proceedings.neurips.cc/paper/2016/file/7634ea65a4e6d9041cfd3f7de18e334a-Paper.pdf
11. Czarnecki, K., Salay, R.: Towards a framework to manage perceptual uncertainty for safe automated driving. In: Gallina, B., Skavhaug, A., Schoitsch, E., Bitsch, F. (eds.) SAFECOMP 2018. LNCS, vol. 11094, pp. 439–445. Springer, Cham (2018). https://doi.org/10.1007/978-3-319-99229-7_37
12. Eaton-Rosen, Z., Bragman, F., Bisdas, S., Ourselin, S., Cardoso, M.J.: Towards safe deep learning: accurately quantifying biomarker uncertainty in neural network predictions. In: Frangi, A.F., Schnabel, J.A., Davatzikos, C., Alberola-López, C., Fichtinger, G. (eds.) MICCAI 2018. LNCS, vol. 11070, pp. 691–699. Springer, Cham (2018). https://doi.org/10.1007/978-3-030-00928-1_78
13. Gehr, T., et al.: AI2: safety and robustness certification of neural networks with abstract interpretation. In: 2018 IEEE Symposium on Security and Privacy (SP), pp. 3–18. IEEE (2018)
14. Gordon, L., et al.: Explainable artificial intelligence for safe intraoperative decision support. JAMA Surg. **154**(11), 1064–1065 (2019)
15. Gu, X., Easwaran, A.: Towards safe machine learning for CPS: infer uncertainty from training data (2019)
16. Guo, C., et al.: On calibration of modern neural networks. In: International Conference on Machine Learning, pp. 1321–1330. PMLR (2017)
17. Hendrycks, D., Dietterich, T.: Benchmarking neural network robustness to common corruptions and perturbations (2019)
18. Holzinger, A., et al.: What do we need to build explainable AI systems for the medical domain? (2017)
19. Isele, D., et al.: Safe reinforcement learning on autonomous vehicles. In: 2018 IEEE/RSJ International Conference on Intelligent Robots and Systems (IROS), pp. 1–6. IEEE (2018)
20. ISO/IEC: Standardization in the area of artificial intelligence. Standard, ISO/IEC, Washington, DC 20036, USA (Creation date 2017). https://www.iso.org/committee/6794475.html
21. Koshiyama, A., et al.: Towards algorithm auditing: a survey on managing legal, ethical and technological risks of AI, ML and associated algorithms. SSRN Electron. J. (2021)
22. Lakshminarayanan, B., et al.: Simple and scalable predictive uncertainty estimation using deep ensembles. In: Proceedings of the 31st International Conference on Neural Information Processing Systems, pp. 6405–6416 (2016)
23. Madhavan, R., et al.: Toward trustworthy and responsible artificial intelligence policy development. IEEE Intell. Syst. **35**(5), 103–108 (2020)
24. Maman, Z.S., et al.: A data analytic framework for physical fatigue management using wearable sensors. Expert Syst. Appl. **155**, 113405 (2020)
25. Mohseni, S., et al.: Practical solutions for machine learning safety in autonomous vehicles. arXiv preprint arXiv:1912.09630 (2019)

26. Mongelli, M., Muselli, M., Ferrari, E.: Achieving zero collision probability in vehicle platooning under cyber attacks via machine learning. In: 2019 4th International Conference on System Reliability and Safety (ICSRS), pp. 41–45. IEEE (2019)
27. Mongelli, M., Ferrari, E., Muselli, M., Fermi, A.: Performance validation of vehicle platooning through intelligible analytics. IET Cyber-Phys. Syst. Theory Appl. **4**(2), 120–127 (2019)
28. Mongelli, M., Muselli, M., Scorzoni, A., Ferrari, E.: Accellerating prism validation of vehicle platooning through machine learning. In: 2019 4th International Conference on System Reliability and Safety (ICSRS), pp. 452–456. IEEE (2019)
29. Maurizio, M., Vanessa, O.: Stability certification of dynamical systems: lyapunov logic learning machine. In: Thampi, S.M., Lloret Mauri, J., Fernando, X., Boppana, R., Geetha, S., Sikora, A. (eds.) Applied Soft Computing and Communication Networks. LNCS, vol. 187, pp. 221–235. (2021). https://doi.org/10.1007/978-981-33-6173-7_15
30. Muselli, M.: Switching neural networks: a new connectionist model for classification (2005)
31. Parodi, S., et al.: Differential diagnosis of pleural mesothelioma using logic learning machine. BMC Bioinform. **16**(9), 1–10 (2015)
32. Parodi, S., et al.: Logic learning machine and standard supervised methods for Hodgkin's lymphoma prognosis using gene expression data and clinical variables. Health Inform. J. **24**(1), 54–65 (2018)
33. Pereira, A., Thomas, C.: Challenges of machine learning applied to safety-critical cyber-physical systems. Mach. Learn. Knowl. Extr. **2**(4), 579–602 (2020)
34. Samek, W., et al.: Explainable artificial intelligence: understanding, visualizing and interpreting deep learning models. ITU J.: ICT Discoveries - Special Issue 1 - The Impact of Artificial Intelligence (AI) on Communication Networks and Services **1**, 1–10 (2017)
35. Saranti, A., Taraghi, B., Ebner, M., Holzinger, A.: Property-based testing for parameter learning of probabilistic graphical models. In: Holzinger, A., Kieseberg, P., Tjoa, A.M., Weippl, E. (eds.) CD-MAKE 2020. LNCS, vol. 12279, pp. 499–515. Springer, Cham (2020). https://doi.org/10.1007/978-3-030-57321-8_28
36. Schwalbe, G., Schels, M.: A survey on methods for the safety assurance of machine learning based systems. In: 10th European Congress on Embedded Real Time Software and Systems (ERTS 2020) (2020)
37. Seshia, S.A., et al.: Formal specification for deep neural networks. In: Lahiri, S.K., Wang, C. (eds.) ATVA 2018. LNCS, vol. 11138, pp. 20–34. Springer, Cham (2018). https://doi.org/10.1007/978-3-030-01090-4_2
38. International Organization for Standardization: Road vehicles safety of the intended functionality PD ISO PAS 21448:2019. Standard, International Organization for Standardization, Geneva, CH, March 2019
39. Sun, Y., et al.: Structural test coverage criteria for deep neural networks. In: 2019 IEEE/ACM 41st International Conference on Software Engineering: Companion Proceedings (ICSE-Companion), pp. 1–23. ACM New York (2019)
40. Varshney, K.R.: Engineering safety in machine learning. In: 2016 Information Theory and Applications Workshop (ITA), pp. 1–5. IEEE (2016)
41. Wiener, Y., El-Yaniv, R.: Agnostic pointwise-competitive selective classification. J. Artif. Int. Res. **52**(1), 179–201 (2015)
42. Williams, N.: The Borg rating of perceived exertion (RPE) scale. Occup. Med. **67**(5), 404–405 (2017)
43. Zhang, X., et al.: DADA: deep adversarial data augmentation for extremely low data regime classification. IEEE Trans. Circuits Syst. Video Technol. 2807–2811 (2019)

Explanatory Pluralism in Explainable AI

Yiheng Yao[✉]

Philosophy-Neuroscience-Psychology Program,
Washington University in St. Louis, St. Louis, MO, USA
yaoyiheng@wustl.edu

Abstract. The increasingly widespread application of AI models motivates increased demand for explanations from a variety of stakeholders. However, this demand is ambiguous because there are many types of 'explanation' with different evaluative criteria. In the spirit of pluralism, I chart a taxonomy of types of explanation and the associated XAI methods that can address them. When we look to expose the inner mechanisms of AI models, we develop Diagnostic-explanations. When we seek to render model output understandable, we produce Explication-explanations. When we wish to form stable generalizations of our models, we produce Expectation-explanations. Finally, when we want to justify the usage of a model, we produce Role-explanations that situate models within their social context. The motivation for such a pluralistic view stems from a consideration of causes as manipulable relationships and the different types of explanations as identifying the relevant points in AI systems we can intervene upon to affect our desired changes. This paper reduces the ambiguity in use of the word 'explanation' in the field of XAI, allowing practitioners and stakeholders a useful template for avoiding equivocation and evaluating XAI methods and putative explanations.

Keywords: Explainable artificial intelligence · Philosophy · Causation · Explanations · Explainability · Interpretability

1 Introduction

There is no doubt that we should be exacting in our demand on explanations for the outputs, functioning, and employment of AI models, given that they are increasingly implicated in decision making that impact humans with potentially undesirable outcomes [31]. However, just what we mean by 'explanations' in the field of Explainable AI (XAI) is currently unclear [24,36]. What is clear is that many different stakeholders have different constraints on the explanations they want from the field [18]. There is a pressing danger that what explains appropriately and sufficiently is lost in translation from stakeholders to practitioners, and vice versa. In other words, even if the General Data Protection Regulation

© IFIP International Federation for Information Processing 2021
Published by Springer Nature Switzerland AG 2021
A. Holzinger et al. (Eds.): CD-MAKE 2021, LNCS 12844, pp. 275–292, 2021.
https://doi.org/10.1007/978-3-030-84060-0_18

(GDPR) strongly enforces[1] that explanations be given when decisions are contested, it is a pyrrhic victory if there are no clear evaluative criteria on the explanations given or worse, that an inappropriate set of evaluative criteria is used to determine which explanation stands as an admissible one.

Therefore, when explanations are requested from AI models, and when explanatory demand is placed on the field of Explainable AI, we should first ask some important questions. Why do we ask for 'explanations' rather than something else to fulfill the desired goal in posing such a request? What objective or purpose is the explanation supposed to serve in a given context? How should we judge whether a given 'explanation' satisfied those objectives or purposes?

Without such clarificatory questions, we run the danger of talking past each other in developmental efforts in XAI and in stakeholder's desiderata for the explanatory products of the field. As astutely noted by Mittelstadt, Russell and Wachter: "many different people ..., are all prepared to agree on the importance of explainable AI. However, very few stop to check what they are agreeing to" [24]. Langer et al. agree that more clarity is required: "Consistent terminology and conceptual clarity for the desiderata are pivotal and there is a need to explicate the various desiderata more precisely" [18]. Indeed, going forward, practitioners would benefit from clarity on the requirements for explanations, and stakeholders would benefit from clarity on the limits of explanatory methods produced by the field which would improve their choice of methods to employ.

Much recent work in XAI investigates just what are the explanatory demands placed onto XAI by way of analyzing social-psychological constraints [23,24, 27], how explanations function in the law [12,14], identifying stakeholders and their desiderata [18], and philosophical treatments of explanatory methods [26, 27]. These reviews correctly identify that explanations have a distinct social dimension as a process rather than purely as a product or text [19,23,26]; that explanations should be contrastive, selective, and non-statistical in their content [23,24]; and that within a social context, explanations of model output do not suffice in isolation [26].

However, talk of explanations in XAI have remained monolithic. In contrast, the stance I would like to present and defend in this paper is one of Explanatory Pluralism in XAI: the notion that there are many different types of explanation requested from the field for which have different effective treatment by means of methods (what we should produce) and different explanatory powers by range of application contexts (where we can use them). The primary contribution of this paper is a taxonomy, as illustrated in Fig. 1, distinguishing the different types of explanation along a Mechanistic-Social axis and a Particular-General axis by identifying the different types of intervention they target. Furthermore, the paper will organize present methods with more specific language introduced using this taxonomy while avoiding the loaded term 'explanation'.

---

[1] Whether a 'right to explain' exists has been debated on the basis of just what explanation is requested by the GDPR [33,37]. This debate in the literature further highlights the urgency of the present discussion to prevent possible equivocation of the different types of explanation.

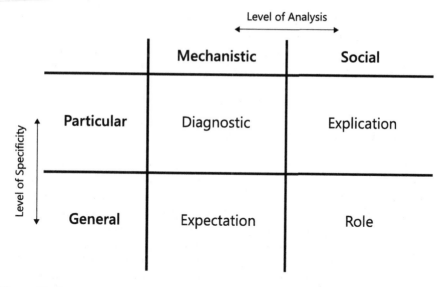

Fig. 1. Evaluative taxonomy proposed by this paper to categorize and distinguish the different types of explanation asked for and produced by XAI.

The idea that there are different explanations requested from XAI is not new [26,36]. However, organization of different explanatory methods have mostly been done in descriptive terms [36]. In this paper, I present a taxonomy based on evaluative terms. XAI methods find membership in the proposed taxonomy by virtue of differences in evaluative criteria rather than differences in descriptive characteristics (when they are assessed – ex-ante/post-hoc, generality of application – agnostic/specific, output format, input data, or problem type [36]). By aligning XAI methods with what interventions they target, the success of each method can then be evaluated on the effectiveness of different interventions. The explanatory pluralistic view is also non-reductive, meaning that each category of explanation thus organized do not subsume other categories even though dependency relations may exist between them. I justify my organization of the taxonomy by appeal to recent work on the nature of scientific explanation in the Philosophy of Science, specifically Woodward's manipulationist account of causation [41], Craver's mechanistic account of scientific explanations [9], and causal relevance [11]. This normative, philosophically grounded taxonomy serves to specify more clearly what the word 'explanation' means in different contexts.

I begin by reviewing in Sect. 2 the diverse explanatory demands for explainability in AI, emphasizing what we are supposed to explain and what we think explanations will help us to achieve. Next in Sect. 3, I provide relevant contemporary philosophical background drawing from the rich literature in Philosophy of Scientific Explanations to motivate the organization in my proposed taxonomy. In Sect. 4, I derive and define the Mechanistic-Social, Particular-General taxonomy illustrated in Fig. 1. Furthermore, in Sect. 5, I take a pragmatic interventionist stance and organize present methods in XAI into each of the four categories identified by the proposed taxonomy, showing how methods in each evaluative category fulfill different explanatory demands. Finally, I conclude with two recommendations and highlight that XAI is not merely a way of looking back and within our models but a way of looking forwards and outwards, a perspective ineliminably involved in the development of truly intelligent systems.

2   Explanatory Demands

By 'explanatory demand' I mean here what is expected of explanations produced by XAI and more broadly, what are the demands placed on the explanatory products of the field (methods which produce explanations). I will anchor my review in the papers by Tim Miller [23], Kieron O'Hara [26], and Langer et al. [18], organized into three broad areas: social-psychological, social-contextual, and functional (exemplified by stakeholder desiderata). Lastly, I will present what is expected of explanations from the law as examined by Doshi-Velez et al. [12] and highlight some regulatory requirements from the recently proposed Harmonized Rules on AI by the EU [8].

Social-Psychological Demand. Drawing from Lombrozo's work on the structure and function of explanations examined as a psychological phenomenon [19], Miller elucidates some key considerations we expect from explanations when they are given to humans [23]: 1) explanations as a social process aim to render something understandable by transferring knowledge between an explainer and explainee; 2) presentation of causes in contrastive terms is preferred; 3) causes cited within an explanation is selective and does not represent the full and complete set of causes; 4) statistical generalizations alone are unsatisfying. In treating explanations as not mere static products but a process that involves social agents, we highlight one important feature of explanations: they elicit understanding (in humans). It is crucial to note that fulfilling this goal sets evaluative criteria that are dependent not upon the content of explanations, whether they do in fact relate to what is explained, but upon how the relevant information is packaged and presented and whether its delivery improves understanding. Put in another way, it is about what makes the light bulb go off in our head, however we reach for the switch[2]. One way of noticing this point is by observing the role of idealized models in science. We do not start teaching with relativity and quantum mechanics but often start by introducing Newtonian physics and constrain our approximate models within some limits such as slow speeds and large sizes. Although such idealized models do not veridically reflect the structure of the world, they lend themselves to better understanding. Of course, we can and should impose the additional constraint that the content of the explanation accurately reflects the underlying causal structure [10,28]. However, the important point is that we have both a factivity criterion and an understanding criterion that can be evaluated independently of one another [27].

---

[2] Craver, 2021, personal communications.

Social-Contextual Demand. When we employ AI models to aid humans in making decisions or to produce outputs that impact humans, we need to situate the AI as part of a larger social context. O'Hara notes that AI models do not have decision-making power in and of themselves. Administrators can choose to intervene upon systems, and how the output is acted upon is distinct from the mechanisms of the AI model that generated it [26]. As such, when we ask for explanations regarding decisions 'made' by AI models, we ought to include relevant details of where such a model is situated in the broader social context surrounding its usage.

Functional Demand. Langer et al. compiles a comprehensive assessment of the different stakeholders who are interested in seeking explanations from AI models [18]: 1) Users seeking *usability* and *trust*; 2) Developers seeking *verification* and *performance*; 3) Affected parties seeking *fairness* and *morality/ethics*; 4) Deployers seeking *acceptance* and *legal compliance*; and 5) Regulators seeking *trustworthiness* and *accountability*. The type of explanation that prove useful to developers of AI models for the purpose of debugging or improving model accuracy would look very different than that which a non-expert user may request for understanding how their personal data is used, precisely because they serve such different purposes. Therefore, it would be insufficient to simply claim that 'explanations' help in all these diverse cases, we need to further specify what type of explanation would help by clarifying the explanandum (what is to be explained).

Legal Demand. Explanations are of value in legal settings for holding AI systems accountable by "exposing the logic behind a decision" and to "ascertain whether certain criteria were used appropriately or inappropriately in case of a dispute" [12]. In their review, Doshi-Velez et al. also note that explanations will be requested only when they "can be acted on in some way" [12], highlighting the cost-benefit trade-off in generating explanations. In addition, the authors note that further explanations may be demanded "even if the inputs and outputs appear proper because of the context in which the decision is made" [12]. Here, three demands on explanations are emphasized. They must: 1) identify contributing factors to the output; 2) identify actionable factors specifically; and 3) attend to the context in which the AI system is deployed to make decisions and take actions. In addition, the proposal for Harmonized Rules on AI in the EU sets additional requirements on employing "high-risk" AI systems intended to be used as "a safety component" [8]. The intent of the proposal echoes that of Article 22 in the GDPR that placed restrictions on automated decisions "which produces legal effects ... or similarly significant affects" on humans subjected to such decisions [7]. In both cases, regulators are interested in identifying AI systems that play a significant role in impacting humans and place additional restrictions on their usage. Furthermore, the newly proposed Harmonized Rules on AI additionally introduce a "Technical Documentation" requirement in Article 11(1) for fielding such "high-risk" AI systems [8]. This document as described in Annex IV includes a comprehensive list of information such as "how the AI system interacts or can be used to interact with hardware or software that is not part of the AI system itself, where applicable" (Annex IV 1(b)), "what the system is designed to optimize for and the relevance of the different parameters" (Annex IV 2(b)), and "metrics used to measure accuracy, robustness, cybersecurity" (Annex IV 2(g)) [8].

2.1 Fulfilling Disparate Explanatory Demands

Explanations are sought for in a multitude of situations, with a diverse set of goals and expectations as reviewed in this section. Considering the importance of explanations in ensuring the responsible usage of AI systems, there is a pressing need to evaluate the quality of explanations given. However, what constitutes as a meaningful explanation differs to the different stakeholders involved. Therefore, we should first acknowledge the plurality of explanations and distinguish between the different types of explanation so we can develop the appropriate evaluative criteria and methods to address the different requests for meaningful explanations. In the next section, I will appeal to recent work in the Philosophy of Science on the nature of scientific explanations to show how we can differentiate between requests for explanations by identifying the relevant level of change in the AI model we wish to affect using the notion of causal relevance.

3   Scientific Explanations

Much has already been said on the nature of explanations, especially what are good explanations in the sciences [42]. One point of agreement between scientific explanations and past work on the nature of explanations in XAI is that explanations should unveil causes [10,23]. However, evaluating the quality of explanations based on the amount of causes they identify or how many why-questions they can answer is insufficient [11]. As previously acknowledged, explanations should further be selective [23,24]. I appeal to recent developments in the Philosophy of Science to state more clearly how we should be selective with our explanations.

3.1   Manipulationist Account of Causation

Firstly, just what is this notion of a 'cause'? The manipulationist account of causation put roughly is that: X causes Y if manipulating X changes the value of Y or its probability distribution. Put in another way, "causal relationships are relationships that are potentially exploitable for purposes of manipulation and control" [41]. Furthermore, Woodward introduces a stability constraint in evaluating which cause is more suitable given some effect Y [40]. Under the stability constraint, causal relationships which "continue to hold under a 'large' range of changes in background circumstances" [40] should be preferable. This may be a driver for the social-psychological demand for explanations presented in contrastive terms. The larger the range of counterfactuals identified under which the causal relationship holds, the more inclined we may be in accepting the identified cause.

3.2   Mechanistic Account of Scientific Explanations

Craver builds upon this notion of causes as manipulable relationships, or points of intervention[3], to develop a mechanistic account of explanations for cognitive neuroscience. In this account, explanations describe mechanisms which are "entities and activities organized such that they exhibit the *explanandum* phenomenon" [9], where entities are the components or parts in a mechanism and activities are causes in the manipulationist sense. Three elements of the mechanistic account will be helpful for explicating different types of explanations in XAI: 1) explanations reveal the relevant causal organization of the explanandum at multiple levels; 2) different explanations given at different levels of realization are non-reductive; and 3) relevant causes are those which make a difference to the effect contrast asked for. In summary, the causal organization revealed by different explanations identify different relevant relationships which can be exploited for purposes of manipulation and control.

Levels of Explanation. Within a mechanism, activities and components in a lower level are organized to realize higher level activities or components [9]. Furthermore, such levels are "loci of stable generalizations" [9] in the sense that the behavior of components within each level are regular and predictable [9]. When we ask for explanations of a mechanism, we can attend to different levels to identify different stable generalizations we are interested in. For example, when we examine an AI model, we may be interested in the behavior of a range of components located at different levels of realization such as the training hyperparameters, model architecture, and optimization function.

Non-reductive. Since there are stable generalizations of mechanisms that are not true of the arrangement of components that realize them [9], there are different causally relevant sets of components at different levels of realization. Explanations of general AI model behavior such as identifying what rules they follow in processing patterns of input features need not necessarily be better substituted with explanations of particular AI model processes that led to an output. The latter may add further details to the former but without situating such details within a higher level, it would be difficult to ascertain similar generalizations of model behavior. An analogy is that to explain the functioning of a program, we need not reduce our explanations to the movement of electrons in the CPU although such movement does realize the program under question at a lower level.

---

[3] In this paper, I sometimes use the term interventions in place of manipulable relationships. The difference between a manipulable relationship and an intervention [41] is a subtle one that does not affect my arguments.

Causal Relevance. The notion of causal relevance stems from the non-reductive nature of levels of explanation as considered above. Causes which are relevant to an explanation should identify "the 'differences that make a difference.'" [11] When we seek explanations, inherent within our request is some class of effect contrast we are attending to. For example, when asking why an AI model classifies images in some way, we may attend to the particular relevance of some subset of features versus others as our contrast class or the distribution of labels over one dataset versus another. To effectively address the request for explanations, we should provide causes relevant to bringing about changes in the requested contrast class. The two ways we answer the question why an image is classified the way it is identify different points of intervention at different levels, so as to change the model behavior in different ways. By attending to particular feature relevance, we target changes in the model's output for a range of similar inputs. By attending to label distribution, we target changes in the model's classification behavior when given different datasets. It is therefore crucial to clarify what is the desired effect contrast so we can provide an appropriate explanation. The notion of an explanation revealing relevant causes at the appropriate level affords us a way to demarcate different types of explanation by identifying different levels of realization, different effect contrasts, and different points in AI systems where we can intervene.

4   Pluralistic Taxonomy

With the need to identify the desired effect contrast at different levels of realization as discussed in Sect. 3.2, I derive my proposed pluralistic taxonomy by augmenting David Marr's famous Three-Levels of Analysis widely applied in cognitive psychology and originally tailored for the biological visual system [20]. Furthermore, drawing inspiration from the taxonomy of Scientific Explanations introduced by Hempel that distinguishes between Particular Facts or General Regularities and Universal Laws or Statistical Laws [15], I arrive at a taxonomy similarly based on a Specific-General axis that additionally considers the augmented levels of analysis along a Mechanistic-Social axis.

4.1   Three Levels of Analysis (Plus One)

Neuroscientist David Marr introduced three levels of analysis to aid with understanding information processing systems [20]: the Computational level (the goal or problem solved); the Algorithmic level (processes and mechanisms used to solve said problem); and the Implementational level (physical substrate used to realize such mechanisms). Mapped onto AI terminology: 1) at the Computational level we can describe our models based on what task it attempts to perform (image classification, text-based summary generation, function minimization, etc.); 2) at the Algorithmic level we can describe what architecture is employed to solve this task (LSTM, RNN, GMM, etc.); and 3) at the Implementational level we can specify what are the hyperparameters that instantiate this

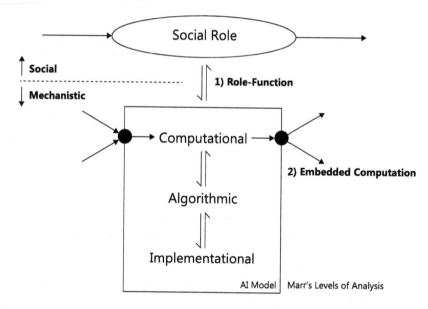

Fig. 2. David Marr's three levels of analysis for information processing systems adapted for AI model employment within social contexts.

particular model and the hardware we use to run it (TPU hours used, Bayesian Optimization value/acquisition functions used, etc.).

However, limiting analysis to the aforementioned three levels of analysis would be insufficient as XAI is specifically interested in types of computation and AI models that are used in some way that influences human decision making, or its outputs impact humans in some other way. For example, we are not typically interested in an isolated NPC (non-playable character) AI within some computer game which may similarly be decomposed into these levels of description and analysis. As such, highlighted in Fig. 2, we need to acknowledge that: 1) the AI models employed realize some Social Role, and 2) the AI model is embedded in additional computation surrounding its usage. Rarely do we have an AI model for which the input is statically specified, and its output directly used [26].

By Social Role, I mean to draw attention to the set of societal expectations surrounding decisions made in the context of application [6]. One may question the authority, ethics, and suitability of the AI model's (or the system's in general) continued employment in such a position that impacts humans or human decision-making. It is one question to ask whether an AI model is functioning as designed and an entirely separate question to ask whether the AI model thus designed could satisfactorily play the role we cast it in. The latter requires that we look outwards to position the AI model within its broader social context and identify whether it satisfies what is expected of such roles they may come to occupy. Granted, part of the difficulty here is that social expectations are typically not explicit[4] [12] and the systems we have for establishing suitable membership in social roles are tailored for human agents[5].

The addition of a Social level to Marr's three levels of analysis emphasizes the point that AI models do not operate in isolation, at least not the ones interesting to XAI. No matter how brightly we illuminate the mechanistic details within the AI model, no matter how transparent our algorithms are [3], we are missing a big chunk of the picture if we restrict discussion to analysis of only the Computational, Algorithmic, and Implementational levels.

4.2   Mechanistic-Social, Particular-General Taxonomy

In addition to the Mechanistic-Social levels of analysis distinction[6], we may also ask for explanations at different levels of specificity much like how Hempel distinguished between explaining particular facts from general regularities [15]. Here, we distinguish between asking questions pertaining to why a particular output was produced, and what types of output tend to be generated. We are also distinguishing between whether a particular social agent can understand outputs or explanations generated by XAI methods, and whether the usage of the AI model under question fits within the broader social context of application.

To be precise in our usage of language and avoid the ambiguous and loaded term 'explanation', each category in this taxonomy introduces a distinct term to disambiguate discourse. When we talk about explanations that identify mechanisms within an AI model contributing to particular outputs, we request for and produce Diagnostic-explanations on the matter (Mechanistic/Particular). When we wish to discern the general regularities of an AI model, we request for Expectation-explanations (Mechanistic/General). When we talk about explanations given to humans, we are requesting for Explication-explanations (Social/Particular). Finally, when we ask for justifications of model usage and seek guidance on regulations and policy, we request for Role-explanations which position an AI model within its context (Social/General).

The advantage of introducing this distinct terminology is two-fold. Firstly, we can keep separate questions which require different XAI methods to address appropriately and develop evaluative metrics and methods within each category independently. Secondly, we can now talk clearly about the relationship between each of these types of explanation and explanatory methods produced by XAI. Furthermore, adopting the view of explanatory pluralism means that we do not place primacy on any one type of explanation but acknowledge that there many types, each suiting a different context or need. For example, it is not the case that a Role-explanation should always be given, as it would do little to determine whether a particular AI model is actually functioning the way it was designed to.

---

[4] Interestingly, there has been research to determine the social norms surrounding trolley-like decision problems in the context of an imminent car crash [4].

[5] Non-human animals are not recognized as legal persons and cannot stand in courts [35]. Can an AI system stand in court as a defendant?

[6] That is not to deny that there may be social mechanisms.

An analogy here is that it is insufficient to ascertain that the person who gave the (incorrect) prescription was a doctor. Rather, we still need to ascertain particular facts of the matter such as whether the doctor made errors in judgment, or employed incorrect diagnostic tools, or whether such tools failed to function correctly which factored into the decision to prescribe the wrong medication. However, it is the case that if we were asking whether it was acceptable that this particular person gave someone else a prescription, we determine whether the person under question is a trained doctor or pharmacologist.

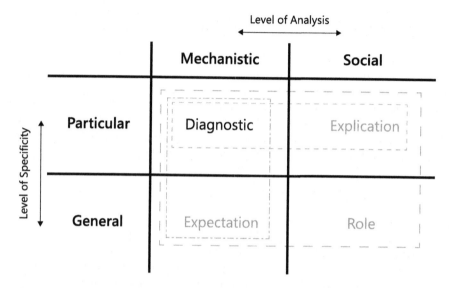

Fig. 3. Highlighting the dependency relations between the different evaluative categories.

This prescription analogy hints at the dependency relations, as highlighted in Fig. 3, between different types of explanation in the proposed taxonomy. When we give explanations that are explicable to human receivers fulfilling the set of social-psychological constraints, we also need to ensure that what we explicate match the mechanisms that produced the object of explication. In other words, as noted by Rudin, there is a worry that explanations produced may not match what the model computes [31]. Therefore, it is important to establish that whatever explanations that are explicable in terms of being contrastive, selective, and non-statistical (criteria noted in Sect. 2) be nonetheless grounded with suitable and accurate Diagnostic-explanations unveiling the relevant mechanisms in the model under scrutiny. Similarly, even if we were to use an Interpretable model with provable bounds which we can generate Expectation-explanations for, we still need to make use of diagnostic methods to verify that the model is functioning correctly. Moreover, simply putting a model for which we have certain bounded expectations on the table does not make its output immediately explicable, although having prior expectations might mean that model outputs lend themselves to easier explication. Expectation-explanations provided still need to fulfill a set of explicability criteria to be understandable to the target audience. Finally, to determine whether a model fits social expectations for the role that it occupies in its social context, we may require that it both be understandable to humans interacting with it and that we can draw generalizations around its function. But an explanation that is both explicable and based on an Interpretable model architecture may, however, still fail to identify and position the model within its social context and thus, fail to be a suitable Role-explanation.

5   Pragmatic Interventionist Stance

We can now position XAI methods within this pluralistic taxonomy by identifying the knobs and levers that we should manipulate to affect our desired effects (fulfillment of desiderata). In other words, the different categories differ in where we intervene upon our system to exact the desired changes. Taken together with the notion of causal relevance, the pragmatic interventionist stance, that explanations help us uncover relevant causes which identify manipulable relationships, affords us a unified way of categorizing XAI approaches.

5.1   Organizing Present XAI Methods

For a more comprehensive review of the methods in XAI, I refer the reader to [36]. In this section, I have chosen some representative examples to illustrate the application of my proposed taxonomy in Fig. 4. Within the category of Diagnostic-explanation are Saliency Maps [25], LIME [29], and Shapley Values [13] which identify particular input features important to affecting the output of models. The Explication-explanation category focuses on techniques to render explanations or model output understandable to humans interacting with the AI model. Such methods may include refining AI model interfaces with Human-Computer-Interaction (HCI) research [34], Google's AI Explanations "What-If" tool [39] to present feature relevance in contrastive terms, or by using the System Causability Scale [17] to measure the extent to which explanations generated were understandable. The Expectation-explanation category includes methods that focus on identifying and building regularities into models [21,22], ensuring robustness against adversarial attacks [5], and avoiding a pattern of output that potentially biases towards inappropriate features [43]. Interpretable models by virtue of their architectural attributes allow us to form certain expectations. For instance, the Neural Additive Model architecture uses a linear combination of neural networks to compute classification [2]. We can expect that a linear combination will combine each input feature in some weighted additive manner rather than have potentially unexpected interactions between features in high dimensions as deep neural networks typically do. Role-explanation emphasize the social context and embedded nature of AI models. By explicitly including humans in the process of decision-making and training, the consideration of Human-In-The-Loop is three-fold: 1) humans may be required to review AI model decisions to

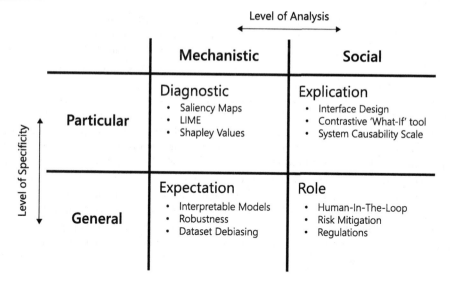

Fig. 4. Selected examples organized under each category in the proposed taxonomy.

comply with regulatory constraints; 2) humans can augment AI models with expertise and skills that AI models currently do not possess [16]; and 3) by including humans within each stage of the AI model, we can better ensure that AI objectives are aligned with human values since such systems open themselves up to more flexible alignment with human preferences [38]. Furthermore, risk mitigation protocols, as required for "high-risk" AI systems under the proposal for Harmonized Rules on AI [8], may identify ways to recover control when the AI system steps outside the boundaries of the role it plays, thereby increasing our trust in the AI system to perform within suitable roles. Finally, the growing body of regulations can help us to clarify what are the roles we envision AI systems can act beneficially within and their associated expectations.

5.2 Descriptive vs Evaluative Taxonomy

XAI methods can be categorized as illustrated in this proposed taxonomy by how we should evaluate them based on the sorts of intervention they identify instead of descriptive characteristics. Since causes identified by explanations should be relevant to the effect contrast we wish to affect [9,11,40], we should ask for XAI methods from the appropriate category of interventions. If we wish to examine changes in the model output, we should intervene at the level of a particular trained model asking for Diagnostic-explanations. If we wish to determine the broad guarantees of a model, then we should intervene at the level of the model architecture and ask for Expectation-explanations. If what we ultimately wish for is human understandability, then we should intervene upon the causes that bring about increased understandability, such as the social-psychological considerations outlined in Sect. 2, and ask for Explication-explanations. Finally, if we wish to better fit the usage of our AI model within its social role, perhaps what we should intervene upon is not the model architecture nor how explicable outputs are, but to involve human controllers and specify their operating procedures or develop regulatory mechanisms surrounding usage of such models and ask for Role-explanations.

Therefore, in addition to disambiguating discourse on different XAI methods, the proposed taxonomy also allows stakeholders to identify a match between their desiderata and the methods that should be employed. For example, if we want to "restore accountability by making errors and causes for unfavorable outcomes detectable and attributable to the involved parties" [18], what we are looking for will be Diagnostic-explanations that identify particular mechanisms in the model contributing to errors as well as Role-explanations that identify the context within which the model was situated. If we wish for users to "calibrate their trust in artificial systems" [18], then we request for Explication-explanations to render model output understandable and Expectation-explanations to identify robustness guarantees.

6   Conclusion

In conclusion, this paper presents an evaluative taxonomy that categorizes XAI methods based on the levels of intervention available and acknowledges the plurality of explanations produced. Furthermore, distinct terminology is introduced for each category to disambiguate the types of explanation we mean: Diagnostic, Expectation, Explication, and Role-explanation. This taxonomy is neither complete nor the only such way we can organize different types of explanation. Rather, this paper makes the point that it is useful for us to differentiate between types of explanation and we should do so on the basis of evaluative criteria rather than descriptive criteria. Additionally, future work is encouraged to develop metrics for evaluating XAI methods in each category. In particular, contributions from the social sciences will be crucial in identifying just what we should look for in Explication-explanations and Role-explanations. Nevertheless, we can now answer some of the clarificatory questions posed in the introduction. Why do we ask for 'explanations'? Because they allow us to identify relevant points of intervention for the desired effect. Furthermore, with the more specific language introduced, we can better distinguish between the evaluative conditions we wish to impose upon explanations requested. This allows stakeholders to more clearly present their objective, purpose and context under which explanations are sought from XAI. I will end with two recommendations for XAI, reemphasizing the point that rather than looking back upon and within our present models, methods developed in XAI can look forward as a way of advancing the field of AI and should look outwards to situate models within their social context.

6.1   Limit of Diagnostics

The first pressing recommendation is to use the more specific term 'AI Model Diagnostics' when we talk about explanatory methods that illuminate mechanisms within AI models. It would be prudent to treat present results from the field of XAI that are mere diagnostic tools as such explicitly to avoid confusion and granting such tools too much authority.

This difference between Diagnostics and full 'explanations' can be illustrated with an analogy to Air Crash Investigations. In the unfortunate case of airplane accidents, the recovery of the plane's Flight Recorder (also known as a black box) is but the first step in forming an investigative report into the accident. The data recorded by flight recorders contain a slice of the plane's flight history, preserving the state of the plane moments before the accident. From this data, investigators may be able to hypothesize what caused the accident by identifying anomalous parameters recorded by the black box. However, in many cases, the causes for airplane accidents do not lie entirely within the plane's state prior to the accident. Rather, the plane exists within the larger context of the flight industry which contains its pilots, maintenance crews, and regulations regarding flight paths and operating procedures. In addition, a key aspect of the final investigative report is to not only identify causes for the accident but recommendations for preventing future accidents from happening [1]. Furthermore, this investigative report also serves to assuage the public of the flight industry's reliability as well as address bereaved families' concerns. In this way, explanations for airplane accidents do not merely contain the causal aspect (which may already exceed the bounds of a plane's black box) but a social aspect of fulfilling the responsibility the flight industry has to its customers.

In a similar fashion, our investigations into AI models must not stop at uncovering what's within the black box (AI models), but look beyond and place the model within its social context. But to do so, we do indeed still require transparency into the inner workings of our AI models in order to render their behavior expectable and explicable. Therefore, a more holistic approach to constructing XAI explanatory products may be necessary by incorporating methods from multiple categories within the proposed taxonomy.

6.2   Ratiocinative AI

The second long-term recommendation is to identify an additional direction XAI can take that is somewhat distinct from the any of the categories defined in the proposed taxonomy: bake into AI models an awareness of its internal processes. In Rosenberg's critical take on connectionism, *Connectionism and Cognition*, he argues that the "mere exercise of a discrimination capacity, however complex, is not yet an example of cognition" [30] and identifies connectionist networks (neural networks) as only capable of mere discrimination by following certain rules. In addition to being rule-conforming (rational), Rosenberg argues in his paper that for truly cognizant systems, they should be rule-aware (ratiocinative) as well. Just so, I believe that for us to eventually develop tools that enable fruitful dialogue between humans and our AI models, we would come to imbue our AI models with an awareness of its internal processes.

What this awareness should be and how it can be implemented is currently unclear. XAI can contribute by not only identifying points of intervention, but eventually allowing us to reflexively surface these interventions back to the AI to develop truly intelligent, ratiocinative systems. Furthermore, present research in Reinforcement Learning agents also paints a promising path for us to achieve this goal. Building on top of its previous successes, DeepMind's recent MuZero agent is able to learn both the rules surrounding permissible actions within its environment as well as an optimal policy to act within this environment [32]. MuZero's awareness of internal rules and policies learned through interacting with the environment is built upon similarly opaque deep neural networks. However, the levels at which we can direct it questions and extract explanations appear to be broader than most other current AI models.

6.3   On Firmer Grounds

In closing, the categorization of many present XAI methods as 'Diagnostics' and admitting a plurality of explanations, thus noting the insufficiency of any single type of explanation, may be viewed as taking a step back. However, by taking this step back to reign in and clarify some of the expectations we have for present XAI methods, we stand on firmer grounds to take the next leap forward in XAI to produce holistic explanations and ensure the responsible usage of AI in society.

Acknowledgments. I thank Carl Craver for many fruitful discussions and helpful comments on multiple drafts of this paper, Jin Huey Lee for feedback on an earlier draft, and my anonymous reviewers for their many insightful comments.

References

1. AAIB: About us (2021). https://www.gov.uk/government/organisations/air-accidents-investigation-branch/about
2. Agarwal, R., Frosst, N., Zhang, X., Caruana, R., Hinton, G.E.: Neural additive models: interpretable machine learning with neural nets (2020)
3. Ananny, M., Crawford, K.: Seeing without knowing: limitations of the transparency ideal and its application to algorithmic accountability. New Media Soc. **20**(3), 973–989 (2016). https://doi.org/10.1177/1461444816676645
4. Awad, E., et al.: The moral machine experiment. Nature **563**, 59–64 (2018). https://doi.org/10.1038/s41586-018-0637-6
5. Blaas, A., Patane, A., Laurenti, L., Cardelli, L., Kwiatkowska, M., Roberts, S.: Adversarial robustness guarantees for classification with Gaussian processes. In: Chiappa, S., Calandra, R. (eds.) Proceedings of the Twenty Third International Conference on Artificial Intelligence and Statistics. Proceedings of Machine Learning Research, vol. 108, pp. 3372–3382. PMLR, 26–28 August 2020. http://proceedings.mlr.press/v108/blaas20a.html
6. Bosak, J.: Social roles. In: Shackelford, T.K., Weekes-Shackelford, V.A. (eds.) Encyclopedia of Evolutionary Psychological Science. Springer, Cham (2018). https://doi.org/10.1007/978-3-319-16999-6_2469-1
7. Council of European Union: Council regulation (EU) no. 2016/679 (2016). https://eur-lex.europa.eu/legal-content/EN/TXT/?uri=CELEX%3A02016R0679-20160504

8. Council of European Union: Proposal for Council Regulation (EU) no. 2021/0106(cod) (2021). https://eur-lex.europa.eu/legal-content/EN/TXT/?uri=CELEX%3A52021PC0206

9. Craver, C.: Explaining the brain: mechanisms and the mosaic unity of neuroscience. Oxford Scholarship Online (2007/2009). https://doi.org/10.1093/acprof:oso/9780199299317.001.0001

10. Craver, C.: The ontic account of scientific explanation. In: Kaiser, M.I., Scholz, O.R., Plenge, D., Hüttemann, A. (eds.) Explanation in the Special Sciences: The Case of Biology and History, pp. 27–52. Springer, Dordrecht (2014). https://doi.org/10.1007/978-94-007-7563-3_2

11. Craver, C., Kaplan, D.: Are more details better? On the norms of completeness for mechanistic explanations. Br. J. Philos. Sci. **71**(1), 287–319 (2020). https://doi.org/10.1093/bjps/axy015

12. Doshi-Velez, F., et al.: Accountability of AI under the law: the role of explanation. Forthcoming (2017). https://doi.org/10.2139/ssrn.3064761

13. Google: AI explanations whitepaper. Technical report (2021). https://storage.googleapis.com/cloud-ai-whitepapers/AI%20Explainability%20Whitepaper.pdf

14. Hacker, P., Krestel, R., Grundmann, S., Naumann, F.: Explainable AI under contract and tort law: legal incentives and technical challenges. Artif. Intell. Law **28**, 415–439 (2020). https://doi.org/10.1007/s10506-020-09260-6

15. Hempel, C.: Aspects of Scientific Explanation and Other Essays in the Philosophy of Science. The Free Press, New York (1965). https://doi.org/10.1086/288305

16. Holzinger, A.: Interactive machine learning for health informatics: when do we need the human-in-the-loop? Brain Inform. **3**, 119–131 (2016). https://doi.org/10.1007/s40708-016-0042-6

17. Holzinger, A., Carrington, A.: Measuring the quality of explanations: the system causability scale (SCS). Künstl. Intell. **34**, 193–198 (2020). https://doi.org/10.1007/s13218-020-00636-z

18. Langer, M., et al.: What do we want from explainable artificial intelligence (XAI)? - a stakeholder perspective on XAI and a conceptual model guiding interdisciplinary XAI research. Artif. Intell. **296**, 103473 (2021). https://doi.org/10.1016/j.artint.2021.103473

19. Lombrozo, T.: The structure and function of explanations. Trends Cogn. Sci. **10**(10), 464–470 (2006). https://doi.org/10.1016/j.tics.2006.08.004

20. Marr, D.: Vision. The MIT Press, Cambridge (1982/2010)

21. Meinke, A., Hein, M.: Towards neural networks that provably know when they don't know. In: International Conference on Learning Representations (2020). https://openreview.net/forum?id=ByxGkySKwH

22. Mell, S., Brown, O.M., Goodwin, J.A., Son, S.: Safe predictors for enforcing input-output specifications. CoRR abs/2001.11062 (2020)

23. Miller, T.: Explanation in artificial intelligence: insights from the social sciences. Artif. Intell. **267**, 1–38 (2019). https://doi.org/10.1016/j.artint.2018.07.007

24. Mittelstadt, B., Russell, C., Wachter, S.: Explaining explanations in AI. In: FAT* 2019, pp. 279–288. Association for Computing Machinery, New York (2019). https://doi.org/10.1145/3287560.3287574

25. Nam, W., Choi, J., Lee, S.: Relative attributing propagation: interpreting the comparative contributions of individual units in deep neural networks. CoRR abs/1904.00605 (2019). http://arxiv.org/abs/1904.00605

26. O'Hara, K.: Explainable AI and the philosophy and practice of explanation. Comput. Law Secur. Rev. **39**, 105474 (2020). https://doi.org/10.1016/j.clsr.2020.105474

27. Páez, A.: The pragmatic turn in explainable artificial intelligence (XAI). Minds Mach. **29**, 441–459 (2019). https://doi.org/10.1007/s11023-019-09502-w

28. Pincock, C.: A defense of truth as a necessary condition on scientific explanation. Erkenntnis (2021). https://doi.org/10.1007/s10670-020-00371-9

29. Ribeiro, M.T., Singh, S., Guestrin, C.: "Why should I trust you?": explaining the predictions of any classifier. CoRR abs/1602.04938 (2016). http://arxiv.org/abs/1602.04938

30. Rosenberg, J.: Connectionism and cognition. In: Haugeland, J. (ed.) Mind Design II, pp. 293–308. The MIT Press, Cambridge (1990)

31. Rudin, C.: Stop explaining black box machine learning models for high stakes decisions and use interpretable models instead. Nat. Mach. Intell. **1**, 206–215 (2019). https://doi.org/10.1038/s42256-019-0048-x

32. Schrittwieser, J., et al.: Mastering Atari, go, chess and shogi by planning with a learned model. Nature **588**, 604–609 (2020). https://doi.org/10.1038/s41586-020-03051-4

33. Selbst, A.D., Powles, J.: Meaningful information and the right to explanation. Int. Data Priv. Law **7**(4), 233–242 (2017). https://doi.org/10.1093/idpl/ipx022

34. Sokol, K., Flach, P.: One explanation does not fit all. Künstl. Intell. **34**, 235–250 (2020). https://doi.org/10.1007/s13218-020-00637-y

35. Staker, A.: Should chimpanzees have standing? The case for pursuing legal personhood for non-human animals. Transnat. Environ. Law **6**(3), 485–507 (2017). https://doi.org/10.1017/S204710251700019X

36. Vilone, G., Longo, L.: Explainable artificial intelligence: a systematic review. CoRR (2020). abs/2006.00093

37. Wachter, S., Mittelstadt, B., Floridi, L.: Why a right to explanation of automated decision-making does not exist in the general data protection regulation. Int. Data Priv. Law **7**(2), 76–99 (2017). https://doi.org/10.1093/idpl/ipx005

38. Wang, G.: Humans in the loop: the design of interactive AI systems. Stanford Human-Centered Artificial Intelligence (2019). https://hai.stanford.edu/news/humans-loop-design-interactive-ai-systems

39. Wexler, J., Pushkarna, M., Bolukbasi, T., Wattenberg, M., Viégas, F., Wilson, J.: The what-if tool: interactive probing of machine learning models. IEEE Trans. Vis. Comput. Graph. **26**(1), 56–65 (2020). https://doi.org/10.1109/TVCG.2019.2934619

40. Woodward, J.: Causation in biology: stability, specificity, and the choice of levels of explanation. Biol. Philos. **25**, 287–318 (2010). https://doi.org/10.1007/s10539-010-9200-z

41. Woodward, J.: Causation and manipulability. In: Zalta, E.N. (ed.) The Stanford Encyclopedia of Philosophy, Winter 2016 edn. Metaphysics Research Lab, Stanford University (2016)

42. Woodward, J.: Scientific explanation. In: Zalta, E.N. (ed.) The Stanford Encyclopedia of Philosophy, Spring 2021 edn. Metaphysics Research Lab, Stanford University (2021)

43. Zhang, G., et al.: Selection bias explorations and debias methods for natural language sentence matching datasets. In: Proceedings of the 57th Annual Meeting of the Association for Computational Linguistics, pp. 4418–4429. Association for Computational Linguistics, Florence, July 2019. https://doi.org/10.18653/v1/P19-1435

On the Trustworthiness of Tree Ensemble Explainability Methods

Angeline Yasodhara$^{(\boxtimes)}$, Azin Asgarian, Diego Huang, and Parinaz Sobhani

Georgian, 2 St Clair Avenue West, Suite 1400, Toronto, ON M4V 1L5, Canada
{angeline,azin,diego,parinaz}@georgian.io
https://georgian.io

Abstract. The recent increase in the deployment of machine learning models in critical domains such as healthcare, criminal justice, and finance has highlighted the need for trustworthy methods that can explain these models to stakeholders. Feature importance methods (e.g. gain and SHAP) are among the most popular explainability methods used to address this need. For any explainability technique to be trustworthy and meaningful, it has to provide an explanation that is accurate and stable. Although the stability of local feature importance methods (explaining individual predictions) has been studied before, there is yet a knowledge gap about the stability of global features importance methods (explanations for the whole model). Additionally, there is no study that evaluates and compares the accuracy of global feature importance methods with respect to feature ordering. In this paper, we evaluate the accuracy and stability of global feature importance methods through comprehensive experiments done on simulations as well as four real-world datasets. We focus on tree-based ensemble methods as they are used widely in industry and measure the accuracy and stability of explanations under two scenarios: 1. when inputs are perturbed 2. when models are perturbed. Our findings provide a comparison of these methods under a variety of settings and shed light on the limitations of global feature importance methods by indicating their lack of accuracy with and without noisy inputs, as well as their lack of stability with respect to: 1. increase in input dimension or noise in the data; 2. perturbations in models initialized by different random seeds or hyperparameter settings.

Keywords: Explainability · Trustworthiness · Tree ensemble

1 Introduction

Owing to the success and promising results achieved in supervised machine learning (ML) paradigm, there has been a growing interest in leveraging ML models in domains such as healthcare [3,30,33], criminal justice [26], and finance [12].

A. Yasodhara and A. Asgarian—Contributed equally to this work.

As ML models become embedded into critical aspects of decision making, their successful adoption depends heavily on how well different stakeholders (e.g. user or developer of ML models) can understand and trust their predictions [4,10,14,20,27]. As a result, there has been a recent surge in making ML models worthy of human trust [31] and researchers have proposed a variety of methods to explain ML models to stakeholders [6]. Among these methods, feature importance methods in particular have received a lot of attention and gained tremendous popularity in industry [6]. The explanations obtained by these methods lie in two categories: 1. local explanations 2. global explanations . Local explanations explain how a particular prediction is derived from the given input data. Global explanations, in contrast, provide a holistic view of what features are important across all predictions. Both explanation methods can be used for the purposes of model debugging, transparency, monitoring and auditing [6]. However, the trustworthiness and applicability of these explanations relies heavily on their accuracy and stability [18].

Previously, Lundberg et al. [22] assess the accuracy of feature importances by comparing them with human attributed importances. Ribeiro et al. [25] limits models to only use ten features from the input. Assuming the models would only pick the top ten important features, he then measures whether the selected features by the model are also captured by feature importances. Although both of these assessments capture whether important features are accurately identified, they do not measure the accuracy with respect to the relative ordering of features. We examine this with and without the presence of noisy inputs and use it to provide a comparison of different global feature importance methods.

In the explainability literature, various definitions are proposed for stability. Alvarez et al. [2] define stability as being stable to local perturbations of the input, or in other words, similar inputs should not lead to significantly different explanations. Hancox-Li provides another definition for stability [18]. He claims that stable explanations reflective of real patterns in the world are those that remain consistent over a set of equally well-performing models. Inspired by these definitions, we consider the following two scenarios to evaluate stability: 1. local perturbations of the input 2. perturbations of the models. We argue that stability with respect to these factors is essential to account for the inherent noisy nature of real-world data and to provide trustworthy explanations.

The stability of local explainability methods under the first scenario has been studied before. For example, Alvarez et al. [2] show LIME [25] and (Kernel) SHAP [22] lack stability for complex black-box models through conducting the following experiments. They slightly perturb the input values and find that the surrogate models and original black box models produce stable output values whereas the explanations provided by LIME and SHAP change drastically in response to the perturbations. Despite these thorough investigations conducted on the stability of local explainability methods, there is yet little understanding about the stability of global explainability methods. With these methods getting embedded into critical aspects of daily life (healthcare, criminal justice, and finance), addressing this knowledge gap becomes crucial to avoid moral and ethical hazards [26].

In this paper, we compare and evaluate the accuracy and stability of global feature explanation methods, gain and SHAP, through comprehensive experiments conducted on synthetic data and four real-world datasets. For this purpose, we use the following tree-based ensemble models as they are widely used in academia and industry: (1) random forest (2) gradient boosting machines [23] and (3) XGBoost [9]. Our findings shed light on the limitations of the global explainability methods and show that they lack accuracy and become unstable when inputs or models are perturbed. For the rest of this paper, we first review the methodologies used in our experiments under Sect. 2. We then describe our experimental setup in Sect. 3. Finally, we present and discuss our findings in Sects. 4 and 5 respectively.

2   Background

Tree ensemble methods are employed widely in research and industry due to their efficiency and effectiveness in modeling complex interactions in the data [7]. The two most common tree ensemble methods are gradient boosting [17] and random forest [8]. In gradient boosting, trees are trained sequentially with upweighting the previously misclassified labels. In contrast, random forest trees are trained in parallel with different subsampling across all trees. We use random forest and gradient boosting machine implemented by sklearn [23], as well as XGBoost, a faster version of gradient boosting that uses second-order gradients [9].

In this study, to compute global feature importances in tree ensemble methods we use gain [16] and SHAP [22], an implementation of the Shapley algorithm. We focus on SHAP instead of LIME [25] as LIME explanations can be fragile due to sampling variance [6] and less resilient against adversarial attacks as shown by [29]. In the following sections, we briefly explain how gain [16] and SHAP [22] are computed.

Gain. For both of the aforementioned tree ensemble methods, sklearn [23] and xgboost [9] libraries provide the implementation to obtain the feature importances based on Hastie's description in the Elements of Statistical Learning [16]. This is also referred to as *gain*. This metric represents the improvements in accuracy or improvements in decreasing uncertainty (or variance) brought by a feature to its branches. At the end, to get a summary of the whole tree ensemble, this measure is averaged across all trees [1,16,19]. In this paper, for the sake of simplicity and consistency we refer to this method as *gain*.

SHAP. SHapley Additive exPlanations (SHAP) [22] has gained a lot of attention in industry as a way to measure feature importance [6]. SHAP is an implementation of Shapley formula that summarizes the contribution of a feature to the overall prediction by approximating the Shapley value presented in the following:

$$\phi_i = \sum_{S \subseteq F \setminus \{i\}} \frac{|S|!(|F| - |S| - 1)!}{|F|!} [f_{S \cup \{i\}}(X_{S \cup \{i\}}) - f_S(X_S)]$$

where ϕ_i is the Shapley value for feature i, S is a subset of all features F that does not include feature i, $f_{S \cup \{i\}}$ is the model trained on features in S and feature i, f_S is the model trained on features in S, and X is the input data.

SHAP inherently calculates local importances, i.e. how each feature contributes to the prediction of a specific input. By averaging the absolute value of these local importances across the training set, one can obtain a global summary of how the feature as a whole contributes to the model. In this paper, we investigate the accuracy and stability of Tree SHAP [21] (a recent extension to Kernel SHAP with faster computation runtime for trees) under various settings. Unlike Kernel SHAP [22] which uses perturbation, Tree SHAP (with tree_path_dependent setting) leverages trees' cover statistics for fast approximation of Shapley values.

3   Experimental Setup

In this section we describe the setup we use to evaluate the accuracy and stability of global feature importance methods.

Datasets. To thoroughly evaluate the accuracy and stability of global feature importances, we conduct our experiments on synthetic data as well as four real-world datasets from various domains.

For synthetic data, we generate 300 training samples with varying number of features (5, 10, 25, 100, and 150 features). We randomly set the features to be either continuous or categorical (each with equal probability). For continuous features, we sample from a uniform distribution between $[0, 1)$. For categorical features, we first randomly sample values like continuous features and we then binarize them based on an independently-sampled threshold selected from $[0, 1)$. Lastly, to obtain the target values, we sum the multiplication of each feature by a randomized set of coefficients (sampled independently per feature between -10 to 10). Then, we categorize the summation values to 1 for values greater than the median and 0 otherwise.

We use the following four real-world datasets in addition to the synthetic data for our stability assessments:

1. Forest Fire: prediction of the amount of burned area resulted from forest fires in the northeast region of Portugal, by using meteorological data, such as coordinates, time, wind, rain, relative humidity, etc. [11].
2. Concrete: prediction of concrete compressive strength given material types, composition, and age [34].
3. Auto MPG: prediction of fuel consumption in miles per gallon (MPG) of cars in the city given its model, horsepower, etc. [24]
4. Company Finance: prediction of whether companies would make a good investment based on their finances.

All datasets except the Company Finance dataset (our proprietary dataset) come from the UCI ML data repository [15] and are parsed with the py_uci package [28]. A summary of these datasets is shown in Table 1.

Table 1. Description of datasets used in this study.

Dataset	Domain	Task type	# Samples	# Features
Synthetic Data		Classification	300	5–150
Forest Fire	Meteorology	Regression	517	12
Concrete	Civil	Regression	1030	8
Auto MPG	Automotive	Regression	406	7
Company Finance	Finance	Classification	2716	892

Experimental Settings. In our experiments, we use random forest and gradient boosting machine implemented by sklearn package [23], as well as XGBoost, an implementation of gradient boosting that uses second-order gradients and has a faster runtime [9]. For each of these models, we run the following experiments:

1. Input perturbation: where the input data are perturbed by adding different levels of noise. Noise is sampled randomly from a normal distribution with mean 0 and standard deviation of: (a) half of the original feature's standard deviation for low noise (b) the original feature's standard deviation for medium noise (c) double of the original feature's standard deviation for high noise.
2. Model perturbation: where the model is perturbed by (a) initializing with a different random seed without hyperparameter tuning, or (b) optimizing hyperparameters [5] (e.g., number of trees, depth of trees, etc.) with a different random seed. In these experiments, we ensure that the predictions of the two models (the original model and the perturbed model) have high correlations, such that of discrepancies in predictions affect the analysis minimally.

We iterate all experiments 50 times with a different random seed, except for the Company Finance dataset. For this dataset, we run the experiments 5 times due to long training time caused by the high number of features. In each iteration of input perturbation experiment, we train two models, one with the original setting (e.g., unperturbed input data) and another with the perturbed setting (e.g., noised input data). In model perturbation experiments, we also train two models in each iteration where we change the random seed of the second model to be different than the first model. For each trained model, we compute gain and SHAP feature importances as described in Sect. 2.

Accuracy Metrics. To evaluate the accuracy of global feature importances, we use simulated data so that the true coefficients (importances) are known. The features are ranked based on the magnitude of their corresponding coefficients used during data generation. We examine the accuracy under the following scenarios: 1. when no noise is added to the input, and 2. when different level of noise is added to the input. We do not consider the model perturbation scenario for this analysis as we are mainly interested in measuring the accuracy of the model's feature importances to the true coefficients.

We evaluate the accuracy of the top features' ranking in the following way:

- First, we rank features based on their coefficients' magnitude, largest magnitude being the most important. Since all features are uniformly sampled from $[0, 1)$, we assume the coefficients' magnitude represent the importances.
- Second, we assess whether these top features are ranked correctly with gain and SHAP feature importances.
- Finally, we count the number of times each top feature is ranked correctly by gain or SHAP feature importances across multiple iterations. If it is ranked incorrectly, there are 2 possible situations: 1. The feature is still considered a top feature by gain or SHAP feature importances,2. The feature is not considered a top feature by gain or SHAP feature importances. We present this count proportionally across the 3 groups (correct, incorrect_but_top, and incorrect) to compare the accuracy of these models on different levels.

Furthermore, to get a sense of feature importances' accuracy across all features, we evaluate the Spearman correlation of gain and SHAP feature importances compared to the coefficients.

Stability Metrics. To evaluate the stability of global feature importances, we consider the following two scenarios: 1. when different levels of noise is added to the input. 2. when models are perturbed by initializing with different random seeds and different hyperparameter settings. We use Spearman correlation to compare feature importances calculated from the 2 models (one unperturbed and the other perturbed), because it is distribution-free unlike parametric tests (e.g., Pearson correlation) [35]. We also report both the Spearman and Pearson correlations between the predicted outputs of the two models trained in each iteration as a sanity check to ensure similar performance.

4   Results

Here, we present our findings from the experiments described in Sect. 3. We first discuss the accuracy of gain and SHAP feature importances in Sect. 4.1. We then dive into the stability of each feature importance method when inputs are perturbed and when models are perturbed in Sect. 4.2. Finally, we present a summary of our findings in Sect. 4.3.

4.1   Accuracy of Gain and SHAP Feature Importances

Table 2 demonstrate the accuracy of gain and SHAP for the top 3 features in synthetic data with a total of 5 features trained with XGBoost. The difference between SHAP and gain proportions are highlighted beneath them. Orange indicates SHAP having a higher proportion and vice versa for blue. Models included in this experiment are highly predictive, with an average area under receiver operating curve (AUROC) of 92.6% with standard deviation of 0.8%.

Surprisingly, we find that the number of features ranked correctly is quite low for both methods even when there is no noise added to the input. For example, the rank #1 feature is correctly ranked approximately 40% of the time by both methods. Despite both SHAP and gain calculating feature importances from the same model, SHAP shows a slightly higher accuracy in ranking top features especially when noise is added into the input.

Table 2. Proportions of correct, incorrect_but_top, and incorrect ranking of the top 3 features on synthetic data (total features: 5) using XGBoost model across all experiment iterations. Proportions in each column add up to 1. Highlighted values indicate the difference between SHAP and gain proportions: orange when SHAP having higher proportion and blue otherwise.

Experiment setting:	No noise added to input						Low noise added to input					
Original feature rank:	1		2		3		1		2		3	
Feature importance method:	gain	shap	gain	shap	gain	shap	gain	shap	gain	shap	gain	shap
correct	0.44	0.5	0.46	0.44	0.32	0.3	0.44	0.46	0.4	0.52	0.3	0.44
	0.06		-0.02		-0.02		0.02		0.12		0.14	
incorrect_but_top	0.26	0.28	0.22	0.26	0.32	0.28	0.24	0.16	0.26	0.2	0.46	0.38
	0.02		0.04		-0.04		-0.08		-0.06		-0.08	
incorrect	0.3	0.22	0.32	0.3	0.36	0.42	0.32	0.38	0.34	0.28	0.24	0.18
	-0.08		-0.02		0.06		0.06		-0.06		-0.06	

To explicitly look at whether the feature importances provides an accurate ranking of all features, we further examine the Spearman correlation between the feature importances and the true coefficients. Figure 1 shows the correlations in a noise-free scenario with increasing number of features. As demonstrated in this Figure, we find that gain and SHAP feature importances do not correlate well with the true coefficients (correlations range from 30–40% and drops to around 20% as the number of features increases). We observe a similar pattern across all other experimental settings (low-noised, medium-noised, or high-noised input).

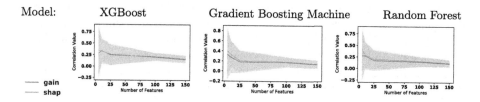

Fig. 1. Spearman correlation of gain and SHAP feature importances (Blue: gain, Orange: SHAP) with the true coefficients with no noise added in simulation. Correlation is quite low across all settings. (Color figure online)

4.2 Stability of Gain and SHAP Feature Importances

In this section we evaluate the stability of feature importances when inputs and models are perturbed. In all of the following experiment settings, the predicted outputs from the perturbed models and the original models are highly correlated (an example for model perturbation is shown in Fig. 2 for synthetic data). This ensures that our models have very similar performance and the results are minimally affected by discrepancies between model predictions.

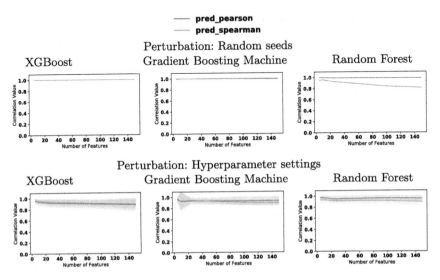

Fig. 2. Correlation of predicted outputs in models trained on synthetic data with model perturbations across different number of features (Blue: Pearson, Orange: Spearman correlation). The predicted output of perturbed models are still highly correlated to those without perturbation. (Color figure online)

Stability of Feature Importances When Inputs are Perturbed. Figure 3 shows us a glimpse of this analysis for low level of noise on synthetic data. From this figure, we see that SHAP is more stable than gain feature importances when we add a small noise to the perturbed input, especially for XGBoost. This uplift between gain and SHAP, however, decreases as noise increases across all models as shown in Fig. 4. We can also see from Fig. 4 that unsurprisingly stability decreases as the level of noise and the number of features increase.

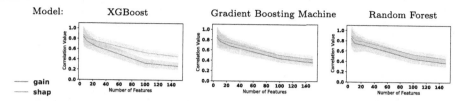

Fig. 3. Correlation of feature importances (Blue: gain, Orange: SHAP) for models trained with low input perturbation on synthetic data. SHAP is more stable across all models although both SHAP and gain both suffer from lack of stability. (Color figure online)

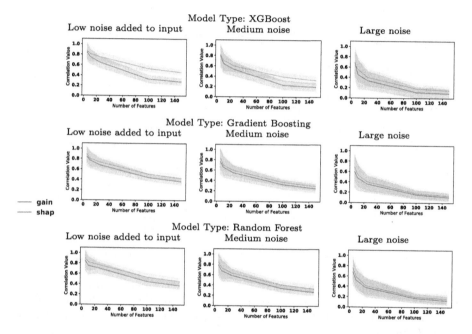

Fig. 4. Correlation of feature importances (Blue: gain, Orange: SHAP) for models trained with input perturbation on synthetic data. SHAP is slightly more stable than gain at low level of noise but are comparable as noise increases. (Color figure online)

As shown in Fig. 5, we see that in real-world datasets when a low noise is injected to the input, the correlations of gain and SHAP feature importances drop very low. For example, in Forest Fire dataset, feature importances correlation averages to around 50% for SHAP while it averages to around 20% for gain. In Company Finance dataset, both gain and SHAP has either 20% correlation or lower. We discover that SHAP is slightly more stable than gain for Forest Fire and Company Finance as can be seen on Fig. 5, although this is not consistent across all datasets. We also observe low correlations with increasing level of noise.

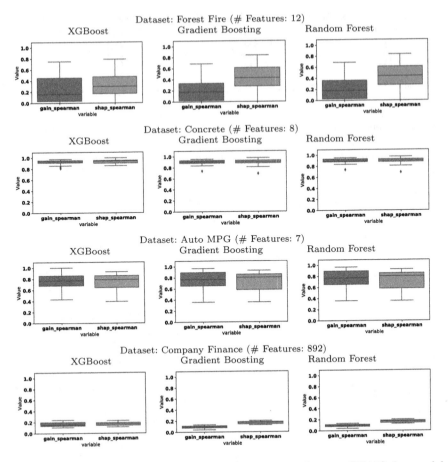

Fig. 5. Correlation of feature importances (Blue: gain, Orange: SHAP) for models trained with input perturbations (low noise) on real-world datasets. SHAP and gain both lack stability overall although SHAP is slightly more stable for certain datasets. (Color figure online)

Stability of Feature Importances When Models are Perturbed. Figure 6 shows the correlation of feature importances when models are perturbed by initializing to a different random seed or by training with different hyperparameter settings. From this figure, we see that the correlation of feature importances is not greatly affected when models are perturbed for small number of features, but it drops significantly (to 80% Spearman correlation for XGBoost and gradient boosting models) as the number of features increases to 150. We find that the correlation of SHAP feature importances is significantly higher compared to gain feature importances, especially in XGBoost trained with different hyperparameter. Although, for gradient boosting machine and random forest, we do not see the same uplift on stability for SHAP. Both gain and SHAP are equally stable for these models.

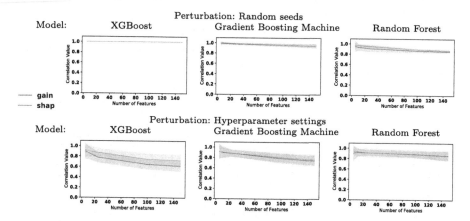

Fig. 6. Correlation of feature importances (Blue: gain, Orange: SHAP) for models trained on synthetic data with model perturbations across different number of features. (Color figure online)

Moreover, we notice a strangely perfect correlation when training XGBoost without hyperparameter optimization but with different random seeds (See Fig. 6, top left). After further investigation, we discover that XGBoost is more deterministic when choosing features even when initialized with different random seeds. The results of our findings are expanded further in Appendix A.

In real world settings, we also notice a decrease in stability for gain and SHAP when models' hyperparameter settings are perturbed (Fig. 7). This is especially bold for Forest Fire dataset. On average, gain feature importances have around 60% Spearman correlation whereas SHAP have around 90% Spearman correlations in this dataset. SHAP tends to be more stable across the different real-world datasets, especially for XGBoost model as shown in Fig. 7, although this uplift is not as apparent in Gradient Boosting Machine and random Forest models.

4.3   Summary of Results

We observe that there is a lack of accuracy with gain and SHAP feature importances even when there is no perturbation involved. In synthetic data with 5 features, the top feature is only ranked correctly around 40% of the time. In addition to lack of accuracy, we also evaluate the lack of stability of these feature importances in various settings. We find that when inputs are perturbed, the correlations drop very low, both in synthetic and real-world datasets. When we perturb the models, especially by using different hyperparameter settings, correlation of feature importances can drop to 70–80%. We find SHAP to be slightly more stable than gain in many cases, but both of their Spearman correlations still reduces to 60% when low noise is added to the input.

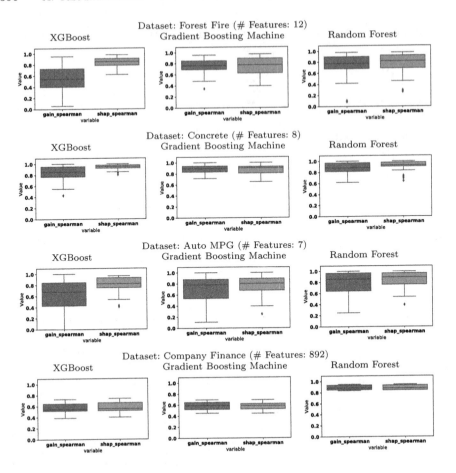

Fig. 7. Correlation of feature importances (Blue: gain, Orange: SHAP) for XGBoost models trained on four real-world datasets with perturbations to the model's hyperparameter settings. SHAP is slightly more stable than gain for XGBoost. (Color figure online)

5 Discussion

We set out to investigate the accuracy and stability of global feature importances for tree-based ensemble methods, such as random forest, gradient boosting machine, and XGBoost. We mainly look at two feature importance methods *gain*, and SHAP. For both of these methods, we evaluate the accuracy in a simulated environment where true coefficients are known with and without noisy inputs. We also evaluate the stability of these methods in two directions, that is 1. when inputs are perturbed, and 2. when model settings are perturbed, either by initializing with a different random seed or by optimizing their hyperparameters with a different random seed.

Accuracy Analysis. We find that SHAP tends to be better at accurately identifying top features compared to gain, although the overall accuracy of both is quite low especially when considering the ordering of all the features.

Stability Analysis. In our experiments, we find that SHAP is either equally or more stable when compared with gain. This is especially interesting as both gain and (Tree) SHAP feature importances investigated here use the innate structure of the trees. The difference lies on the fact that gain measures the feature's contribution to accuracy improvements or decreasing of uncertainty/variance whereas SHAP measures the feature's contribution to the predicted output.

Future Work. There has been recent work on extending Shapley values to other cooperative game theory algorithms, such as the core [32]. We will investigate this approach when a public implementation of this algorithm becomes available. In this study, we mostly focus on the stability of global features importance across the same model trained with perturbed hyperparameters/random seeds or inputs. Dong and Rudin recently suggest the idea of using a variable cloud importance, capturing the many good (but not necessarily the same) explanations coming from a group of models with almost equal performance [13]. In our future work, we will investigate the stability and usability of this methodology. We will also extend our analysis to new scenarios and datasets.

Conclusion. We investigate the accuracy and stability of global feature importances for tree ensemble methods. We find that even though SHAP in many cases can be more stable than gain feature importance, both methods still have limitations in terms of accuracy and stability and more work needs to be done to make them trustworthy. We hope that our paper will continue propel the discussion for trustworthy global feature importances and for the community to investigate this more thoroughly.

Appendix

A   Determinism of XGBoost Feature Importances

In this experiment, we simulate 1000 samples with 10 redundant features where each feature is equally important in predicting the target. Figure 8 shows the distribution of the default feature importance in random forests, gradient boosting, and XGBoost across 30 iterations with different random seeds. As shown on the bottom left, XGBoost always assigns all importance to the first feature it saw no matter the random seed. When we shuffle the order of the features, we are able to break down this pattern (shown on bottom right). This is why on Fig. 6, there is a perfect correlation of importance for XGBoost initialized with different random seeds. With shuffled features, we still find SHAP to be more stable for XGBoost overall, although the correlation still decreases with higher number of features (See Fig. 9).

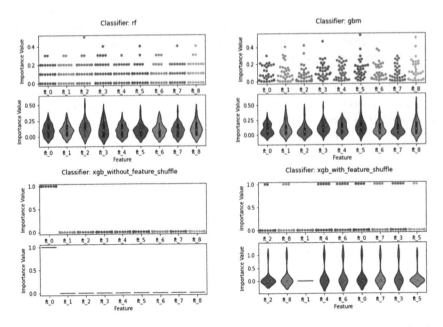

Fig. 8. These plots show the distribution of feature importances across 10 redundant features for random forest (top left), gradient boosting (top right), XGBoost (bottom left), and XGBoost with feature shuffling (bottom right). XGBoost by its implementation is more deterministic compared to other methods at assigning feature importance. For the same hyperparameter with different seeds, when the features are redundant, it will always pick the first feature in order. With feature shuffling though, we are able to break this pattern a little bit.

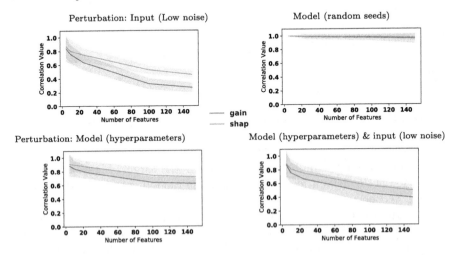

Fig. 9. SHAP is more stable overall for XGBoost with shuffled features as can be seen on the plots above across input perturbation (low noise) experiments, model perturbations and both. Each row represents a different set of experiments with Spearman correlations of the default feature importance (Blue) and SHAP feature importance (Orange). (Color figure online)

References

1. Abu-Rmileh, A.: Be careful when interpreting your features importance in xgboost!, February 2019. https://towardsdatascience.com/be-careful-when-interpreting-your-features-importance-in-xgboost-6e16132588e7
2. Alvarez-Melis, D., Jaakkola, T.S.: On the robustness of interpretability methods. arXiv preprint arXiv:1806.08049 (2018)
3. Asgarian, A., et al.: A hybrid instance-based transfer learning method. arXiv preprint arXiv:1812.01063 (2018)
4. Asgarian, A., et al.: Limitations and biases in facial landmark detection D an empirical study on older adults with dementia. In: CVPR Workshops, pp. 28–36 (2019)
5. Bergstra, J., Yamins, D., Cox, D.: Making a science of model search: hyperparameter optimization in hundreds of dimensions for vision architectures. In: International Conference on Machine Learning, pp. 115–123 (2013)
6. Bhatt, U., et al.: Explainable machine learning in deployment. In: Proceedings of the 2020 Conference on Fairness, Accountability, and Transparency, pp. 648–657 (2020)
7. Biau, G., Scornet, E.: A random forest guided tour. TEST **25**(2), 197–227 (2016)
8. Breiman, L.: Random forests. Mach. Learn. **45**(1), 5–32 (2001)
9. Chen, T., Guestrin, C.: XGBoost: a scalable tree boosting system. In: Proceedings of the 22nd ACM SIGKDD International Conference on Knowledge Discovery and Data Mining KDD 206, pp. 785–794. ACM, New York (2016)
10. Christodoulakis, C., Asgarian, A., Easterbrook, S.: Barriers to adoption of information technology in healthcare. In: Proceedings of the 27th Annual International Conference on Computer Science and Software Engineering, pp. 66–75 (2017)
11. Cortez, P., Morais, A.: A data mining approach to predict forest fires using meteorological data (2007)
12. Dixon, M.F., Halperin, I., Bilokon, P.: Machine Learning in Finance. Springer, Cham (2020). https://doi.org/10.1007/978-3-030-41068-1
13. Dong, J., Rudin, C.: Variable importance clouds: a way to explore variable importance for the set of good models. arXiv preprint arXiv:1901.03209 (2019)
14. Doshi-Velez, F., Kim, B.: Towards a rigorous science of interpretable machine learning. arXiv preprint arXiv:1702.08608 (2017)
15. Dua, D., Graff, C.: UCI machine learning repository (2017). http://archive.ics.uci.edu/ml
16. Friedman, J., Hastie, T., Tibshirani, R.: The Elements of Statistical Learning. Springer Series in Statistics, vol. 1. Springer, New York (2001). https://doi.org/10.1007/978-0-387-21606-5
17. Friedman, J.H.: Stochastic gradient boosting. Comput. Statist. Data Anal. **38**(4), 367–378 (2002)
18. Hancox-Li, L.: Robustness in machine learning explanations: does it matter? In: Proceedings of the 2020 Conference on Fairness, Accountability, and Transparency, pp. 640–647 (2020)
19. Lewinson, E.: Explaining feature importance by example of a random forest, April 2020. https://towardsdatascience.com/explaining-feature-importance-by-example-of-a-random-forest-d9166011959e
20. Lipton, Z.C.: The mythos of model interpretability. Queue **16**(3), 31–57 (2018)
21. Lundberg, S.M., et al.: From local explanations to global understanding with explainable AI for trees. Nat. Mach. Intell. **2**(1), 2522–5839 (2020)

22. Lundberg, S.M., Lee, S.I.: A unified approach to interpreting model predictions. In: Advances in Neural Information Processing Systems 30, pp. 4765–4774. Curran Associates, Inc. (2017)
23. Pedregosa, F., et al.: Scikit-learn: machine learning in Python. J. Mach. Learn. Res. **12**, 2825–2830 (2011)
24. Quinlan, J.R.: Combining instance-based and model-based learning. In: Proceedings of the 10th International Conference on Machine Learning, pp. 236–243 (1993)
25. Ribeiro, M.T., Singh, S., Guestrin, C.: "Why should I trust you?" explaining the predictions of any classifier. In: Proceedings of the 22nd ACM SIGKDD International Conference on Knowledge Discovery and Data Mining, pp. 1135–1144 (2016)
26. Rudin, C.: Stop explaining black box machine learning models for high stakes decisions and use interpretable models instead. Nat. Mach. Intell. **1**(5), 206–215 (2019)
27. Selbst, A.D., Barocas, S.: The intuitive appeal of explainable machines. Fordham L. Rev. **87**, 1085 (2018)
28. Skafte, N.: py_uci (2019). https://github.com/SkafteNicki/py_uci
29. Slack, D., Hilgard, S., Jia, E., Singh, S., Lakkaraju, H.: Fooling lime and shap: adversarial attacks on post hoc explanation methods. In: Proceedings of the AAAI/ACM Conference on AI, Ethics, and Society, pp. 180–186 (2020)
30. Spann, A., Yasodhara, A., Kang, J., Watt, K., Wang, B., Goldenberg, A., Bhat, M.: Applying machine learning in liver disease and transplantation: a comprehensive review. Hepatology **71**(3), 1093–1105 (2020)
31. Wiens, J., et al.: Do no harm: a roadmap for responsible machine learning for health care. Nat. Med. **25**(9), 1337–1340 (2019)
32. Yan, T., Procaccia, A.D.: If you like shapley then you'll love the core (2020)
33. Yasodhara, A., Dong, V., Azhie, A., Goldenberg, A., Bhat, M.: Identifying modifiable predictors of long-term survival in liver transplant recipients with diabetes mellitus using machine learning. Liver Transpl. **27**(4), 536–547 (2021)
34. Yeh, I.C.: Modeling of strength of high-performance concrete using artificial neural networks. Cem. Concr. Res. **28**(12), 1797–1808 (1998)
35. Zwillinger, D., Kokoska, S.: CRC Standard Probability and Statistics Tables and Formulae. CRC Press, New York (1999)

Human-in-the-Loop Model Explanation via Verbatim Boundary Identification in Generated Neighborhoods

Xianlong Zeng[(✉)], Fanghao Song, Zhongen Li, Krerkkiat Chusap, and Chang Liu[(✉)]

School of Electrical Engineering and Computer Engineering, Ohio University, Athens, OH 45701, USA
{xz926813,liuc}@ohio.edu

Abstract. The black-box nature of machine learning models limits their use in case-critical applications, raising faithful and ethical concerns that lead to trust crises. One possible way to mitigate this issue is to understand how a (mispredicted) decision is carved out from the decision boundary. This paper presents a human-in-the-loop approach to explain machine learning models using verbatim neighborhood manifestation. Contrary to most of the current eXplainable Artificial Intelligence (XAI) systems, which provide hit-or-miss approximate explanations, our approach generates the local decision boundary of the given instance and enables human intelligence to conclude the model behavior. Our method can be divided into three stages: 1) a neighborhood generation stage, which generates instances based on the given sample; 2) a classification stage, which yields classifications on the generated instances to carve out the local decision boundary and delineate the model behavior; and 3) a human-in-the-loop stage, which involves human to refine and explore the neighborhood of interest. In the generation stage, a generative model is used to generate the plausible synthetic neighbors around the given instance. After the classification stage, the classified neighbor instances provide a multifaceted understanding of the model behavior. Three intervention points are provided in the human-in-the-loop stage, enabling humans to leverage their own intelligence to interpret the model behavior. Several experiments on two datasets are conducted, and the experimental results demonstrate the potential of our proposed approach for boosting human understanding of the complex machine learning model.

Keywords: Explainable artificial intelligence · Method classification · Human-in-the-loop · Deep learning

1 Introduction

Machine learning models are typically designed and fine-tuned for optimal accuracy, which often results in layers of weights that are difficult to explain or

Published by Springer Nature Switzerland AG 2021
A. Holzinger et al. (Eds.): CD-MAKE 2021, LNCS 12844, pp. 309–327, 2021.
https://doi.org/10.1007/978-3-030-84060-0_20

understand. In the meantime, recent successes of machine learning systems have attracted adoption from more end-users, who need to better understand the model in order to trust or properly use such machine learning systems. To make these two ends meet, researchers and practitioners alike have adopted several approaches, including 1) using approximate models just for explanation [2]; 2) linear local explanation for complex global models (e.g. LIME [9]); 3) example-based explanation by finding and showing most influential training data points [5]. These approaches all have their own merits, but none of them deliver everything needed by end-users [10].

The fundamental limitation of these approaches is that they assume that 1) certain aspects of machine learning systems, especially complex deep neural networks, cannot be understood by human beings, and 2) typical human users can only understand simple concepts such as linear systems.

We have an opportunity to improve on previous attempts with two assumptions. First, human users are intelligent, just not in the same way as machines. Humans can identify patterns intelligently but may not be able to scale up to thousands of data points easily. Second, machine learning systems are built to reflect actual physical systems that follow logical and physical rules. What worked well most likely can be explained, even though the explanation could be complex. What cannot be explained most likely is not a good reflection of the underlying physical properties.

We intend to make improvements in this area by 1) presenting various aspects of the actual model through verbatim model manifestation (instead of trying to approximate the models), and 2) identifying and generating a manageable number of data points to present to users in the local context of the point-of-interest, so that human users can use their own intelligence to understand what the actual model is trying to do within a limited scope that is manageable by a human being.

With this intuition, we aim to design an approach to facilitate human users' understanding of machine learning models through 1) verbatim manifestation of certain aspects of the underlying machine learning systems and 2) contextualized visualization of carefully curated or generated data points that facilitates human understanding. In other words, we try to build a bridge between machine and human intelligence to address machine learning models' explainability problems. Furthermore, we observe that a typical human user does not need to understand the complete machine learning model to gain confidence in the results from the model. The user only needs to understand the rationale behind the decision related to the current task.

In this paper, we present a three-stage human-in-the-loop XAI system, a high-level illustration of which is depicted in Fig. 1. For a given (mispredicted) point-of-interest, our framework tries to carve out its local decision boundary and delineate the model behavior through a neighborhood manifestation. Our framework leverages variational autoencoders (VAE) to generate neighborhood examples that cross the decision boundary. Human users are involved in exploring the neighborhood through three carefully designed intervention points. These intervention points help human users limit the neighborhood's scope and enable them to gain insights from the model behavior. The source code of our work is public available on GitHub: https://github.com/drchangliu/xai.

The main contributions of our work are:

- We proposed a novel human-in-the-loop framework that could mitigate the trust crisis between human users and machine learning models.
- Several case studies are presented to illustrate the potential of our approach to facilitating human understanding of complex machine learning models.
- A general framework to depict the local decision boundary around the (mis-predicted) instance-of-interest.

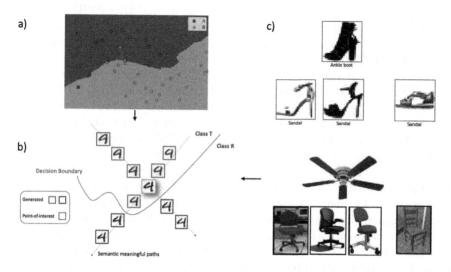

Fig. 1. A high-level illustration of our proposed framework. a) For a (mispredicted) point-of-interest (red x) and a trained machine learning model, b) our framework tries to carve out the local decision boundary and delineate the model behavior through a manageable neighborhood manifestation. c) Images of *sandals* and *ankle boot* from the fashionMNIST dataset that cause confusion to a classifier. Human users can understand the classification errors by seeing the context that some sandals have boot-shape heels. Another classification error is from the Caltech 101 dataset. Trust crisis can be mitigated given the context that some chairs have fan-shaped bases. (Color figure online)

2 Related Work

Machine learning researchers and practitioners have always used techniques and tools to better understand machine learning models. In this section, we examine a few state-of-the-art tools that are publicly accessible in an attempt to shed some light on how they can help software engineers adopt machine learning components.

To understand the information flow of a deep network, Ancona et al. [1] has studied the problem of assigning contributions to each input feature of a network. Such methods are known as *attribution methods*, which can be divided into two categories: perturbation-based and backpropagation-based. The perturbation-based methods, such as Occlusion [18], LIME [9] and Shapely value [2], change the input features and measure the difference between the new output and the original output, while backpropagation-based methods compute the attributions for all input features through the network. Backpropagation-based methods include the feature-wise manner and the layer-wise manner. Feature-wise approaches includes Gradient*Input [13] and Integrated Gradients [15]). Layer-wise approaches includes Layer-wise Relevance Propagation [3], Class activation maps [4,11,14,16] and DeepLIFT [12].

Among these related research efforts, LIME [9] and DEEPVID [17] are the two most relevant methods as compared to our framework. LIME, proposed by Ribeiro et al., was an approach that was able to explain the predictions of any model [9]. LIME utilized a locally interpretable model to interpret the black-box model's prediction results and constructed the relationship between the local sample features and the prediction results. Explanations from LIME do not exactly reflect the underlying model. LIME describes the prediction outcomes obtained even with different complex models, such as Random Forest, Support Vector Machine, Bagged Trees, or Naive Bayes. LIME can handle different input data types, including tabular data, image data, or text data.

DEEPVID, proposed by Wang et al., was a visual analytics system that leverages knowledge distillation and generative modeling to generate a visual interpretation for image classifiers [17]. Given an image of interest, DEEPVID applied a generative model to generate samples near it. These generated samples were used to train a local interpretable model to explain how the original model makes the decision. The difference between our approach and DEEPVID is that, instead of utilizing interpretable models such as linear regression to provide interpretation, our approach visualizes boundary examples directly. End-users can then leverage their human intelligence to interpret the model decision.

DeepDIG [6,7], developed by Karmi et al., was a framework that used to characterize the decision boundary for deep neural networks. The main contribution can be divided into two parts. The first part is to generating borderline instances that are near the decision boundary. This part is completed in three steps, the first and second steps are used to generate adversarial instances by Autoencoder. The third step is used to generate the borderline instances based on the binary search and adversarial instances produced after step one and step two. The second contribution is related to the characterization that is used to measure the decision boundary complexity in the input space and embedding space. The input space complexity is calculated by the generated borderline instances from the first contribution. The embedding space complexity is measured by developing a linear Support Vector Machine (SVM) model.

3   The Proposed Human-in-the-Loop Framework

Given a trained machine learning model and a (mispredicted) point-of-interest, we intend to generate a neighborhood that can enable a better human understanding of the model. The generated neighborhood needs to satisfy three critical criteria:

- The instances in the neighborhood need to be semantically close to the point-of-interest.
- The decision boundary is at least partially visible within the neighborhood.
- The neighborhood needs to maintain the number of instances in a manageable size so that human users can gain insight from it.

To generate a neighborhood that can satisfy the above three criteria, we propose the human-in-the-loop framework that contains three stages, as shown in Fig. 2. In the first stage, a neighborhood is generated based on the given sample through a trained generative model. In the second stage, the pre-trained machine learning model is used to yield classification on the generated instances to carve out the local decision boundary and delineate the model behavior. Next, three intervention points are provided to enable human users for a throughout exploration for gaining insights. In the following section, we explain each stage in detail.

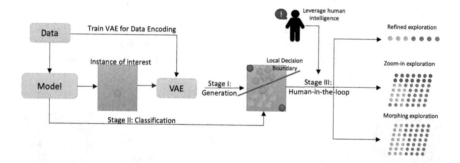

Fig. 2. The proposed human-in-the-loop framework. It contains three stages. In stage (I), a neighborhood is generated based on the given sample through a trained variational autoencoder. In stage (II), the pre-trained machine learning model is used to yield classification on the generated instances to carve out the local decision boundary and delineate the model behavior. In stage (III), human users are enabled with three intervention points to explore the neighborhood: a) refined multifacet path exploration, b) "zoom-in" & "zoom-out" area exploration, and c) boundary-crossing morphing exploration.

3.1   Stage (I): Neighborhood Generation

Stage one can be described as a stochastic process that generates neighbors from the given point-of-interest. There are two approaches to accomplish such a procedure: Variational Auto-Encoders (VAEs) and Generative Adversarial Networks (GANs). Both of these two generative methods assume an underlying latent space that is mapped to the original data space through a deterministic parameterized function. The generative model often consists of an encoder that can map the given data into the latent space, and a decoder that can decode the latent space vector back to the original space. In this work, we adopt VAE as the generative model because of its more straightforward model structure.

As shown in Fig. 3, we train an encoder-decoder CNN-VAE with ten latent dimensions on the MNIST dataset to learn the underlying latent distribution. A hyper-parameter *step-length* is applied to each latent space via linear interpolation to generate the perturbed latent vectors. The perturbed latent vectors are then fed through the decoder to generate neighbors around the point-of-interest.

More formally, a VAE model that consists of encoder $q_\theta(z|x)$ and decoder $q_\phi(x|z)$ are trained on the dataset X, where $X = \{(x_1, y_1), (x_2, y_2), ..., (x_n, y_n)\}$, $x_i \in R^D$ and $y_i \in [1, c]$. The VAE is trained with the negative log-likelihood with regularizer. The loss function l_i for data instance x_i is:

$$- E_{z-q_\theta(z|x_i)}[log_{p_\theta}(x_i|z)] + KL(q_\theta(z|x_i)||p(z)), \tag{1}$$

where $z \in R^d$ denotes the *d-dimension* embedding space learned by the VAE encoder.

Utilized by the trained VAE, examples near the point-of-interest can be generated and form the neighborhood. A hyper-parameter *step-length* needs to be chosen to determine the border of the neighborhood. In practice, we set *step-length* equal to one as the default value.

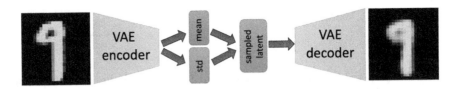

Fig. 3. The architecture of our selected generative model, i.e., a Variational AutoEncoder (VAE)

3.2   Stage (II): Neighborhood Classification

To identify and visualize the local decision boundary, the given trained machine learning model is applied to the generated instances. The classification results are highlighted with different colors so that the model behavior can be delineated. We call this classification results as classified neighborhood. A classified neighborhood is one where every data point within the neighborhood has been classified by the model-under-investigation so that the decision boundary is identified and visualized verbatim. Because the actual model is used, this is a verbatim manifestation of the model decision boundary within the neighbourhood. In practice, a larger value of *step-length* is recommended to ensure a decision boundary with clear difference between the opposite sides. In our experiments, we set the *step-length* to 1.

3.3   Stage (III): Human-in-the-Loop Exploration

Three intervention points are provided in our human-in-the-loop stage. Specifically,

- a refinement intervention point that provides a multifacet refined neighborhood exploration.
- a "zoom-in" & "zoom-out" intervention point that enables human users to take a closer look at the certain region of interest.
- a morphing intervention point that selects two examples from each side of the decision boundary and creates a visualization path.

For the first intervention point, human users are enabled to identify the dimensions of interest, i.e., specific dimensions from the d-dimensional latent space. Next, we allow the human to adjust the hyper-parameter *step-length* along the selected latent dimension for exploration. A larger value of the *step-length* will enrich the semantic variation, while a smaller value can provide a more concentrated result. The *step-length* serves as a "tuning knob" to adjust traversal speed in the latent space, which helps human users to understand how a prediction is carved out from specific changes.

Human users are allowed to identify two hidden dimensions of interest for the second intervention point and construct a morphing matrix based on these two-dimension spaces. Allowing the morphing of two dimensions simultaneously can provide a richer context around the point-of-interest. The second intervention point acts as a "zoom-in" & "zoom-out" effect to assist human users in gathering insights from the generated examples.

For the third intervention point, a few instances that are semantically close to the given point-of-interest at two sides of the decision boundary are provided. Next, a morphing path between the two instances are created and the path passes through the point-of-interest. The algorithm for identifying the nearest neighbor and creating the morphing path is shown in Algorithm 1. Such morphing traverses data manifold while crossing the decision boundary, which can delineate the model behavior and explain how and why a particular image is relevant to the prediction.

Algorithm 1: Pseudocode for the proposed method

Given: Dataset (X, Y)
Given: Classifier $F()$ to be interpreted
Given: Pretrained VAE: $(VAE\text{-}enc, VAE\text{-}dec)$
Given: Data instance of interest (x_i, y_i), where $y_i = c_1$, but mispredicted $F(x_i) = c_2$

1: $enc\text{-}x_i = VAE\text{-}enc(x_i)$
2: **for** $(x_j, y_j) \in (X, Y), y_j = c_1$ **do**
3:     $enc\text{-}x_j = VAE\text{-}enc(x_j)$
4:     update x_j s.t. $\|enc\text{-}x_j - enc\text{-}x_i\|_{L1}$ is smallest
5: **end for**
6: **for** $(x_k, y_k) \in (X, Y), y_k = c_2$ **do**
7:     $enc\text{-}x_k = VAE\text{-}enc(x_k)$
8:     update x_k s.t. $\|enc\text{-}x_k - enc\text{-}x_i\|_{L1}$ is smallest
9: **end for**
10: interval$=(enc\text{-}x_k - enc\text{-}x_i)/$num-neighbors
11: neighbors$=[]$
12: labels$=[]$
13: **for** i=0, i$\leq$num-neighbors; i++ **do**
14:     neigh $= enc\text{-}x_i \pm$interval
15:     neighbors.append(neigh)
16:     labels.append(F(neigh))
17: **end for**
18: Visualize(neighbors, labels)

---

4   Experiment Setup

To verify the effectiveness of our proposed framework, we conduct several experiments on two datasets. Section 4.1 describes the datasets and the trained machine learning model architectures. Section 4.2 presents the detailed experimental settings for our framework.

4.1   Dataset and Trained Machine Learning Architecture

We investigate the proposed framework against two datasets, MNIST and FashionMNIST. The MNIST dataset is a large database of handwritten digits, while FashionMNIST is a dataset of Zalando's article images. The images in these datasets are 28×28 grayscale images associated with a label of 10 classes. Both MNIST and FasionMNIST are commonly used for training various image processing machine learning models. The details of the datasets and the chosen model performance are shown in Table 1.

4.2   Our Proposed Framework Settings

In this subsection, we describe the training detail of each stage. Stage (I) utilizes an autoencoder that is pre-trained on the dataset to generate the neighborhood

Table 1. Description of the investigated datasets.

	MNIST	FashionMNIST
# of training examples	60,000	60,000
# of testing examples	10,000	10,000
# of output classes	10	10
Original data space (i.e., # of dimension)	784	784
Test accuracy of the chosen model	94.1	92.5

based on the given point-of-interest. Table 2 demonstrates the hyper-parameters of the pre-trained autoencoder for both datasets. Since MNIST contains simpler data points than FashionMNIST, we use a 10-dimensional latent space to represent the images in MNIST, while a 20-dimensional latent space for FashionMNIST.

Table 2. Description of variational autoencoder models used in Stage (I) and classifiers that need to be explained. The model architecture, activation function, and the number of hidden layers are shown accordingly.

	VAE	Classifier
MNIST	$CNV(32,64,\ 64),\ ReLU,\ 10$	$Linear(20,10),\ ReLU$
FashionMNIST	$CNV(32,64,\ 64),\ ReLU,\ 20$	$Linear(20,10),\ ReLU$

5   Result

This section will first apply our proposed framework to the MNIST dataset and illustrate how our framework works by providing multiple examples. Then, we apply our method to the FashionMNIST dataset. The examples we presented here demonstrate our framework's potential for improving human understanding of the black-box machine learning models. Note that due to the page limits we only present a handful case studies on two datasets. We also apply our framework on other datasets such as 3-D point cloud data. More interesting examples can be found in our GitHub Page.

5.1   MNIST

A CNN model trained on the MNIST dataset for digit classification is selected and yields a 94.1% accuracy on the testing dataset. A mispredicted example is chosen for the case study. Figure 4 and Fig. 5 show the selected mispredicted point-of-interest and the stage (I) and stage (II) process. As shown in Fig. 4, the neighborhood of the point-of-interest is generated in grey-scale. The examples in the neighborhood satisfied the criteria in Sect. 3 as they are all semantically close to the original data point. The classified neighborhood is shown in Fig. 5. The colors refer to the classification results. We observe that despite being classified to the same label, images close to the decision boundary have higher fidelity. This observation is consistent with our intuition that the model is more likely mispredicting samples near the decision boundary. One can also draw a similar conclusion by visually examining the classified neighborhood: examples near the decision boundary often have an ambiguous shape that sometimes confuses machine learning models. Through stage (I) and stage (II), our framework generates examples that delineate the model behavior by depicting the local decision boundary.

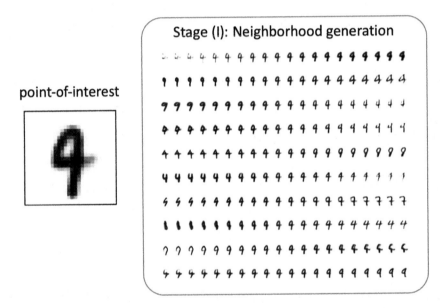

Fig. 4. Stage (I) of our framework. In Stage (I), the neighborhood of the point-of-interest is generated. The examples in the neighborhood satisfied the criteria in Sect. 3 as they are all semantically close to the original data point.

After getting the classified neighborhood that carves the local decision boundary around the point-of-interest, human users could be invited to explore the neighborhood using their own intelligence. Figure 6, Fig. 7, Fig. 8 and Fig. 9 illustrate the three possible human-in-the-loop exploration strategies. From Fig. 6, one can observe that at stage (III-a) there exist three interesting ways of morphing between *digit-4* and *digit-9*. Therefore, human users can gain insights by investigating the relevant features that have been changed along the process of *digit-4* morphing to *digit-9*. In this example, the three identified morphing paths revealed three related features: 1) the tartness of the circle, 2) the size of the

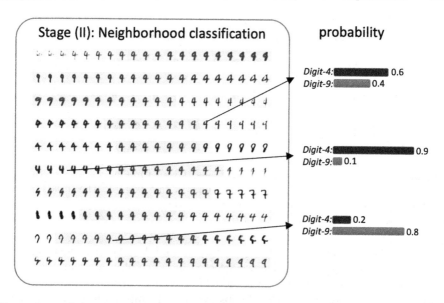

Fig. 5. Stage (II) of our framework. In Stage (II), the generated neighborhood is classified with the given trained machine learning model. Purple color indicates the image is classified as *digit-4*, orange color indicates the image is classified as *digit-9* and all other classification results are marked as color grey. We also observe that despite being classified to the same label, images close to the decision boundary have higher fidelity. (Color figure online)

circle and, 3) the straightness of the line. Next, human users can combine two paths for a "zoom-in" & "zoom-out" investigation. Combining two paths allows human users to gather richer information related to the decision boundary. As shown in Fig. 8 and Fig. 7, two possible combinations are chosen and presented, and the *step-length* are adjusted for the "zoom-in" effect and the "zoom-out" effect. From the denser region manifestation, one might conclude that 1) an "open-circle" at the top could help the given predictor correctly identify a *digit-4*, and 2) lines with roundness instead of tartness could mislead the predictor to mispredict a *digit-4* to *digit-9*. Such conclusions could help human users better understand how the model behaves in a certain region.

Figure 9 demonstrates the result generated by our third intervention point. As shown in the Figure, a *digit-4* is mispredicted as *digit-9*. By examining the morphing from the nearest digit 4 (in purple) to the nearest digit 9 (in orange), the circled area can be identified by human intelligence as one of the explanations for the misprediction.

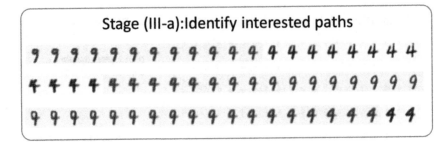

Fig. 6. Stage (III-a) of the framework. In Stage (III-a), three paths are identified and the morphing is highlighted with different colors. In this example, the three identified morphing paths revealed three related features: 1) the tartness of the circle, 2) the size of the circle and, 3) the straightness of the line.

Two other examples are shown in Fig. 10 and Fig. 11. The local decision boundary of the model near the two selected instance-of-interest are displayed, end-users can better understood the model behavior by visually examining these samples. In these two cases, we could observe that the mispredictions are likely to be caused by the circle areas in the image's top-left region. Note that human users can leverage their own intelligence to generate their own understanding with respect to the model behavior. Our framework only provides the intervention points that bridge the gaps between human minds and the black-box nature of machine learning models.

5.2 FashionMNIST

We provide another experiment using FashionMNIST dataset. In Fig. 12, a sandal is mispredicted as an ankle boot (in green) by a pre-trained CNN. Without the context that some sandals are boot-shaped, it would be difficult to understand the cause of this error. We select this mispredicted image as an item-of-interest and apply the trained VAE to extract its latent vector. Next, we explore the latent space around the extracted latent vector and generate a manageable number of neighbor images. The trained CNN is then applied to classify the generated images. The decision boundary can be observed as the classified label is morphing from sandal (in purple) to ankle boot (in orange). By visually displaying the neighborhood and the decision boundary (the area that purple turns into orange), the end-user can observe the smooth transition between sandal and ankle boot. Human users can easily draw the conclusion that the circled areas might cause the misprediction, i.e., if a boot-shaped image with blank space at the circled areas, it is likely the image will be classified as ankle boot.

6   Workflow of Human Users of the Proposed Framework

This study aims to improve explainability of machine learning models in a human-centric fashion. In this section, we present how a human user or a software engineer can leverage our framework to understand why a given ML model misclassifies a data point. There are three human intervention points.

Step length set to 1 (small)

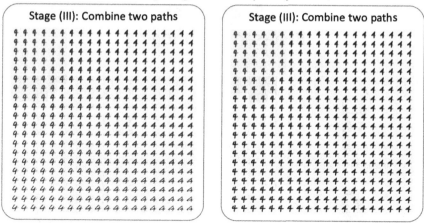

Fig. 7. Stage (III-b) of the framework. In Stage (III-b), the combination of two paths is presented to achieve a "zoom-in" effect for better carving out the model behavior. From this denser region manifestation, one might conclude that 1) an "open-circle" at the top could help the given predicter correctly identify a *digit-4*, and 2) lines with roundness instead of tartness could mislead the predictor to mispredict a *digit-4* to *digit-9*.

Step length set to 3 (large)

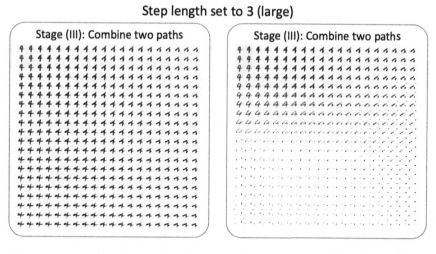

Fig. 8. Stage (III-b) of the framework. In Stage (III-b), the combination of two paths is presented to achieve a "zoom-out" effect for better carving out the model behavior.

6.1   Identifying the Point-of-Interest

First, the human user identifies a mispredicted point-of-interest, which software engineers routinely encounter as they debug software systems with ML components.

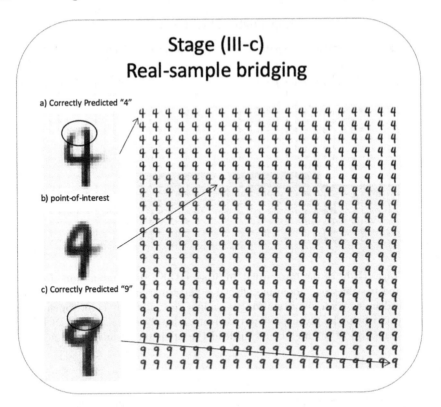

Fig. 9. Stage (III-c) of the framework. In this stage, two nearest data samples from the original dataset are selected to bridge the gaps between the point-of-interest and real samples on two sides of the decision boundary. (Color figure online)

6.2 Identifying Interesting Dimensions and Appropriate Step Lengths

Second, the key question from a user's perspective is: how and why a particular region of the point of interest is relevant to the prediction. That is where human users can again contribute by identifying the most interesting dimensions of semantic changes. Our framework leverages a powerful generative model, variational autoencoders, to generate a neighborhood of closely related data points. The generated neighborhood displays a progressive set of plausible variations of the point-of-interest and visualizes the semantic changes across all directions. The human user can use his common sense judgement to identify more interesting dimensions and more appropriate step lengths of changes on these dimensions so that changes in neighboring data-points are perceivable but not too dramatic.

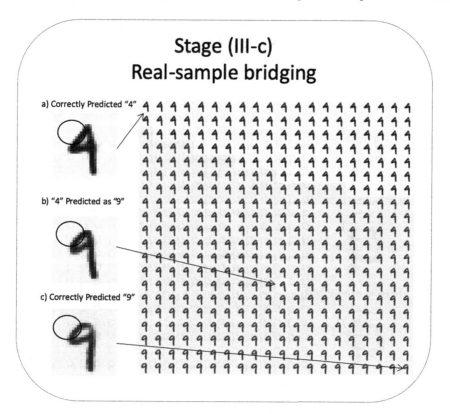

Fig. 10. Stage (III-c) of the framework. In this stage, two nearest data samples from the original dataset are selected to bridge the gaps between the point-of-interest and real samples on two sides of the decision boundary.

6.3 Selecting Two Most Revealing Dimensions to Generate a Matrix for Decision Boundary Visualization

Third, human users then select two most revealing dimensions so that a matrix of data-points can be generated to visualize the efforts of gradual changes on both dimensions. This matrix represents the neighborhood of interest. All generated data-points in the neighborhood are passed through the actual model-under-investigation so that the decision boundary is identified and visualized verbatim. Human users can gain knowledge and insights by walking through the classified instances and examining the decision boundary.

These three intervention points provide helpful exploration tools to help human users see, select, and manipulate the neighborhood of the data-point-of-interest and the decision boundary within it and therefore better understand the behavior of the underlying model.

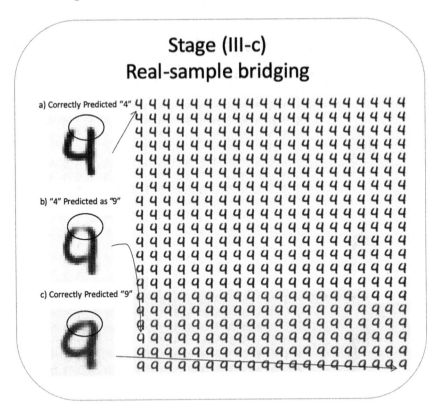

Fig. 11. Stage (III-c) of the framework. In this stage, two nearest data samples from the original dataset are selected to bridge the gaps between the point-of-interest and real samples on two sides of the decision boundary.

7   Discussion, Limitations and Future Works

This paper proposes a human-in-the-loop framework to improve human understanding of the black-box machine learning models locally through verbatim neighborhood manifestation.

However, the proposed method is limited in several ways. First, the neighborhood is generated based on the reconstructed data point. We lack a quantitative measure of the fidelity of the generated neighborhood to the original samples. Though the generated samples are derived from the VAE that was directly trained on the original dataset, some details are lost. Second, we adopt a standard VAE to encode the data point into latent space. Moving in such a latent space typically affects several factors of variation at once, and different directions interfere with each other [8]. This entanglement effect poses challenges for interpreting these directions' semantic meaning and, therefore, hinders human users from understanding the machine learning models.

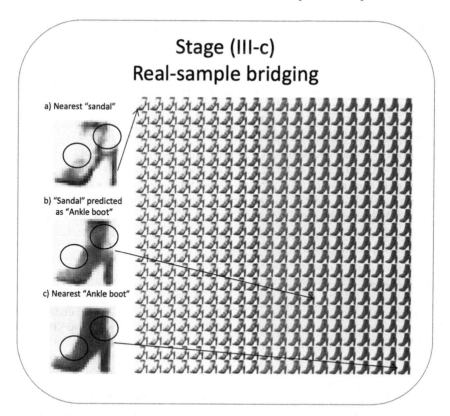

Fig. 12. A sandal is mispredicted as Ankle boot in FashionMNIST dataset. Without the context that some sandal have boot-shaped, it would be difficult to understand the cause of this error. The neighborhood manifestation provided by our framework enable human users to explore the context environment thus gain understanding of this type of mistakes.

Each of the limitations mentioned above points to a potential direction for future work. We want to quantify the fidelity of the generated data through metrics such as mean-absolute-error or binary-cross-entropy. For the second limitation, we are considering leveraging disentangle-VAE to generate neighborhoods along with semantic meaningful directions. We are also interested in learning a set of latent space directions inducing orthogonal transformations that are easy to distinguish from each other and offer robust semantic manipulations in the neighborhood manifestation. These future works introduce exciting challenges for bridging the gaps between the black-box nature of machine learning models and human understanding.

8   Conclusion

Machine learning models are mainly being developed and fine-tuned for optimal accuracy, while understanding these models has not attracted much attention. Existing XAI models focus on providing approximate hit-or-miss explanations, which do not involve humans in explaining and neglect human intelligence. We propose a human-in-the-loop explanation framework that reframes the explanation problem as a human-interactive problem to tackle this limitation. Our approach utilizes a generative model to enrich the (mispredicted) point-of-interest neighborhood and crave out the local decision boundary by highlighting the model prediction results. We provide three human-involved exploration intervention points that assist human users to leverage their own understanding of the model behavior. We conducted case studies on two datasets, and the experimental results demonstrate the potential of our framework for building a bridge between machine and human intelligence.

References

1. Ancona, M., Ceolini, E., Öztireli, C., Gross, M.: Towards better understanding of gradient-based attribution methods for deep neural networks. arXiv preprint arXiv:1711.06104 (2017)
2. Ancona, M., Öztireli, C., Gross, M.: Explaining deep neural networks with a polynomial time algorithm for shapley values approximation. arXiv preprint arXiv:1903.10992 (2019)
3. Bach, S., Binder, A., Montavon, G., Klauschen, F., Müller, K.R., Samek, W.: On pixel-wise explanations for non-linear classifier decisions by layer-wise relevance propagation. PLoS ONE **10**(7), e0130140 (2015)
4. Chattopadhay, A., Sarkar, A., Howlader, P., Balasubramanian, V.N.: Grad-cam++: generalized gradient-based visual explanations for deep convolutional networks. In: 2018 IEEE Winter Conference on Applications of Computer Vision (WACV), pp. 839–847. IEEE (2018)
5. Kabra, M., Robie, A., Branson, K.: Understanding classifier errors by examining influential neighbors. In: Proceedings of the IEEE Conference on Computer Vision and Pattern Recognition, pp. 3917–3925 (2015)
6. Karimi, H., Derr, T., Tang, J.: Characterizing the decision boundary of deep neural networks. arXiv preprint arXiv:1912.11460 (2019)
7. Karimi, H., Tang, J.: Decision boundary of deep neural networks: challenges and opportunities. In: Proceedings of the 13th International Conference on Web Search and Data Mining, pp. 919–920 (2020)
8. Mathieu, E., Rainforth, T., Siddharth, N., Teh, Y.W.: Disentangling disentanglement in variational autoencoders. In: International Conference on Machine Learning, pp. 4402–4412. PMLR (2019)
9. Ribeiro, M.T., Singh, S., Guestrin, C.: "why should I trust you?": explaining the predictions of any classifier. In: Proceedings of the 22nd ACM SIGKDD International Conference on Knowledge Discovery and Data Mining, San Francisco, CA, USA, 13–17 August 2016, pp. 1135–1144 (2016)
10. Rudin, C.: Stop explaining black box machine learning models for high stakes decisions and use interpretable models instead. Nat. Mach. Intell. **1**(5), 206–215 (2019)

11. Selvaraju, R.R., Cogswell, M., Das, A., Vedantam, R., Parikh, D., Batra, D.: Grad-cam: visual explanations from deep networks via gradient-based localization. In: Proceedings of the IEEE International Conference on Computer Vision, pp. 618–626 (2017)

12. Shrikumar, A., Greenside, P., Kundaje, A.: Learning important features through propagating activation differences. arXiv preprint arXiv:1704.02685 (2017)

13. Shrikumar, A., Greenside, P., Shcherbina, A., Kundaje, A.: Not just a black box: learning important features through propagating activation differences. arXiv preprint arXiv:1605.01713 (2016)

14. Simonyan, K., Vedaldi, A., Zisserman, A.: Deep inside convolutional networks: visualising image classification models and saliency maps. CoRR abs/1312.6034 (2014)

15. Sundararajan, M., Taly, A., Yan, Q.: Axiomatic attribution for deep networks. arXiv preprint arXiv:1703.01365 (2017)

16. Wang, H., et al.: Score-cam: score-weighted visual explanations for convolutional neural networks. In: Proceedings of the IEEE/CVF Conference on Computer Vision and Pattern Recognition Workshops, pp. 24–25 (2020)

17. Wang, J., Gou, L., Zhang, W., Yang, H., Shen, H.W.: Deepvid: deep visual interpretation and diagnosis for image classifiers via knowledge distillation. IEEE Trans. Visual Comput. Graphics **25**(6), 2168–2180 (2019)

18. Zeiler, M.D., Fergus, R.: Visualizing and understanding convolutional networks. In: Fleet, D., Pajdla, T., Schiele, B., Tuytelaars, T. (eds.) ECCV 2014. LNCS, vol. 8689, pp. 818–833. Springer, Cham (2014). https://doi.org/10.1007/978-3-319-10590-1_53

MAIRE - A Model-Agnostic Interpretable Rule Extraction Procedure for Explaining Classifiers

Rajat Sharma, Nikhil Reddy[(✉)], Vidhya Kamakshi, Narayanan C. Krishnan, and Shweta Jain

Indian Institute of Technology Ropar, Rupnagar, India
{2015csb1026,2018csm1011,2017csz0005,ckn,shwetajain}@iitrpr.ac.in

Abstract. The paper introduces a novel framework for extracting model-agnostic human interpretable rules to explain a classifier's output. The human interpretable rule is defined as an axis-aligned hyper-cuboid containing the instance for which the classification decision has to be explained. The proposed procedure finds the largest (high *coverage*) axis-aligned hyper-cuboid such that a high percentage of the instances in the hyper-cuboid have the same class label as the instance being explained (high *precision*). Novel approximations to the coverage and precision measures in terms of the parameters of the hyper-cuboid are defined. They are maximized using gradient-based optimizers. The quality of the approximations is rigorously analyzed theoretically and experimentally. Heuristics for simplifying the generated explanations for achieving better interpretability and a greedy selection algorithm that combines the local explanations for creating global explanations for the model covering a large part of the instance space are also proposed. The framework is model agnostic, can be applied to any arbitrary classifier, and all types of attributes (including continuous, ordered, and unordered discrete). The wide-scale applicability of the framework is validated on a variety of synthetic and real-world datasets from different domains (tabular, text, and image).

Keywords: Interpretable machine learning · Explainable models · Rule based explanations

1 Introduction

The working of classic machine learning models such as simple decision trees, linear regression can be easily interpreted by analyzing the parameters of the model. But, for want of higher accuracy or better generalization performance, complex classifiers such as deep neural networks, support vector machines, and decision forests are being employed. However, improved performance comes at the cost of reduced human interpretability. Recent research focuses on explaining

© IFIP International Federation for Information Processing 2021
Published by Springer Nature Switzerland AG 2021
A. Holzinger et al. (Eds.): CD-MAKE 2021, LNCS 12844, pp. 329–349, 2021.
https://doi.org/10.1007/978-3-030-84060-0_21

the working of these complex black-box models, thereby bridging the accuracy-interpretability trade-off and making them useful and trustworthy.

An explainable approach that can work irrespective of the underlying black-box model is desirable. Such approaches are referred to as model agnostic approaches in the literature [8,13,16,17]. Short and concise rules are highly human interpretable [8,12]. Hence we would like to develop a model-agnostic explainable approach that is capable of providing the human interpretable rules for any black-box machine learning model. A major challenge in designing such approaches lies in preserving *faithfulness* to the black-box. An explanation method is said to be faithful to a black-box model if it identifies features that are truly important for the working of the model.

A popular method of explaining the black-box model's working is by assigning ranks to the features relative to the importance the black-box model gives to a feature. This feature rank is easy to understand but is not complete. The ranking approach does not capture the class discriminative information based on the range of values. In other words, if a specific range of values for a feature results in classification to a class and outside the range corresponds to a different class, such a mechanism would not be revealed by feature ranking approaches [4,6]. For an explanation based on feature ranking approach to be complete, we need a measure to say how relevant is the feature value to a particular prediction. The sensitivity of the changes to the output of the model due to small changes in the feature values must also be captured. In a nutshell, feature ranking by itself is an incomplete explanation. Various factors like the importance of a feature, tolerable range of values to get the same prediction, the influence of a feature value towards a prediction, is to be additionally considered along with the feature rank to provide a complete explanation to the working of the black-box model.

Another class of methods, called rule-based methods, provide intuitive explanations. Decision trees, decision lists that provide rules in the form of if-then-else statements in a hierarchical fashion, come under this category. These are global explanation models that aim to explain the working of the model in the whole instance space. Though the explanations are intuitive, it is not always simple to comprehend. If the hierarchical structure of if-then-else statements grows into longer chains, it is difficult to comprehend, and the interpretability suffers [12]. It is to be noted that these methods partition the instance space based on the attribute values. Longer chains of if-then-else statements would mean small partitions created in the instance space. This further complicates the scenario as the rules are less generalizable.

Anchors [17], a recent approach overcomes the limitations of feature ranking and rule-based approaches. It builds on the observation that a complex tree of rules encompasses many simple trees. Hence instead of attempting to build a tree that spans the entire instance space and provides an explanation of the black-box model globally, it is better to provide a local explanation spanning a smaller partition of the whole instance space. The precision and coverage metrics defined in [17] help to preserve the desired properties of posthoc explanations. But a limitation of the Anchors approach is that it is applicable only for discrete attributed datasets. In the case of continuous-valued attributes, Anchors can be applied after the continuous values are mapped onto a discrete values set only.

Binning is employed to discretize the continuous-valued attributes by identifying a threshold to create bins. The binning threshold plays a crucial role and it may not be always possible to obtain the tighter bounds on the range of tolerable values for a prediction. Thus, an unsuitable binning threshold may lead to loss of subtle class discriminant information that may otherwise be present in the original continuous-valued attributes.

The proposed framework MAIRE is a non-trivial extension of Anchors that is applicable across any attribute type - continuous, discrete (ordered or unordered). A sample explanation generated from our approach is shown in the Table 1.

Table 1. Example MAIRE explanations obtained for the Adult, Abalone and German credit datasets

	If		Predict	Coverage	Precision
Adult	17 < Age ≤ 43 Education = High School grad 0.00 < Capital-Loss ≤ 1291.44		≤ 50K	0.35	0.95
Abalone	Sex = F 0.07 < Length ≤ 0.48 0.05 < Diameter ≤ 0.37		≤9 Rings	0.25	0.94
German credit	Housing = own 20 < Duration ≤ 25 38 < Age ≤ 54		good	0.28	0.94

2 Related Work

The significant efforts towards improving the explainability aspect of machine learning models can be broadly categorized into three directions.

Model agnostic methods are like 'meta-learning' approaches that are capable of explaining the behavior of any black-box classifier. LIME[16] approximates the working of the black box classifier in a local neighborhood by fitting a linear model on the black box predictions for the neighbors. Anchors[17] finds the decision rule for black-box model prediction such that the rule anchors the prediction adequately as governed by the precision and coverage metrics. Both LIME and Anchor generate global explanations but apply only to discrete-valued datasets. MAIRE overcomes this constraint by its novel optimization framework.

Learning To Explain (L2X) [6] does instance wise feature selection by maximizing the mutual information between the subset of features and the target variable. SHAP [15] uses Shapley values to predict the importance of features towards a prediction. Both L2X and SHAP use feature ranking approach, which is accurate in text classification. However, feature ranking may not always be optimal as feature values may play an important role in distinguishing between two classes. MAIRE, on the other hand, explains in terms of a range of values of an attribute. LORE [8] explains the black-box model by extracting rules using a decision tree applicable in a local neighborhood generated by a genetic algorithm. This method is applicable for low dimensional datasets only as with high dimensional datasets; the decision tree may grow complex, reducing interpretability. LLORE [9] uses an autoencoder to perform dimensionality reduction so that LORE [8] can be applied in the reduced dimensional space. Dimensionality reduction may lead to loss of information and should be avoided. Further, LLORE [9] can be used only for images and an extension to handle text data has only been mentioned as a future possibility. Our proposed approach MAIRE can be applied to different domains (text, tabular, image) and does not require any modification to the dimensionality of the feature space, thus preserving all the information.

The proposed work is close to that of Lakkaraju et al. [13] in the broader sense from the perspective of explanation generation in terms of rules as per attribute ranges. But their explanation generation algorithm requires value ranges to be provided, and explanation is in terms of the rules explaining how the black-box model works in the subspace defined by the given attribute values. This flexibility may be beneficial for the tabular datasets, where the value range for attributes shall make sense to end-users. While the approach in [13] is model agnostic like MAIRE, the extensive experimentation has been carried out only on tabular datasets. This need to give attribute value ranges for explanation generation is challenging in case of images or textual datasets where the attributes the black-box model works on may be different from how humans perceive the data. Our proposed approach MAIRE does not have this requirement and hence is readily applied to explain black-box models working on data from different domains.

Model specific explainable methods are designed to explain the working of a single or a class of models. Approaches like Guided Back Propagation [21], CAM [23], and its extension [18] are applicable to architectures involving Convolutional Neural Networks only. Specifically for deep learning architectures, an attribution based technique, DeepLIFT [19] provides a set of rules to assign contribution scores to every unit of a deep neural network. In contrast, the MAIRE framework explains the output of any black-box model.

Models explainable by design consist of methods that propose new explainable classifiers that are trained from scratch. Interpretable CNN [22] uses mutual information to learn interpretable parts that are filtered through predefined templates. A self-explaining architecture involving an autoencoder that determines representative prototypes clustered around inputs in a latent space was proposed by Li et al. [14]. In another approach, the convolutional layer feature maps are used as latent representations that helps to localize regions of the image that are similar to the prototypes [5, 10]. Models that explain the output in terms of human interpretable rules have also been proposed [2, 12]. However, these models cannot be applied to an already deployed model.

3   Methodology

3.1   Problem Statement

Let $\{\mathbf{x}_n, y_n\}_{n=1}^N$ be a set of N training examples, where $\mathbf{x}_n \in \mathbb{R}^D$ is a data point and $y_n \in \mathcal{Y}$ is the associated label. For simplicity, let us assume that all the attributes are continuous and are normalized to the range $[0, 1]$ and that the classification task is binary. The MAIRE framework can be easily extended for discrete attributes and multi-class classification. Given a query data point $\mathbf{x}_q'$ and a classifier $f : \mathbb{R}^D \to \mathcal{Y}$, our objective is to explain the decision of the classifier at $\mathbf{x}_q'$ i.e. $f(\mathbf{x}_q')$. Prior literature suggests that explanations in the form of rules defined on the values of the attributes are human interpretable [12,17]. A simple way to define these rules for continuous attributes is in terms of range on the values. Thus, we define an explanation as $\mathbf{E} = \{\mathbf{l}, \mathbf{u}\}$, where $\mathbf{l}, \mathbf{u} \in \mathbb{R}^D$ represent the lower and upper bounds of intervals such that $l_i \leq x_{qi}' \leq u_i$, $\forall i \in \{1, \dots, d\}$.

The Cartesian product of these intervals represents a hyper-cuboid denoted by $S(\mathbf{l}, \mathbf{u})$. This is illustrated as a rectangle for the 2D dataset presented in Fig. 1(a). Our objective is to find an explanation that has high coverage and satisfying a certain threshold on precision. Coverage of an explanation $\mathbf{E}$, $Cov(\mathbf{l}, \mathbf{u})$, is defined as the fraction of data points that lie within the hyper-cuboid,

$$Cov(\mathbf{l}, \mathbf{u}) = \frac{1}{N} \sum_{i=1}^N \mathbf{1}(\mathbf{x}_i \in S(\mathbf{l}, \mathbf{u})) \tag{1}$$

where $\mathbf{1}(A)$ is the indicator function that takes the value 1 if the argument A is true. High coverage means that more data points are explained using the hyper-cuboid.

Precision, $Pre(\mathbf{l}, \mathbf{u})$, is defined as the fraction of training instances that lie within the hyper-cuboid representing the explanation $\mathbf{E}$ and whose classifier predictions match with the classifier prediction of the query point,

$$Pre(\mathbf{l}, \mathbf{u}) = \frac{\sum_{i=1}^N \mathbf{1}(f(\mathbf{x}_i) = f(\mathbf{x}_q') \text{ and } \mathbf{x}_i \in S(\mathbf{l}, \mathbf{u}))}{\sum_{i=1}^N \mathbf{1}(\mathbf{x}_i \in S(\mathbf{l}, \mathbf{u}))} \tag{2}$$

The MAIRE framework allows for a user to define a minimum value P for the precision of an explanation $Pre(\mathbf{l}, \mathbf{u})$. Thus the overall objective of the framework is to find an explanation (or the hyper-cuboid) that maximizes the coverage, while ensuring that the precision of the estimated explanation does not fall below the threshold P i.e.,

$$\underset{\{\mathbf{l}, \mathbf{u}\}}{\operatorname{argmax}} \quad Cov(\mathbf{l}, \mathbf{u}) \quad \text{s.t.} \quad Pre(\mathbf{l}, \mathbf{u}) \geq P. \tag{3}$$

The above optimization problem is challenging to solve due to the involvement of the indicator function. For a binary classification setting, with $P = 1$, this problem becomes the bichromatic rectangle problem, a widely studied combinatorial problem in computational geometry. Bichromatic rectangle problem involves computing a rectangle containing maximum number of red points and no blue points amongst the set containing n red points and m blue points in $d-$dimensional space. Most of the results in this area hold for $2D$ [1,3]. Further, the problem is NP-hard for arbitrary dimension [7]. Solving the above problem even approximately is therefore important for many applications.

We propose a novel method to transform the coverage and precision functions into differentiable approximations (with non-zero gradients), thereby making it easier to optimize using gradient-based methods.

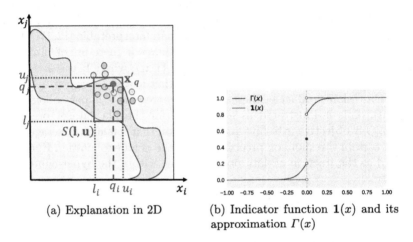

(a) Explanation in 2D

(b) Indicator function $1(x)$ and its approximation $\Gamma(x)$

Fig. 1. [Best viewed in color] Illustration of the explanation and the approximation to the indicator function

3.2   Approximations to Coverage and Precision

We first define the function Γ, which is an approximation to the indicator function, as $\Gamma(z) = c_1\sigma(c_2 z)+c_3(\text{sgn}(z)c_4+c_5$ where c_1, c_2, c_3, c_4, and c_5 are constants that determine the quality of the approximation, σ is the Sigmoid function, and $\text{sgn}(z)$ is the Signum function. The constant c_1 is chosen to scale down the sigmoid function so that, $c_1\sigma(c_2 z)$ takes values in a small range (effectively modeling the horizontal lines of the indicator function, but still retaining non-zero gradients). The constant c_2 has a high value to model the steep increase at $z = 0$ while making $\sigma(c_2 z)$ flatter for $z < 0$ and $z > 0$. The constants c_4 and c_5 are chosen so that $(sgn(z)c_4 + c_5)$ is 1 when $z > 0$, 0.5 when $x = 0$ and 0 otherwise. c_3 is chosen to provide a step at $z = 0$ such that $\Gamma(z) \in (0,1)$. This makes $\Gamma(z)$ piece-wise differentiable with non-zero gradients. The behavior of $\Gamma(z)$ is illustrated in Fig. 1(b). Note that, $\Gamma(z) = c_1\sigma(c_2 z)$ if $z < 0$, $\Gamma(z) = c_1\sigma(c_2 z)+c_3$ if $z > 0$, and $\Gamma(z) = 0.5$ if $z = 0$.

We can now approximate $1(x_1 > x_2)$ as $G(x_1, x_2) = \Gamma(x_1 - x_2)$ and $1(x_1 \geq x_2)$ as $GE(x_1, x_2) = \Gamma(x_1 - x_2 + c_l)$, where c_l is a constant that has a low value (close to 0). The approximation to the function $1(x_1 \text{ and } x_2 \dots \text{ and } x_m)$ for the logical operator 'and' is defined as $A(x_1, x_2, \ldots, x_m) = \Gamma(\frac{1}{m} \sum_{i=1}^{m} x_i - c_h)$, where c_h is a constant that has a high value (close to 1).

Let us define the functions $a_{2j-1}(\mathbf{l}, \mathbf{u}, \mathbf{x}) = G(x_j, l_j)$ and $a_{2j}(\mathbf{l}, \mathbf{u}, \mathbf{x}) = GE(u_j, x_j)$ with $j \in \{1, \ldots, D\}$. Then the approximation to the indicator function $\mathbf{1}(\mathbf{x} \in S(\mathbf{l}, \mathbf{u}))$ can be defined as $h(\mathbf{l}, \mathbf{u}, \mathbf{x}) = A(a_1(\mathbf{l}, \mathbf{u}, \mathbf{x}), a_2(\mathbf{l}, \mathbf{u}, \mathbf{x}), \ldots, a_{2D}(\mathbf{l}, \mathbf{u}, \mathbf{x}))$ Note that $h(\mathbf{l}, \mathbf{u}, \mathbf{x})$ should take a value close to 1 if the point $\mathbf{x}$ lies inside the hyper-cuboid $S(\mathbf{l}, \mathbf{u})$, else should take a value close to 0. We can now define approximate coverage and approximate precision as:

$$\hat{Cov}(\mathbf{l}, \mathbf{u}) = \frac{1}{N} \sum_{i=1}^{N} h(\mathbf{l}, \mathbf{u}, \mathbf{x}_i)$$

$$\hat{Pre}(\mathbf{l}, \mathbf{u}) = \frac{\sum_{i=1}^{N} h(\mathbf{l}, \mathbf{u}, \mathbf{x}_i)(1 - (f(\mathbf{x}_i) - f(\mathbf{x}'_q))^2)}{\sum_{i=1}^{N} h(\mathbf{l}, \mathbf{u}, \mathbf{x}_i)}$$

3.3 Accuracy of the Approximation

In this section, we theoretically bound the accuracy of our approximation functions $\hat{Cov}$ and $\hat{Pre}$. The accuracy of the approximation depends on the values of the constants in the definition of Γ. Note that, by definition $c_4 = c_5 = 0.5$ and $c_3 = 1 - c_1$. Thus we need to tune the parameters c_1, c_h, c_l, and c_2. Before we bound $\hat{Cov}$ and $\hat{Pre}$, we would like to make the following observation for the function $\Gamma(x)$ which is defined as $\Gamma(x) = c_1\sigma(c_2 x) + c_3(sgn(x)c_4 + c_5)$.

Observation 1. *When $c_4 = c_5 = 0.5$ and $c_3 = 1 - c_1$, we have:*

- *If $x > 0$, $\Gamma(x) = c_1\sigma(c_2 x) + c_3$*
- *If $x < 0$, $\Gamma(x) = c_1\sigma(c_2 x)$*
- *If $x = 0$, $\Gamma(x) = 0.5$*

We first begin with bounding the term $h(l, u, x)$. The following Lemma shows $h(l, u, x)$ is a good enough approximation for the indicator function for any poing $x \in \mathbb{R}^d$.

Lemma 1. *Let $c = \frac{c_1}{2}$ and $c_h > 1 - c$. If $c < \frac{1}{4D}$, we have $\forall x_i$:*

- *If $l_j < x_{ij} \leq u_j$ $\forall j = \{1, 2, \ldots, D\}$, we have:*

$$h(l, u, x) \leq 1 \text{ and}$$
$$h(l, u, x) \geq 1 - c$$

i.e. for all points lying inside the hypercuboid, function $h(\cdot)$ is very close to 1.

– If $\exists k, m$ with $k + m \geq 1$, such that $x_{ij} \leq l_j$ for k attributes or $x_{ij} > u_j$ for m attributes then:

$$h(l, u, x) \leq c \text{ and}$$
$$h(l, u, x) \geq 0$$

i.e. for all points lying outside the hypercuboid, function $h(\cdot)$ is very close to 0.

Proof. The proof considers four cases depending on the number of attributes of a data point that lie between the lower bound and upper bound of the hyper-rectangle.

Case 1: $\forall j \in \{1, 2, \ldots, D\}, l_j < x_{ij} \leq u_j$.

$$h(l, u, x_i) = \Gamma \left(\frac{\sum_{j=1}^{D} \Gamma (x_{ij} - l_j) + \sum_{j=1}^{D} \Gamma (u_j - x_{ij} + c_l)}{2D} - c_h \right)$$

$$= \Gamma \left(\frac{\sum_{j=1}^{D} (c_1 \sigma (c_2 (x_{ij} - l_j)) + c_3)}{2D} + \frac{\sum_{j=1}^{D} (c_1 \sigma (c_2 (u_j - x_{ij} + c_l)) + c_3)}{2D} - c_h \right)$$

$$\text{(From Observation 1)}$$

Let, $t = c_1 \dfrac{\sum_{j=1}^{D} \sigma(c_2(x_{ij} - l_j)) + \sum_{j=1}^{D} \sigma(c_2(u_j - x_{ij} + c_l))}{2D} + c_3 - c_h$, then using the fact that $\sigma(x) \geq 0.5$ if $x > 0$, we have:

$$t \geq \frac{c_1}{2} + c_3 - c_h$$
$$\geq 1 - \frac{c_1}{2} - c_h$$
$$> 0 \qquad \qquad \text{(if } c_h + \frac{c_1}{2} < 1)$$

Thus, if $c_h + \frac{c_1}{2} < 1$, we have $t > 0$. Thus, we get, $h(l, u, x_i) = \Gamma(t) = c_1 \sigma(c_2 t) + c_3$ from Observation 1. Since, $t > 0$, $c_2 t > 0$ for any $c_2 > 0$, we have, $h(l, u, x_i) \geq \frac{c_1}{2} + c_3 \geq 1 - \frac{c_1}{2}$. Also, $h(l, u, x_i) = c_1 \sigma(c_2 t) + c_3 \leq c_1 + c_3 \leq 1$.

Case 2: Let us assume that $\exists k$ such that $x_{ij} \leq l_j$ for k attributes i.e. point lie outside or on the lower bound of hypercuboid for $k \geq 1$ attributes and $\exists m$ such that $x_{ij} > u_j$ for $m \geq 1$ attributes. Out of k attributes, let k_1 attributes have $x_{ij} = l_j$ and $k - k_1$ attributes $x_{ij} < l_j$. Then, we have:

– For all k_1 attributes: $\Gamma(x_{ij} - l_j) = 0.5$
– For $k - k_1$ attributes: $\Gamma(x_{ij} - l_j) = c_1 \sigma(c_2(x_{ij} - l_j)) \leq c_1$
– For $D - k$ attributes: $\Gamma(x_{ij} - l_j) = c_1 \sigma(c_2(x_{ij} - l_j)) + c_3 \leq 1$
– For all m attributes: $\Gamma(u_j - x_{ij} + c_l) = c_1 \sigma(c_2(u_j - x_{ij} + c_l)) \leq c_1$
– For $D - m$ attributes: $\Gamma(u_j - x_{ij} + c_l) = c_1 \sigma(c_2(u_j - x_{ij} + c_l)) + c_3 \leq 1$

We get,

$$h(l, u, x_i)$$

$$= \Gamma \left(\frac{0.5k_1 + \sum_{j=1}^{k-k_1} c_1\sigma(c_2(x_{ij} - l_j))}{2D} + \frac{\sum_{j=k+1}^{D}((c_1\sigma(c_2(x_{ij} - l_j))) + c_3)}{2D} \right.$$

$$\left. + \frac{\sum_{j=1}^{m} c_1\sigma(c_2(u_j - x_{ij} + c_l))}{2D} + \frac{\sum_{j=m+1}^{D} c_1\sigma(c_2(u_j - x_{ij} + c_l) + c_3)}{2D} - c_h \right)$$

Let, $h(l, u, x_i) = \Gamma(t)$ i.e. consider the entire term in Γ expression to be t then:

$$t \leq \frac{0.5k_1 + (k - k_1)0.5c_1 + (D - k)(c_1 + c_3)}{2D} + \frac{0.5mc_1 + (D - m)(c_1 + c_3)}{2D} - c_h$$

$$t \leq \frac{0.5k_1(1 - c_1) + 0.5kc_1 + 0.5mc_1 + 2D - k - m}{2D} - c_h$$

$$(1 \leq k + m \leq 2D, \ k_1 \leq D, \text{ and } 0 < c_1 < 1)$$

$$\leq \frac{0.5D(1 - c_1) + 0.5c_1 D}{2D} + \frac{2D - 1}{2D} - c_h$$

$$\leq \frac{1}{4D} + \frac{2D - 1}{2D} - c_h \leq \frac{4D - 1}{4D} - c_h$$

Thus, when $c_h > \frac{4D-1}{4D}$, then we get $t < 0$. In this case, we have: $h(l, u, x_i) = \Gamma(t) = c_1\sigma(c_2 t) \leq \frac{c_1}{2}$. From Case 1, we have $\frac{c_1}{2} < 1 - c_h$. Substituting $c_h > \frac{4D-1}{4D}$, we get, $\frac{c_1}{2} < \frac{1}{4D}$. Thus, if any of the attribute of the example lies outside the boundary, we get $h(l, u, x_i) \leq \frac{1}{4D}$ and if all the attributes lie inside the boundary, we get $h(l, u, x_i) \geq \frac{4D-1}{4D}$.

Thus, choosing c_1 and c_h according to the lemma ensures that h is a good approximation to the indicator function $\mathbb{1}(x \in S(l, u))$. Further, we can arrive at the following important result that bounds the difference between Cov and the corresponding approximation $\hat{Cov}$.

Theorem 1. If $c_1 < \frac{1}{2D}$ and $c_h > \frac{4D-1}{4D}$, then $\left(\frac{4D-1}{4D}\right) Cov \leq \hat{Cov} \leq \frac{1}{4D} + \left(\frac{4D-1}{4D}\right) Cov$

Proof. Let the actual coverage from the hypercuboid (l, u) be $\frac{k}{N}$ i.e. $\sum_{i=1}^{N} \mathbb{I}(x_i \in S(l, u)) = k$. Then:

$$\hat{Cov} = \frac{1}{N} \sum_{i=1}^{N} h(l, u, x_i)$$

$$= \frac{1}{N} \sum_{x_i \in S(l,u)} h(l, u, x_i) + \frac{1}{N} \sum_{x_i \notin S(l,u)} h(l, u, x_i)$$

$$\geq \frac{1}{N} k(1 - c) \geq Cov \left(\frac{4D - 1}{4D} \right)$$

$$\hat{Cov} = \frac{1}{N} \sum_{x_i \in S(l,u)} h(l, u, x_i) + \frac{1}{N} \sum_{x_i \notin S(l,u)} h(l, u, x_i)$$

$$\leq \frac{k}{N} + \frac{N-k}{N} c \leq c + Cov\,(1-c) \leq \frac{1}{4D} + Cov\left(\frac{4D-1}{4D}\right)$$

The above result is interesting not only because it bounds the approximate coverage in terms of true coverage but it also suggests that as the features (dimension) increases, approximate coverage becomes closer to the true coverage. We also verify this from our experiments in Table 2.

We also have additional result for the bounds on the approximate precision.

Theorem 2. $\hat{Pre} \leq Pre\left(1 + \frac{1}{\hat{Cov}}\left(\frac{4D}{4D-1}\right)\right)$. Thus, when algorithm returns a hypercuboid with $\hat{Pre} \geq P$ then $Pre \geq \frac{1}{\left(1 + \frac{1}{\hat{Cov}}\left(\frac{4D}{4D-1}\right)\right)}P$

Proof. Let, k points be inside the hyper-cuboid, out of k points, q points satisfy $f(x_i) = f(x_q)$ and m points satisfy $f(x_i) = f(x_q)$ in total.

$$\hat{Pre} = \frac{\sum_{\substack{x_i = x_q \\ x_i \in S(l,u)}} h(l, u, x_i) + \sum_{\substack{x_i = x_q \\ x_i \notin S(l,u)}} h(l, u, x_i)}{\sum_{x_i \in S(l,u)} h(l, u, x_i) + \sum_{x_i \notin S(l,u)} h(l, u, x_i)}$$

$$\leq \frac{q + (m-q)c}{(1-c)k}$$

$$\leq Pre + \frac{Pre}{Cov}\left(\frac{4D}{4D-1}\right)$$

In summary, when the dimension of the dataset increases, $c_1 \approx 0$ and $c_h \approx 1$, and the difference between analytical coverage and the corresponding approximation tends to 0. Even though the theoretical bounds depend on the dimensionality of the data, we conclude from the experiments that the values $c_1 = 0.4, c_2 = 15, c_3 = 0.6, c_4 = 0.5, c_5 = 0.5, c_l = 0.02$, and $c_h = 0.8$ work well for a wide variety of datasets and do not have to be tuned for a new dataset. We use these values for all the experiments performed in the paper.

4   Optimization to Estimate the Explanation

Our next objective is to formulate the optimization problem in the MAIRE framework for estimating the explanation. In the simplest case, we want to find the optimal values for the parameters l and u that maximize coverage while maintaining a minimum precision P. The user sets this lower bound on precision. It is assumed that any value of $Pre(\mathbf{l}, \mathbf{u})$ above P is acceptable. If the current value of precision is greater than the threshold, we would only like to maximize the coverage. If the analytical precision becomes less than the specified threshold, then in addition to maximizing coverage, the MAIRE framework also maximizes the precision $\hat{Pre}$. This component is weighted by a constant factor λ_1 to signify the importance of increasing the precision at the cost of reducing the coverage. If λ_1 is high, $\hat{Pre}$ will increase whenever the precision is less than the threshold. Thus the overall objective function $\mathcal{L}(\mathbf{l}, \mathbf{u})$ is defined as:

$$\mathcal{L}(\mathbf{l}, \mathbf{u}) = \hat{Cov}(\mathbf{l}, \mathbf{u}) + \lambda_1 \hat{Pre}(\mathbf{l}, \mathbf{u})(1 + \text{sgn}(P - Pre(\mathbf{l}, \mathbf{u})))$$

Note that the analytical precision value $Pre(\mathbf{l}, \mathbf{u})$ is only required to activate the approximation term. The objective function $\mathcal{L}(\mathbf{l}, \mathbf{u})$ is maximized subject to two constraints. First, the lower and upper bound vectors $\mathbf{l}$ and $\mathbf{u}$ need to be in $[0, 1]^D$. As $\Gamma(z)$, the approximator to the indicator function, never truly achieves a zero gradient, if the explanation is unbounded, then the optimization procedure might never converge as the explanation could keep expanding in all directions indefinitely. This constraint is implemented by clipping the values of $\mathbf{l}$ and $\mathbf{u}$ at 0 and 1 respectively after every iteration. The second constraint is that the explanation finally generated must contain the query instance: $l_j \le x'_{qj} \le u_j, \forall j = 1, \ldots, D$. The optimizer focuses on these constraints, only when they are not satisfied. When the constraints are satisfied, the optimizer only maximizes the coverage. This is achieved by using the ReLU function on the difference $l_j - x'_{qj}$ and $x'_{qj} - u_j$. These constraints are added to the final optimization function with a weighting constant λ_2 (can be viewed as the Lagrange multiplier used for constrained optimization) that signifies the penalty on the objective when the constraint is not satisfied. Thus, the final objective function that is maximized with respect to the parameters $\mathbf{l}$ and $\mathbf{u}$ is defined as

$$\underset{\mathbf{l}, \mathbf{u}}{\arg\max} \, \mathcal{L}(\mathbf{l}, \mathbf{u}) - \lambda_2 \left(\sum_{j=1}^{D} ReLU(l_j - x'_{qj}) + \sum_{j=1}^{D} ReLU(x'_{qj} - u_j) \right) \quad (4)$$

Adam optimizer [11] with default parameter values is used for this non-linear and non-convex optimization.

4.1 Greedy Attribute Elimination for Human Interpretability

The explanations created might still be too large (containing non-trivial bounds for many dimensions) for a human to understand. We reduce the size of the generated explanations using a greedy elimination procedure to improve human interpretability. An attribute whose removal results in a maximum increase in the coverage while retaining precision above the user-defined threshold is eliminated. If no such attribute exists, then the attribute whose removal reduces the precision by the minimum extent is excluded from the explanation. Attributes are removed greedily at least for $D - K$ times, where K is the maximum number of attributes that can be part of an explanation as set by the user. Note that once we get the hypercuboid $S(l, u)$, the greedy selection of a single feature will take $O(D)$ time because computing coverage and precision for a given hypercuboid with one feature removal will take constant time to compute.

4.2 Local to Global Explanations

To gain a broader understanding of how the classification model works on the entire dataset, we would need to generate multiple explanations for a comprehensive set of instances. This is an infeasible task due to the significant computational complexity. Instead of creating a global explanation by combining local explanations of randomly selected instances, we identify an optimal set of local explanations that can approximate the global behavior of the classifier.

The process of creating a global explanation is started by considering a moderately sized subset of the training set (chosen randomly). Local explanations are generated using the MAIRE framework for all the instances in this set. A subset of these explanations is selected greedily, such that every new local explanation added to the global explanation leads to the maximum increase in the overall coverage of the global explanation. We call this procedure Maximum Symmetric Difference (MSD Select) as the local explanation that results in the maximum symmetric difference with the current estimate of the global explanation is added to the global explanation.

The global explanation can be viewed as a new rule-based classifier $f'(\mathbf{x})$. Given a test data point, several local explanations that are part of $f'(\mathbf{x})$ can be applied to predict the class label. We propose to use the majority class label among the applicable explanations for generating the class label.

4.3 Extension to Discrete Attributes

The MAIRE framework is directly applicable on ordered discrete attributes. The final explanation is a set of consecutive discrete values. The generated explanation is slightly modified for ordered discrete attributes by changing l_i to the smallest discrete value that is greater than or equal to l_i and changing u_i to the largest discrete value that is lesser than or equal to u_i. This modification does not affect coverage or precision and improves readability. In the case of a categorical attribute (unordered), finding intervals is not meaningful. We instead convert all categorical attributes to their equivalent one-hot encoding. The transformed boolean representation is treated as ordered discrete attributes. If an explanation contains both the values of a boolean attribute, the corresponding attribute is dropped from the explanation. If only the value one is selected, then the value of the unordered attribute in $\mathbf{x}'_q$ is included in the explanation. Due to the enforcement of the second constraint, selection of only 0 is not possible as $\mathbf{x}'_q$ has the value 1 for the corresponding boolean attribute.

5 Experiments and Results

The source code for the method is publicly accessible[1]. The MAIRE framework is tested on a wide variety of real-world datasets.

---

[1] https://github.com/maire-code/code-submission.

5.1 Tabular Datasets

We conducted experiments to study the quality of the approximations to coverage and precision using the tabular datasets. Explanations for 100 randomly sampled data points for each of the datasets were computed. The true coverage and precision were determined for each explanation as well as the values for the corresponding approximations. The mean squared error between the true and approximate values averaged over 100 data points for the three datasets is presented in Table 2. It can be noticed that difference in the true values and the corresponding approximations is not significant. Further this difference reduces as the number of attributes increases supporting our theoretical analysis. The German credit dataset has the highest number of attributes (20), followed by Adult (14), and Abalone (8) data sets.

The MAIRE framework is evaluated on three tabular datasets - Adult, Abalone, and German credit datasets. A three-layer neural network (containing 150, 100, and 50 nodes in each layer with ReLU activation) serves as the black-box model (though any classifier can serve the purpose). The datasets are divided into train and test splits according to the ratio of 3:1. The neural network is trained for 100 epochs. The test accuracy of the black box model for the Adult, Abalone, and German credit datasets is 81.52%, 87.76%, and 79.28%, respectively. A sample of the explanations generated by MAIRE for the three datasets is presented in Table 1.

We compare the quality of the global explanations extracted from MAIRE against other model-agnostic rule-based explanation methods capable of composing global explanations, namely; LIME and Anchors. LIME and Anchor are applicable only on discrete datasets. Hence, for a fair comparison, we have used the same discretized version of the dataset across all the models, including MAIRE. The precision threshold is set at 0.95 for all the datasets. We compare the sub-modular pick (SP) versions of LIME and Anchor against the MSD Select of MAIRE. Coverage over unseen test instances in the global explanation is used as the metric for comparison. Figure 3a–c presents the results averaged over five trials on the three tabular datasets. MSD-MAIRE consistently performs better than SP-LIME and SP-Anchor, achieving the maximum coverage using a lesser number of explanations. Thus MSD-MAIRE has higher coverage at the same precision threshold. It is also observed that SP-LIME performs better than SP-Anchors on the German-credit dataset.

We further conduct experiments on the original non-discretized version of the tabular datasets only using MAIRE. We compare the global explanation created by MSD-MAIRE against randomly selected local explanations - RP-MAIRE. The results on the adult dataset are presented in Fig. 3d. Similar trends were observed for the Abalone and German-credits datasets.

Table 2. Mean Square difference between Cov and $\hat{Cov}$, Pre and $\hat{Pre}$ for adult, abalone and German credit datasets averaged over 100 data points.

	Adult	Abalone	German credit
MSD coverage	0.0015	0.0004	8.552e−05
MSD precision	0.3217	0.1265	0.0985

Figures 2(a–c) compare the change in precision as the local explanations are incrementally added to the global explanation for the three tabular datasets. It is observed that the proposed framework results in a minimal reduction in precision consistently across the three datasets. The observation is consistent with the mechanism to create a global explanation ensuring a minimum reduction in precision. LIME shows the maximum decrease in precision.

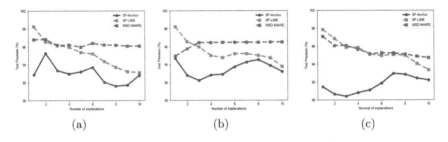

(a)   (b)   (c)

Fig. 2. [Best viewed in color] Change in test precision as a function of number of local explanations included in the global explanation Test Coverage for (a) Adult (b) Abalone (c) German-Credit datasets for SP-LIME, SP-Anchors and MSD-MAIRE.

Figures 3(a–c) compares the performance of the MAIRE framework for both the discretized and non-discretized versions of the tabular datasets. We observe that the coverage of the global explanation for MSD-MAIRE for both versions of the datasets is comparable for Adult and Abalone datasets. However, we notice a significant improvement in the performance of MAIRE on the discretized version of the German-Credit dataset that requires further investigation.

5.2   Text Datasets

The MAIRE framework is evaluated on two text datasets- IMDB movie reviews and a reduced 20-Newsgroups dataset (containing the data belonging to the four classes - medicine, graphics, Christian, and atheism). We illustrate the model-agnostic capability of the MAIRE framework by training a decision forest classifier for the IMDB dataset and a deep learning classifier for the Newsgroup dataset. The datasets are divided into train and test splits in the ratio of 4:1. A bag of words representation was used to characterize the reviews and documents.

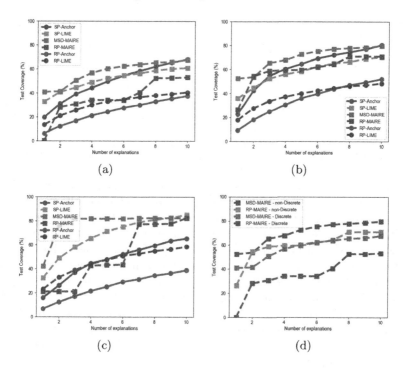

Fig. 3. [Best viewed in color] Change in test coverage as a function of number of local explanations included in the global explanation Test Coverage for (a) Adult (b) Abalone (c) German-Credit Data sets (d) Comparison of discretized vs non-discretized version of the datasets for RP-MAIRE and MSD-MAIRE

In the case of IMDB movie reviews, we considered a random forest with 500 trees as our black-box model to be explained. In the case of 20-Newsgroup dataset, we have only considered output labels 'medicine', 'graphics', 'Christian', and 'atheism' as it is not feasible to present 20 labels to a human subject. We use a four-layer network consisting of two hidden layers with 512 nodes each, having ReLU activation, and dropout probability set to 0.3 among layers, and softmax activation at the output layer as the base classifier. The model is trained for 30 epochs using Adam optimizer. The test classification accuracy on the two datasets is 87.3% and 81.2%, respectively.

We use ten data points (three medicine, three atheism, two graphics, two Christian) and generated explanations for each review using 5 different approaches mentioned in the paper. For generating the MAIRE explanation, the review was converted into a bag of words vector, and the sample points for computing $Cov, \hat{Cov}, Pre,$ and $\hat{Pre}$ were taken by randomly flipping bits in the bag of words. The words are ranked based on the effect they have on the classification using Greedy Attribute Elimination.

We conduct human subject experiments on the explanations for ten random test instances for each of the datasets to compare MAIRE against other

Table 3. Human accuracy of various model agnostic approaches on IMDB and Newsgroup datasets.

Method	LIME	SHAP	Anchor	L2X	MAIRE
IMDB	0.66	0.56	0.62	0.70	0.67
Newsgroup	0.66	0.64	0.70	0.69	0.75

model-agnostic approaches, namely; LIME and Anchors and feature ranking approaches, namely; L2X, and SHAP. We employ the experimental protocol of Chen et al., [6] for computing human accuracy. We assume that the explanations, in terms of the keywords (maximum of 10), convey sufficient information about the sentiment or class label of the document. We ask human subjects to infer the sentiment or the class label of the text when provided with only the explanations. The explanations from the different models and the various instances of a dataset are randomized. The final label for each document is averaged over the results of 25 human annotators. We measure the accuracy of the label predicted by the human annotator against the output of the model. The subjects are also allowed to label an explanation "can not infer" if the explanation is not sufficiently informative. We use the *Human Accuracy* metric for comparing the different approaches and treat the instances labeled as "can not infer" as misclassified instances.

The results are reported in Table 3. The human judgment given only ten words aligns best with the model prediction when the words are chosen from L2X and MAIRE for the IMDB and Newsgroup datasets, respectively. While on the binary classification dataset (IMDB), L2X is better than MAIRE by around 3%, on the more challenging 4-way classification dataset (Newsgroup) MAIRE leads over L2X by 6%. Overall the result indicates the competitiveness of MAIRE against other feature ranking approaches. It is also evident that MAIRE has significantly higher human accuracy over the other model-agnostic approaches LIME and Anchors. Table 4 shows exemplars for various models.

5.3   Image Datasets

We use the MAIRE framework to explain the classification results of the VGG16 model [20]. For explaining the model output, the image is segmented into superpixels and each superpixel is treated as a Boolean attribute. x'_q is taken to be a vector of 1 indicating the presence of all superpixels in the image. Sample points for calculating Cov, $\hat{Cov}$, Pre and $\hat{Pre}$ are computed by flipping the bits of x'_q randomly (i.e. randomly removing some superpixels). In the final explanation, the superpixels that covered both the values $\{0, 1\}$ of the corresponding Boolean attributes are removed as these superpixels do not affect the decision of the classifier. Figures 4(a–c) show the explanation generated by MAIRE and the heat map of the explanation (generated by ordering the superpixels chosen in the local explanation using Greedy Attribute Elimination) for a sample image (beagle). The VGG model has high confidence in its prediction for this image.

Table 4. Sample explanations for documents in the IMDB and Newsgroup dataset.

Review: I have to say that this miniseries was the best interpretation of the beloved novel "Jane Eyre". Both Dalton and Clarke are very believable as Rochester and Jane. I've seen other versions, but none compare to this one. The best one for me. I could never imagine anyone else playing these characters ever again. The last time I saw this one was in 1984 when I was only 13. At that time, I was a bookworm and I had just read Charlotte Bronte's novel. I was completely enchanted by this miniseries and I remember not missing any of the episodes. I'd like to see it because it's so good. :-)
LIME: best, completely, believable, 13, say, just, imagine, good, read, remember
SHAP: beloved, none, interpretation, good, novel, missing, remember, best, read, imagine
Anchors: believable remember, best, novel
L2X: imagine, interpretation, best good, novel, just, remember, read, characters, believable
MAIRE: enchanted, best, interpretation, remember, good, believable, novel, imagine, beloved, completely
Document: In article 47974@sdcc12.ucsd.edu— wsun@jeeves.ucsd.edu (Fiberman) writes: Is erythromycin effective in treating pneumonia? It depends on the cause of the pneumonia. For treating bacterial pneumonia in young otherwise-healthy non-smokers, erythromycin is usually considered the antibiotic of choice, since it covers the two most-common pathogens: strep pneumoniae and mycoplasma pneumoniae.
LIME: cause, treating, edu, common, effective, healthy, usually, antibiotic, bacterial, non
SHAP: healthy, writes, common, cause, effective, young, pneumoniae, choice, treating, cover
Anchors: pneumonia, healthy, antibiotic
L2X: cause, treating, antibiotic, edu, young, covers, bacterial, pathogens, choice, considered
MAIRE: common, bacterial, covers, young, pathogens, healthy, usually, smokers, cause, pneumoniae

We first validate the performance of the MAIRE framework by measuring the classifier confidence when random superpixels are removed (RSR) and when the superpixels picked by the MAIRE framework for the explanation are removed (MSR) from the original image. The results of this experiment are presented in Fig. 4(d). It is observed that the decrease in the classifier confidence on the removal of superpixels picked by the MAIRE framework is significantly larger than randomly selecting a superpixel. This illustrates that the MAIRE framework does indeed select the superpixels that have a big impact on the classifier.

In the second experiment, we only pick the superpixels selected by the greedy algorithm in the MAIRE framework. We iteratively remove the selected superpixels in the decreasing order of importance as estimated by the greedy algorithm, while also computing the classifier confidence. Our hypothesis is that if the greedy algorithm does indeed pick only important superpixels, then we would expect a sharp drop in the classifier confidence when the initial set of

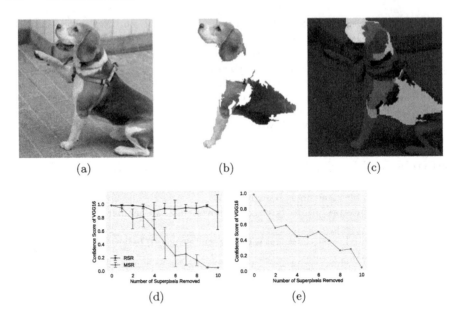

Fig. 4. [Best viewed in color] Results on the beagle image (a) Original Image (b) Explanation Generated (c) Heat Map (d) Confidence Score as more number of Superpixels are Removed (RSR = Random Superpixels Removed, MSR = MAIRE Superpixels Removed) (e) Confidence Score as more number of Superpixels are Removed (the removal order is from most important to least as given by Greedy Attribute Elimination)

superpixels are removed from the image. Figure 4(e) presents the results for the beagle image. We observe that by removing the top 4 superpixels selected by the MAIRE framework, the classifier confidence drops to less than 0.5. The Fig. 5 shows the explanation generated by the MAIRE framework and heat map of the explanation (generated by ordering the superpixels chosen in the local explanation using Greedy Attribute Elimination) for the bluetick image. The VGG model has high confidence in its prediction for this image as well. The decrease in the classifier confidence (Fig. 5(d)) with the removal of superpixels picked by the MAIRE framework is more significant than randomly selecting a superpixel. This also illustrates that the MAIRE framework does indeed select the superpixels having a significant impact on the classifier. We also observe that by removing the top 2 superpixels selected by the MAIRE framework, the classifier confidence drops to less than 0.5 for the bluetick image. It is interesting to note that the images in Figs. 5(b) and 5(c) show that the MAIRE framework selected superpixels mostly from the background in the bluetick image. Surprisingly, the VGG16 model classified the image, containing only the superpixels selected by the MAIRE framework for the bluetick image, correctly with the confidence of 0.953. Further, when we remove the superpixel containing the background snow, the VGG16 classifier confidence drops to 0.007. This indicates

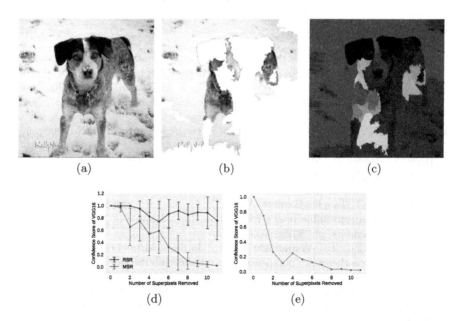

Fig. 5. [Best viewed in color] Results on the bluetick image (a) Original Image (b) Explanation Generated (c)Heat Map (d) Confidence Score as more number of Superpixels are Removed (RSR = Random Superpixels Removed, MSR = MAIRE Superpixels Removed) (e) Confidence Score as more number of Superpixels are Removed (the removal order is from most important to least as given by Greedy Attribute Elimination)

that the VGG16 network is focusing on perhaps incorrect regions of the image. The MAIRE framework is effective at detecting such wrong correlations learned by the machine learning model.

6   Summary

In this paper, we propose a novel model-agnostic interpretable rule extraction (MAIRE) framework for explaining the decisions of black-box classifiers. The framework quantifies the goodness of the explanations using coverage and precision. We propose novel differentiable approximations to these measures that are then optimized using the gradient-based optimizer. The flexible framework can be applied to any classifier for a wide variety of datasets. We test the framework on multiple datasets (tabular, text, and image) and show that the generated explanations are competitive to state-of-the-art approaches.

References

1. Acharyya, A., De, M., Nandy, S.C., Pandit, S.: Variations of largest rectangle recognition amidst a bichromatic point set. Discrete Appl. Math. **286**, 35–50 (2019)
2. Angelino, E., Larus-Stone, N., Alabi, D., Seltzer, M., Rudin, C.: Learning certifiably optimal rule lists for categorical data. J. Mach. Learn. Res. **18**(1), 8753–8830 (2017)
3. Armaselu, B., Daescu, O.: Maximum area rectangle separating red and blue points. arXiv preprint arXiv:1706.03268 (2017)
4. Bach, S., Binder, A., Montavon, G., Klauschen, F., Müller, K.R., Samek, W.: On pixel-wise explanations for non-linear classifier decisions by layer-wise relevance propagation. PloS One **10**(7), e0130140 (2015)
5. Chen, C., Li, O., Tao, D., Barnett, A., Rudin, C., Su, J.K.: This looks like that: deep learning for interpretable image recognition. In: Advances in Neural Information Processing Systems, pp. 8928–8939 (2019)
6. Chen, J., Song, L., Wainwright, M.J., Jordan, M.I.: Learning to explain: an information-theoretic perspective on model interpretation. arXiv preprint arXiv:1802.07814 (2018)
7. Eckstein, J., Hammer, P.L., Liu, Y., Nediak, M., Simeone, B.: The maximum box problem and its application to data analysis. Comput. Optim. Appl. **23**(3), 285–298 (2002)
8. Guidotti, R., Monreale, A., Giannotti, F., Pedreschi, D., Ruggieri, S., Turini, F.: Factual and counterfactual explanations for black box decision making. IEEE Intell. Syst. **34**, 14–23 (2019)
9. Guidotti, R., Monreale, A., Matwin, S., Pedreschi, D.: Black box explanation by learning image exemplars in the latent feature space. ECML, PKDD (2019)
10. Hase, P., Chen, C., Li, O., Rudin, C.: Interpretable image recognition with hierarchical prototypes. In: Proceedings of the AAAI Conference on Human Computation and Crowdsourcing, vol. 7, pp. 32–40 (2019)
11. Kingma, D.P., Ba, J.: Adam: a method for stochastic optimization. CoRR (2015)
12. Lakkaraju, H., Bach, S.H., Leskovec, J.: Interpretable decision sets: a joint framework for description and prediction. In: Proceedings of the 22nd ACM SIGKDD International Conference on Knowledge Discovery and Data Mining, pp. 1675–1684 (2016)
13. Lakkaraju, H., Kamar, E., Caruana, R., Leskovec, J.: Faithful and customizable explanations of black box models. In: Proceedings of the 2019 AAAI/ACM Conference on AI, Ethics, and Society, pp. 131–138 (2019)
14. Li, O., Liu, H., Chen, C., Rudin, C.: Deep learning for case-based reasoning through prototypes: a neural network that explains its predictions. In: Thirty-Second AAAI Conference on Artificial Intelligence (2018)
15. Lundberg, S.M., Lee, S.I.: A unified approach to interpreting model predictions. In: Advances in Neural Information Processing Systems, pp. 4765–4774 (2017)
16. Ribeiro, M.T., Singh, S., Guestrin, C.: Why should i trust you?: explaining the predictions of any classifier. In: Proceedings of the ACM International Conference on Knowledge Discovery and Data Mining, pp. 1135–1144 (2016)
17. Ribeiro, M.T., Singh, S., Guestrin, C.: Anchors: high-precision model-agnostic explanations. In: Proceedings of the AAAI Conference on Artificial Intelligence (2018)

18. Selvaraju, R.R., Cogswell, M., Das, A., Vedantam, R., Parikh, D., Batra, D.: Grad-CAM: visual explanations from deep networks via gradient-based localization. In: Proceedings of the IEEE International Conference on Computer Vision, pp. 618–626 (2017)
19. Shrikumar, A., Greenside, P., Kundaje, A.: Learning important features through propagating activation differences. In: Proceedings of the 34th International Conference on Machine Learning, volu. 70, pp. 3145–3153. JMLR. org (2017)
20. Simonyan, K., Zisserman, A.: Very deep convolutional networks for large-scale image recognition. arXiv preprint arXiv:1409.1556 (2014)
21. Springenberg, J.T., Dosovitskiy, A., Brox, T., Riedmiller, M.: Striving for simplicity: the all convolutional net. arXiv preprint arXiv:1412.6806 (2014)
22. Zhang, Q., Nian Wu, Y., Zhu, S.C.: Interpretable convolutional neural networks. In: Proceedings of the IEEE Conference on Computer Vision and Pattern Recognition, pp. 8827–8836 (2018)
23. Zhou, B., Khosla, A., Lapedriza, A., Oliva, A., Torralba, A.: Learning deep features for discriminative localization. In: Proceedings of the IEEE Conference on Computer Vision and Pattern Recognition, pp. 2921–2929 (2016)

Transparent Ensembles for Covid-19 Prognosis

Guido Bologna$^{(\boxtimes)}$ (iD)

University of Applied Sciences and Arts of Western Switzerland,
Rue de la Prairie 4, 1202 Geneva, Switzerland
Guido.Bologna@hesge.ch, Guido.Bologna@unige.ch

Abstract. A natural method aiming at explaining the answers of a black-box model is by means of propositional rules. Nevertheless, rule extraction from ensembles of Machine Learning models was rarely achieved. Moreover, experiments in this context have rarely been evaluated by cross-validation trials. Based on stratified tenfold cross-validation, we performed experiments with several ensemble models on Covid-19 prognostic data. Specifically, we compared the characteristics of the propositional rules generated from: Random Forests; Shallow Trees trained by Gradient Boosting; Decision Stumps trained by several variants of Boosting; and ensembles of transparent neural networks trained by Bagging. The Discretized Interpretable Multi Layer Perceptron (DIMLP) allowed us to generate rules from all the used ensembles by transforming Decision Trees into DIMLPs. Our rule extraction technique simply determines whether an axis-parallel hyperplane is discriminative or not, with a greedy algorithm that progressively removes rule antecedents. Rules extracted from Decision Stumps trained by modest Adaboost were the simplest with the highest fidelity. Our best average predictive accuracy result was equal to 96.5%. Finally, we described a particular ruleset extracted from an ensemble of Decision Stumps and it turned out that the rule antecedents seem to be plausible with respect to several recent works related to the Covid-19 virus.

Keywords: Ensembles · Model transparency · Rule extraction · Covid-19.

1 Introduction

Numerous projects aiming at making black-box models transparent are now emerging, especially in deep learning. In medicine, it is difficult to accept a successful model without being able to explain its answers. A natural method close to human thinking consists in explaining the answers of a model by means of propositional rules [13]. For black-box models such as neural networks, many techniques have been introduced. Specifically, all rule extraction algorithms applied to Multi Layer Perceptrons (*MLP*) have been categorized by the taxonomy introduced by Andrews et al. [1].

© IFIP International Federation for Information Processing 2021
Published by Springer Nature Switzerland AG 2021
A. Holzinger et al. (Eds.): CD-MAKE 2021, LNCS 12844, pp. 351–364, 2021.
https://doi.org/10.1007/978-3-030-84060-0_22

Among the few transparent Machine Learning models, Decision Trees (DTs) represent a valuable alternative to connectionist models, mainly when deep models are unnecessary. For structured data, Random Forests (RF) [6] are very often among the best classification models. Nevertheless, since RFs represent a combination of a large number of trees, rule extraction becomes difficult [23]. Another successful strategy is the boosting of weak learners, such as ensembles of shallow decision trees [12]. But again, with these ensembles, we encounter the same problem of transparency as with RFs. A number of representative techniques in this context are reported in [18].

In this work, we propose to generate rules from ensembles of DTs, such as RFs and Boosted shallow trees. We accomplish this task by inserting into transparent MLPs all the rules generated from each DT of an ensemble. In practice, any propositional rule can be transformed into a (small) MLP with weight values depending on rule antecedents. In this work, the transparent MLPs are DIMLP networks (*Discretized Interpretable Multi Layer Perceptron*). From them, we are able to generate rules from both single DIMLPs and DIMLP ensembles [4]. Hence, a ruleset representing a DT ensemble is generated from a DIMLP ensemble. The main novelty of this work is the application of our rule extraction technique proposed in [2] to RFs and Gradient Boosting GB of shallow trees.

Only a few authors proposed rule extraction algorithms from neural network ensembles. Zhou et al. introduced the *REFNE* technique (Rule Extraction from Neural Network Ensemble) [27], which utilizes the trained ensembles to generate samples and then extracted symbolic rules from those samples. For Johansson, rule extraction from ensembles is an optimization problem to solve with genetic algorithms in which a trade-off between accuracy and comprehensibility is taken into account [17]. Hara and Hayashi proposed the two-MLP ensembles [11] by using the "Recursive-Rule eXtraction" (*Re-RX*) algorithm [22]. Re-RX utilizes C4.5 decision trees and back-propagation to train MLPs recursively.

A novel classification problem on the prognosis of Covid-19 was recently presented in [26]. Based on stratified ten-fold cross-validation, we perform experiments with the Covid-19 related dataset. Here, we are specially interested to characterize the complexity of the rules produced by the applied models. It is worth noting that the proposed rule extraction technique can be applied to any classification problem. Therefore, our contribution is much more oriented toward a general algorithm, rather than the application itself. In this latter case, the involvement of some subject-matter expert would be necessary to fully understand the perceived usefulness of the resulting propositional rules. In the following sections we present the models, the experiments, followed by the conclusion.

2   Models

In this section we present Decision Trees (DTs) and their transformation into transparent MLPs. By transparency, we mean the production of propositional rules from the models. Specifically, the format of a rule is given as: "if tests on antecedents are true then class K"; where "tests on antecedents" are in the form

$x_i \leq t_i$ or $x_i \geq t_i$; with x_i as an input attribute and t_i as a real number. Class K designates a class among several possible classes.

Ensemble training is based on meta-learning algorithms, such as Bagging [5] and Boosting [9]. Bagging and Boosting use resampling techniques. For the former, assuming a training set of size N, bagging selects for each classifier included in an ensemble N samples drawn with replacement from the original training set. In such a manner, some diversity in each individual classifier proves beneficial for the combined set of classifiers. With Boosting, the samples of each classifier are chosen according to the probabilities defined for each sample in the original training set. After the first classifier has been trained, the probability of sample selection in a new training set is increased for all unlearned samples and decreased for the others.

2.1   Ensembles of Binary Decision Trees

A binary DT possesses nodes and branches. Each node represents a predicate with respect to an attribute and a threshold. Specifically, a node corresponds to a rule antecedent. A node without successors is denoted as a leaf. Any path from the root to a leaf represents a propositional rule. Thus, a binary tree is considered transparent by construction. Several examples of tree induction algorithms are reported in [8, 20].

In this work, we use among others, small trees with a unique node also denoted as Decision Stumps (DS). These particular trees represent "weak" learners, because their power of expression is very limited. With the use of Boosting techniques [9], ensembles of weak learners become strong classifiers [21]. We use four Boosting algorithms to train shallow trees:

- Adaboost [9];
- Gentle Adaboost [10];
- Modest Adaboost [24];
- Gradient Boosting [12].

Random Forests (RF) are also ensembles of DTs [6]. They are trained by Bagging; moreover, each tree can be constrained to select a small proportion of the available attributes. As a consequence, the training phase is very fast for two reasons:

- each tree of an ensemble has to determine a new split from a limited number of attributes;
- each tree is independent from the others; hence, the induction of the trees is usually performed with parallel hardware.

2.2   Transformation of Trees into Transparent MLPs

The key idea behind rule extraction from ensembles of DTs being their transformation into ensembles of transparent MLPs, we first describe how to transform a rule antecedent into an MLP. Then, we generalize to rules with many antecedents.

Figure 1 illustrates an MLP that represents a symbolic rule with a unique antecedent. Any neuron in the middle or output layer receives a signal, which is the result of a weighted sum of inputs and weights. Then, an activation function is applied; in the middle layer it is a step function given as:

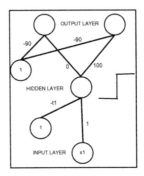

Fig. 1. A transparent MLP coding a propositional rule with a unique antecedent ($x_1 > t1$).

$$t(x) = \begin{cases} 1 \text{ if } x > 0; \\ 0 \text{ otherwise.} \end{cases} \tag{1}$$

In the output layer we have a sigmoid function given as:

$$\sigma(x) = \frac{1}{1 + \exp(-x)}. \tag{2}$$

Therefore, the MLP represented in Fig. 1 represents the following propositional rule:

- $(x_1 > t_1) \rightarrow C_2$; with C_2 designating the second class coded by vector $(0, 1)$.

It is worth noting that if $x_1 \leq t_1$ then the output will be a vector of two components with their values very close to zero.

Figure 2 shows an MLP that represents a propositional rule with two antecedents:

- $(x_1 > t_1)$ AND $(x_2 \leq t_2) \rightarrow C_1$; with C_1 designating the first class coded by vector $(1, 0)$.

Generally, the number of rule antecedents in a rule is unconstrained. To correctly code an arbitrary number of antecedents we must ensure that the weight of the bias neuron between the middle layer and the output layer is adequate relative to the number of antecedents. Specifically, for a unique antecedent if a constant K of high value ($K = 100$) allows to transmit the fact that an antecedent is

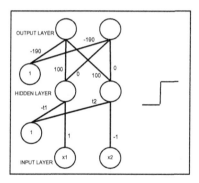

Fig. 2. A transparent MLP coding a propositional rule with two antecedents $(x_1 > t_1)$ AND $(x_2 \le t_2)$.

true, then the bias weight is equal to $-9K/10$. With p antecedents, K is replaced by pK. In this way, any rule generated from the root to a leaf of a DT is inserted into a transparent MLP.

Figure 3 represents the same transparent MLP as Fig. 1, but with an additional layer having an Identity activation function that is useful to encode the weighting of a rule (coefficient w at the top right). Indeed, with boosted DTs we have to take into account the weighting of each tree.

2.3   DIMLP Networks

The transparent MLPs presented in the previous paragraphs represent special cases of *Discretized Interpretable Multi Layer Perceptrons* (DIMLPs) [4]. DIMLP differs from standard Multi Layer Perceptrons in the connectivity between the input layer and the first hidden layer. Specifically, any hidden neuron receives only a connection from an input neuron and the bias neuron, as shown in Fig. 3. Furthermore, the activation function for the first hidden layer is a staircase function that approximates the sigmoid function. Note also that the staircase function generalizes the step function. Above the first hidden layer, neurons are fully connected.

Since rule antecedents correspond to axis-parallel hyperplanes, the rule extraction technique simply determines whether a hyperplane is discriminative or not, depending on the weight values of the neurons above the first hidden layer. A greedy algorithm that generates unordered rules progressively removes antecedents and rules. The fidelity of the generated ruleset, which is the degree of matching between network classifications and rules' classifications is equal to 100%, with respect to the training set. More details on the rule extraction technique can be found in [3]. Finally, Rule extraction from DIMLP ensembles can also be performed, since an ensemble of DIMLP networks can be viewed as a single DIMLP with one more hidden layer [4].

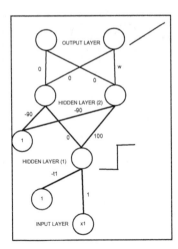

Fig. 3. A transparent MLP coding a propositional rule with a unique antecedent ($x_1 > t_1$). The activation function of the output layer is the Identity, with coefficient w coding the rule weight.

3   Experiments

The used dataset is related to blood samples from 375 infected patients by the Covid-19 virus in the region of Wuhan, China [26]. The medical data of all patients were collected between January and February 2020. The number of attributes for each patient is 76, among which for 74 attributes we calculated the average/min/max values during the patient's hospital stay ($74 * 3 = 222$). Thus, the input vectors have a size of 224 components. The whole dataset describes 201 patients who recovered from COVID-19, while the remaining 174 died. A description of the attributes is reported in [26].

3.1   Models and Learning Parameters

Our experiments are based on 10 repetitions of stratified tenfold cross validation trials. The Covid-19 dataset was normalized by Gaussian normalization. We trained the following models:

- Random Forests (RF);
- Shallow trees trained by Gradient Boosting (GB);
- DIMLP ensembles trained by Bagging;
- Boosted Decision Stumps (DS) trained by Adaboost.
- Boosted Decision Stumps trained by gentle Adaboost;
- Boosted Decision Stumps trained by modest Adaboost;

In all the ensembles, the number of trained classifiers was fixed to: 25; 50; 100; and 150. For RF we used default learning parameters defined in the *Scikit*

Learn library [19]. Note that the depth of a single tree is unconstrained. By contrary, with GB the default maximal depth is three. Finally, Decision Stumps are DTs with a unique node.

For DIMLP ensembles we used default learning parameters (see [4]). The default number of neurons in the first hidden layer is equal to the number of input neurons and the number of neurons in the second hidden layer is empirically defined in order to obtain a number of weight connections that is less than the number of training samples. Here, we used ten neurons.

3.2 Random Forest Results

In the next tables, columns from left to right designate the average values of:

- Train accuracy;
- Accuracy on the testing set (predictive accuracy);
- Fidelity, which is the degree of matching between the rules and the model (on the testing set); specifically, with P samples in the testing set and Q samples for which the classification of the rules correspond to the classification of an ensemble, fidelity is Q/P.
- Accuracy of the rules on the testing set;
- Accuracy of the rules when rules and model agree on the testing set;
- Number of extracted rules and number of rule antecedents.

For RFs, Table 1 shows in the last column that the average complexity of the generated rulesets increases with the number of trees in an ensemble. Moreover, the average predictive accuracy and the average predictive accuracy of the rules when rules and ensembles agree are relatively stable. Finally, as we will see in the results provided by the other models, the average complexity of the rulesets generated by RF is the highest (last column of the Tables).

Table 1. Average results obtained by Random Forests. The last column depicts the average number of rules and the average number of rule antecedents.

#Trees	Train acc.	Test acc.	Fidelity	Test acc. (r1)	Test acc. (r2)	#Rules/#Ant
25	**100.0**	95.8	**96.8**	**94.3**	96.5	**30.5/3.7**
50	100.0	**96.0**	96.4	93.8	**96.6**	36.3/3.9
100	100.0	95.8	95.6	93.0	96.5	43.2/4.1
150	100.0	95.9	96.0	93.2	96.4	44.6/4.2

3.3 Gradient Boosting Results

Table 2 illustrates the results obtained by GB of shallow trees. The average complexities of the obtained rulesets is lower than that given by RFs. This is probably due to the complexity of a shallow tree, which is lower than that of a typical RF tree without depth constraints. We also observe that average fidelity decreases with the increasing average complexity of extracted rulesets. Finally, the average predictive accuracy of the models (third column) decreases slowly after using 50 trees in an ensemble.

Table 2. Average results obtained by Gradient Boosting.

#Trees	Train acc.	Test acc.	Fidelity	Test acc. (r1)	Test acc. (r2)	#Rules/#Ant
25	99.5	95.0	**97.7**	94.6	95.9	**14.0/3.0**
50	99.5	**95.9**	97.3	94.3	96.3	16.5/3.1
100	**100.0**	95.8	97.0	**94.9**	96.7	26.5/3.4
150	100.0	95.7	96.8	94.8	**96.8**	27.6/3.4

3.4   Results Obtained by DIMLPs Trained by Bagging

The average fidelity of the rulesets is the lowest among all the used models (see Table 3). Furthermore, it is worth noting that the difference between the average testing accuracy and the average testing accuracy of the rules is the highest. Finally, the average predictive accuracy of the rules is also the lowest, although the average predictive accuracy of the rules when models and rules agree is acceptable (sixth column).

Table 3. Average results obtained by DIMLPs trained by Bagging.

#DIMLP	Train acc.	Test acc.	Fidelity	Test acc. (r1)	Test acc. (r2)	#Rules/#Ant
25	98.2	95.3	95.3	92.1	95.9	**26.4/3.6**
50	98.6	**95.5**	95.2	92.3	96.1	28.6/3.7
100	**98.7**	95.5	**95.7**	**92.8**	**96.2**	28.1/3.7
150	98.7	95.5	95.6	92.6	96.1	27.3/3.7

3.5   Results of Boosted Decision Stumps

Tables 4, 5, and 6 depict the results obtained by Boosted DSs trained with Adaboost, gentle Adaboost and modest Adaboost, respectively. With respect to average complexity, Adaboost and gentle Adaboost provide similar results, while modest Adaboost involves lower values. Furthermore, all boosted DSs tend to generate rulesets of higher average complexity, as the number of trees in an ensemble increases.

Table 4. Average results obtained by Decision Stumps trained by Adaboost.

#Trees	Train acc.	Test acc.	Fidelity	Test acc. (r1)	Test acc. (r2)	#Rules/#Ant
25	99.7	**96.0**	97.7	94.9	96.5	**14.3/3.0**
50	**100.0**	95.9	97.5	95.2	**96.7**	18.8/3.1
100	100.0	96.0	**97.8**	95.1	96.5	21.5/3.3
100	100.0	95.7	97.4	**95.3**	96.7	22.7/3.4

The highest average predictive accuracy was obtained by gentle Adaboost with 96.5% (±0.6). Moreover, the highest average predictive accuracy provided by the rulesets was equal to 95.5% (±0.4). Finally, the best average predictive accuracy when the rulesets agree with their corresponding models (sixth column of all Tables) was equal to 96.9% (±0.5).

Table 5. Average results obtained by Decision Stumps trained by gentle Adaboost.

#Trees	Train acc.	Test acc.	Fidelity	Test acc. (r1)	Test acc. (r2)	#Rules/#Ant
25	99.5	**96.5**	**97.7**	95.1	**96.9**	**14.1/3.1**
50	**100.0**	96.4	97.7	95.3	96.9	18.8/3.2
100	100.0	95.9	97.7	**95.5**	96.8	21.5/3.3
150	100.0	95.7	96.8	94.8	96.8	27.6/3.4

Compared to the other ensembles, DSs tend to provide rulesets with the best average fidelities. For instance, with modest Adaboost (see Table 6) the best average fidelity reached 98.0% (±0.7).

Table 6. Average results obtained by Decision Stumps trained by modest Adaboost.

#Trees	Train acc.	Test acc.	Fidelity	Test acc. (r1)	Test acc. (r2)	#Rules/#Ant
25	97.0	94.3	97.9	93.3	94.8	**9.4/2.5**
50	98.2	95.3	97.3	93.9	95.9	10.5/2.7
100	99.1	96.2	97.6	95.0	**96.8**	12.2/2.9
150	**99.2**	**96.4**	**98.0**	**95.2**	96.7	12.2/2.9

3.6   An Example of Generated Ruleset

Table 7 presents a summary of the results obtained by taking into account for each model the highest average accuracy on the test set. Figure 4 illustrates a ruleset generated from an ensemble of DSs trained by modest Adaboost during cross-validation trials. Its accuracy on the training set is 99.1% and 100% on the testing set. By mere chance the accuracy on the testing set is higher than that obtained on the training set. The class *NEG* designates patients who have recovered from the Covid-19 virus, while the class *POS* indicates patients who did not recover. In addition, the number of samples covered by the rules for the training and testing sets is shown in parentheses. Note also that any sample can be covered by several rules.

Table 7. Summary of the results obtained when the highest average predictive accuracy is obtained for each different model. First row indicates Gradient Boosting, then Random Forests, DIMLPs, shallow trees trained by Adaboost (AB), Gentle Adaboost (GAB) and modest Adaboost (MAB). Standard deviations are given between brackets.

Model (#Trees)	Train acc.	Test acc.	Fidelity	Test acc. (r1)	Test acc. (r2)	#Rules/#Ant
GB (50)	**100.0** (0.0)	96.0 (0.3)	96.4 (0.8)	93.8 (0.5)	96.6 (0.4)	36.3 (1.1)/3.9 (0.1)
RF (50)	99.5 (0.0)	95.9 (0.7)	97.3 (0.8)	94.3 (0.9)	96.3 (0.5)	16.5 (2.1) /3.1 (0.1)
DIMLP (50)	98.6 (0.1)	95.5 (0.5)	95.2 (1.1)	92.3 (1.2)	96.1 (0.6)	28.6 (2.1) /3.7 (0.1)
AB (25)	99.7 (0.0)	96.0 (0.8)	97.7 (0.6)	94.9 (0.8)	96.5 (0.6)	14.3 (0.6)/3.0 (0.1)
GAB (25)	99.5 (0.0)	**96.5** (0.6)	97.7 (0.7)	95.1 (0.8)	**96.9** (0.5)	14.1 (0.5)/3.1 (0.1)
MAB (150)	99.2 (0.0)	96.4 (0.5)	**98.0** (0.7)	**95.2** (0.6)	96.7 (0.6)	**12.2** (0.3)/**2.9** (0.0)

In general, the higher the number of trees and their depth, the higher the risk of overfitting. Here, with decision stumps the depth parameter is equal to one, which is the minimal value. Moreover, cross-validation results obtained by modest Adaboost show that the average predictive accuracy increases progressively as the number of trees in an ensemble increases from 25 trees (94.3% predictive accuracy ±0.8) to 150 trees (96.4% ±0.4). In case of overfitting, we would observe that the average predictive accuracy would start to decrease at some point, which is not the case here.

The thresholds of the antecedents shown in Fig. 4 are related to normalised values. For instance, the first rule means that if the average measure of monocytes during the patient stay at the hospital are greater than a given threshold and if the average measure of Lactic Dehydrogenase (LDH) is below a given threshold and finally if the minimal value of Hypersensitive C reactive Protein (HcRP) is less than a given threshold then a patient recovers from the virus. High levels of LDH could indicate severe disease or multiple organ failure, while a high level of HcRP in the blood is a marker of inflammation. Hence, the first rule requires low levels of LDH/HcRP and a sufficient number of monocytes to be able to fight against an aggressor.

The sixth rule says that if the average LDH is above a given threshold and if the average prothrombin activity is below another threshold then a patient is

```
R1:  (avg_monocytes > -0.507)
     (avg_LDH < -0.356)
     (min_HcRP < -0.393) --> NEG (146/23)

R2:  (avg_lymphocyte > -0.360)
     (avg_LDH < -0.356)
     (min_glucose < -0.519) --> NEG (121/24)

R3:  (avg_LDH > -0.356)
     (min_glucose > -0.724)
     (max_PlateletCount < 0.655) --> POS (111/12)

R4:  (age > -0.749)
     (avg_ProthrAct < -0.210)
     (avg_lymphocyte < -0.360) --> POS (111/11)

R5:  (avg_ProthrAct > -0.210)
     (min_procalcitonin < -0.245)
     (min_glucose < -0.519) --> NEG (110/22)

R6:  (avg_ProthrAct < -0.210)
     (avg_LDH > -0.356) --> POS (106/10)

R7:  (age < 0.223)
     (avg_ProthrAct > -0.210)
     (min_procalcitonin < -0.245)
     (min_HcRP < -0.393) --> NEG (97/19)

R8:  (avg_monocytes < -0.323)
     (min_HcRP > -0.393) --> POS (94/8)

R9:  (min_glucose > -0.519)
     (min_HcRP > -0.393) --> POS (82/10)

R10: (age < -0.749)
     (avg_LDH < -0.356) --> NEG (73/12)

R11: (avg_HCT1 > -0.178)
     (avg_monocytes > -0.323)
     (avg_lymphocyte < -0.294)
     (max_PlateletCount < 0.655) --> POS (14/4)
```

Fig. 4. An example of ruleset extracted from an ensemble of Decision Stumps trained by modest Adabost. In brackets after the class is the number of samples covered in the training set and the testing set, respectively.

unable to recover. Prothrombin helps blood to clot. Since prothrombin activity is measured in seconds, if the activity time is too low, one could be at risk for a blood clot. The other rules involve other attributes, such as:

- Average measure of lymphocytes (avg_lymphocyte);
- Minimal level of glucose (min_glucose);
- Maximal platelet count (max_PlateletCount);
- Minimal level of procalcitonin (min_procalcitonin);

– Age;
– Average measure of Highly sensitive troponin I (avg_HCT1).

We may wonder about the plausibility of these attributes from the extracted ruleset (Fig. 4). For instance, lymphocytes exceeding a certain threshold tend to fight the virus (second rule) and are not sufficient to recover below this same threshold (fourth rule). This is conceivable, since lymphocytes are white blood cells that combat the virus.

The value of the minimal glucose attribute is present in two negative rules (second rule and fifth rule). It favours healing, when it is below a certain threshold, but tends to be fatal above another threshold (third and ninth rule). Interestingly, a recent meta-analysis provides evidence that severe COVID-19 is associated with increased blood glucose [7]; hence, our rules are credible.

In the third and the last rule we have the maximal platelet count attribute. Here, values below a given threshold do not allow patients to recover. This fact is corroborated with a meta-analysis in an article [16] stating that the non-survivors had a much lower platelet count than the survivors.

The minimal procalcitonin attribute is present in the fifth and seventh rule. In a recent work, it was stated that procalcitonin levels were over eight times higher in critical patients than in moderate patients [15]. Therefore in our rules, low levels of minimal procalcitonin supporting the recovery of patients is plausible.

The highly sensitive cardiac troponinI (HCT1) is in the eleventh rule. A recent study demonstrated that the risk of in-hospital death among COVID-19 patients with cardiac injury can be predicted by the peak levels of HCT1 [25]. We are not sure of the plausibility of the eleventh rule, since we do not know whether the patients had cardiac injury. Finally, patient age is well known to be an important factor of prognosis and it is a rule antecedent in three rules.

3.7   Related Work

In [26] the authors determined a small number of informative features with a prognostic predictive value. The following three attributes were chosen: lactic dehydrogenase (LDH); lymphocytes; and high-sensitivity C-reactive protein (HcRP). Note also that these three attributes are present in the ruleset described above.

At this point, a question arising was whether these three attributes would allow us to obtain equivalent performance by cross-validation. Hence, we selected these small set of attributes and performed tenfold cross-validation experiments. In Table 8, we illustrate the predictive accuracy results obtained by ensembles of DTs. The average predictive accuracy of these ensembles was higher with the use of all attributes, rather than a small subset.

Table 8. Average predictive accuracy obtained by ensemble of DTs with the use three attributes out of 76 (average predictive accuracy with the 76 attributes is given between brackets).

#Trees	Adaboost	Gentle Adaboost	Modest Adaboost	Random Forests	Grad. Boost
25	94.2 (96.0)	92.4 (**96.5**)	93.4 (94.3)	94.2 (95.8)	93.6 (95.0)
50	93.7 (95.9)	93.1 (96.4)	93.4 (95.3)	94.4 (96.0)	94.0 (95.9)
100	93.8 (96.0)	92.9 (95.9)	93.4 (96.2)	**94.5** (95.8)	94.1 (95.8)
150	93.2 (95.7)	93.1 (95.7)	93.4 (96.4)	94.4 (95.9)	94.2 (95.7)

4   Conclusion

In this work, we extracted unordered rules from ensembles of Random Forests, Boosted Shallow Trees and DIMLPs. On a classification problem related to Covid-19 prognosis, Decision Stumps trained by modest Adaboost produced the less complex rulesets with good predictive accuracy. Generally, rule extraction from ensembles of Machine Learning models was rarely tackled. Our best result on average predictive accuracy was equal to 96.5% ($\pm$0.6) and the average predictive accuracy of the rules was equal to 95.5% ($\pm$0.4). Moreover, the average predictive accuracy of the rules when rules and models agree was 96.9% ($\pm$0.5). Afterward, we described a particular ruleset extracted from an ensemble of Decision Stumps and it turned out that the rule antecedents seem to be plausible. We would like to encourage researchers to work with transparent models, since they potentially help to discover new knowledge. Random forests, decision trees and transparent neural networks are of considerable interest for the future, because of their interpretability. Another important approach based on these models would be to insert a human into the loop who can provide conceptual knowledge, experience and contextual understanding, which no AI algorithm can provide today [14].

References

1. Andrews, R., Diederich, J., Tickle, A.B.: Survey and critique of techniques for extracting rules from trained artificial neural networks. Knowl.-Based Syst. **8**(6), 373–389 (1995)
2. Bologna, G.: A study on rule extraction from several combined neural networks. Int. J. Neural Syst. **11**(03), 247–255 (2001)
3. Bologna, G.: A model for single and multiple knowledge based networks. Artif. Intell. Med. **28**(2), 141–163 (2003)
4. Bologna, G.: Is it worth generating rules from neural network ensembles? J. Appl. Log. **2**(3), 325–348 (2004)
5. Breiman, L.: Bagging predictors. Mach. Learn. **24**(2), 123–140 (1996)
6. Breiman, L.: Random forests. Mach. Learn. **45**(1), 5–32 (2001)
7. Chen, J., Wu, C., Wang, X., Yu, J., Sun, Z.: The impact of covid-19 on blood glucose: a systematic review and meta-analysis. Front. Endocrinol. **11** (2020)
8. Cohen, W.W.: Fast effective rule induction. In: Proceedings of the Twelfth International Conference on Machine Learning, pp. 115–123 (1995)

9. Freund, Y., Schapire, R.E.: A desicion-theoretic generalization of on-line learning and an application to boosting. In: Vitainyi, P. (eds.) Computational Learning Theory. EuroCOLT 1995. Lecture Notes in Computer Science (Lecture Notes in Artificial Intelligence), vol. 904. Springer, Berlin, Heidelberg (1995). https://doi.org/10.1007/3-540-59119-2_166

10. Friedman, J., Hastie, T., Tibshirani, R., et al.: Additive logistic regression: a statistical view of boosting (with discussion and a rejoinder by the authors). Ann. Stat. **28**(2), 337–407 (2000)

11. Hara, A., Hayashi, Y.: Ensemble neural network rule extraction using re-rx algorithm. In: The 2012 International Joint Conference on Neural Networks (IJCNN), pp. 1–6. IEEE (2012)

12. Hastie, T., Tibshirani, R., Friedman, J.: The Elements of Statistical Learning: Data Mining, Inference, and Prediction. Springer Science & Business Media (2009)

13. Holzinger, A., Biemann, C., Pattichis, C.S., Kell, D.B.: What do we need to build explainable AI systems for the medical domain? arXiv:1712.09923 (2017)

14. Holzinger, A., Malle, B., Saranti, A., Pfeifer, B.: Towards multi-modal causability with graph neural networks enabling information fusion for explainable AI. Inf. Fusion **71**, 28–37 (2021)

15. Hu, R., Han, C., Pei, S., Yin, M., Chen, X.: Procalcitonin levels in covid-19 patients. Int. J. Antimicrob. Agents **56**(2), 106051 (2020)

16. Jiang, S.Q., Huang, Q.F., Xie, W.M., Lv, C., Quan, X.Q.: The association between severe covid-19 and low platelet count: evidence from 31 observational studies involving 7613 participants. Br. J. Haematol. **190**(1), e29–e33 (2020)

17. Johansson, U.: Obtaining accurate and comprehensible data mining models: an evolutionary approach. Linköping University, Department of Computer and Information Science (2007)

18. Mashayekhi, M., Gras, R.: Rule extraction from decision trees ensembles: new algorithms based on heuristic search and sparse group lasso methods. Int. J. Inf. Technol. Dec. Making 1–21 (2017)

19. Pedregosa, F., et al.: Scikit-learn: machine learning in python. J. Mach. Learn. Res. **12**, 2825–2830 (2011)

20. Quinlan, J.R.: C4.5: Programs for machine learning. morgan kaufmann publishers, inc., 1993. Mach. Learn. **16**(3), 235–240 (1994)

21. Schapire, R.E.: A brief introduction to boosting. Ijcai. **99**, 1401–1406 (1999)

22. Setiono, R., Baesens, B., Mues, C.: Recursive neural network rule extraction for data with mixed attributes. IEEE Trans. Neural Netw. **19**(2), 299–307 (2008)

23. Van Assche, A., Blockeel, H.: Seeing the forest through the trees: learning a comprehensible model from an ensemble. In: Kok, J.N., Koronacki, J., Mantaras, R.L., Matwin, S., Mladenic, D., Skowron, A. (eds.) Machine Learning: ECML 2007. ECML 2007. Lecture Notes in Computer Science, vol. 4701. Springer, Berlin, Heidelberg (2007). https://doi.org/10.1007/978-3-540-74958-5_39

24. Vezhnevets, A., Vezhnevets, V.: Modest adaboost-teaching adaboost to generalize better. In: Proceedings of the International Conference on Computer Graphics in Europe and Asia, vol. 12, pp. 987–997. Computer Graphics in Russia (2005)

25. Wang, Y., et al.: The peak levels of highly sensitive troponin i predicts in-hospital mortality in covid-19 patients with cardiac injury: a retrospective study. Eur. Heart J. Acute Cardiovasc. Care **10**(1), 6–15 (2021)

26. Yan, L., et al.: An interpretable mortality prediction model for covid-19 patients. Nat. Mach. Intell. **2**(5), 283–288 (2020)

27. Zhou, Z.H., Jiang, Y., Chen, S.F.: Extracting symbolic rules from trained neural network ensembles. Artif. Intell. Commun. **16**(1), 3–16 (2003)

Author Index

Printed in the United States
by Baker & Taylor Publisher Services